T0221160

THE FAULT LINES OF FARM POLICY

THE FAULT LINES OF FARM POLICY

A LEGISLATIVE AND POLITICAL HISTORY OF THE FARM BILL

JONATHAN COPPESS

University of Nebraska Press

LINCOLN & LONDON

Library of Congress Cataloging-in-Publication Data
Names: Coppess, Jonathan, author.
Title: The fault lines of farm policy: a legislative and
political history of the farm bill / Jonathan Coppess.
Description: Lincoln: University of Nebraska Press, 2018. |
Includes bibliographical references and index.
Identifiers: LCCN 2018022855
ISBN 9781496205124 (hardback)
ISBN 9781496212528 (epub)
ISBN 9781496212535 (mobi)
ISBN 9781496212542 (pdf)
Subjects: LCSH: United States. Agricultural Act of
2014—Legislative history. | Agricultural laws and
legislation—Legislative history—United States. | Agricul-
tural laws and legislation—United States—History—20th
century. | Agricultural policy—United States. | Agricul-
tural policy—United States—History—20th century. |
BISAC: POLITICAL SCIENCE / Government / National. |
TECHNOLOGY & ENGINEERING / Agriculture / General.
| HISTORY / United States / General.
Classification: LCC KF1681.A3282014 A15 2018
| DDC 343.7307/6—dc23 LC record available at
https://lccn.loc.gov/2018022855

Set in Minion Pro by E. Cuddy.

To Susan, Abigail, and Warren, for their love, support, and patience.

And in memory of my grandfather, Sam Coppess, a farmer and politician. I can picture him reading this book in one sitting and declaring, "It was all right."

CONTENTS

PREFACE

I can trace the beginnings of this book to the fall of 2012 and the quiet of a Congress in recess for campaign season. It began with questions about policy, politics, and history—an attempt to better understand what had occupied so much time and energy. Initially, in fact, I had no intention of writing a book. But what began as an attempt to answer questions progressed over the course of five years into one. The following serves as an explanation for my writing this book but also as full disclosure regarding my role in a small part of the history.

I have been a participant in the process. I had the privilege of serving in the U.S. Senate and at the U.S. Department of Agriculture (USDA) from 2006 to 2013. During that time I worked on the 2008 farm bill for then-senator Ben Nelson (D-NE), as well as the energy bill that created the second renewable fuels standard. I went on to a lead role in implementing the 2008 farm bill as administrator of the Farm Service Agency. From early 2011 to late 2013, I served as chief counsel to the Senate Committee on Agricultural, Nutrition, and Forestry for then-chairwoman Senator Debbie Stabenow (D-MI). In that role I was very involved in the bill that would become the Agricultural Act of 2014.

After more than a year of intense negotiations, the 2012 farm bill sat in limbo; the Senate had passed a bill, but House leadership refused to bring it up for debate. Recess and the campaign afforded me time to think about the issues more deeply, to delve into questions put aside during the heat of the work. The most prominent issue was the intense regional fight over farm programs, and in particular, using prices fixed in statute to trigger payments. Thus, my first steps down this path involved basic research about the origins of target or reference prices. These prices had been at the heart of the dispute among farm interests, senators, and representatives from the three main commodity production regions, the South, Midwest, and Great Plains. This research effort was short-lived, however, as Congress returned after the elections for a lame-duck session that was followed by the difficult 2013 debate. The work of negotiations and debate reengaged, the research was set aside.

In the fall of 2013, I embarked on a new career path at the University of Illinois. The farm bill had not been completed. It remained, in fact, in some doubt in a Congress on the brink of a federal government shutdown. At Illinois I began the adjustment to life outside the Beltway, at some remove from the debates and goings-on of Congress. I also began to reflect further on the history of farm policy, picking up the research from the prior fall. I was encouraged to develop a class on the subject, which further drove my work—research morphing into my attempts at teaching undergraduate students. I also began co-teaching a short summer course on the topic at the Vermont Law School.

Research to better answer my questions combined with the need for material for class. I was helped most by three excellent books on the topic: Bill Winders's *The Politics of Food Supply: U.S. Agricultural Policy in the World Economy*; John Mark Hansen's *Gaining Access: Congress and the Farm Lobby, 1919–1981*; and Murray R. Benedict's *Farm Policies of the United States, 1790–1950: A Study of Their Origins and Development*. As demonstrated by the bibliography, these were among many great resources. They were integral in helping me dive deeper into the history of the policy and the larger political history that surrounded congressional efforts. I have returned to them time and again throughout my research, writing, and teaching efforts.

There are also a wealth of articles and a few books on farm programs and policy by agricultural economists. These include, among others, great works by Carl Zulauf, David Orden, Willard Cochrane, Don and Robert Paarlberg, Edward and Frederick Schapsmeier, Daniel Sumner, and Robert Spitze. I found these helpful for economic analysis of the policies but also for snapshots of the times in which the bills were written and debated. I was, however, looking for a different perspective on, and understanding of, the policies and programs. Thus, I next turned to a concept from my background in law known as legislative history.

Legislative history can be used by courts to help interpret statutory language. It involves reviewing debates on the House and Senate floors, committee reports, testimony at hearings, and other public records. While in legal circles its use for interpreting statutes is a matter of some controversy, I found it incredibly helpful in writing this book. I read the congressional floor debates for each of the major farm bills and other public records, such as the bills, committee reports, and hearing testimony. I used those records to help me understand what issues were prevalent at the times the bills were written and the key matters that were debated. They helped me develop a more complete sense of what politics were involved and what members of Congress were seeking to accomplish. I found that the legislative histories added substantially to the economic, political, and general histories. Thus, these materials form the core of this book.

Maybe somewhat ironically, the two recent farm bills, especially the 2014 farm bill, were the toughest for me to write about. These were the bills for which I had direct involvement and extensive personal knowledge. I acknowledge that I cannot be considered a neutral observer of the 2008 and 2014 farm bills. I worked to write their histories, however, using the same research methods used for all previous bills. I relied on the *Congressional Record* for floor debates and on committee reports and materials from hearings. I also supplemented this record substantially with the public record created by Keith Good on FarmPolicy.com and with news sources such as David Rogers's extensive reporting on the 2014 farm bill for Politico.com. My experiences informed my writing and the concluding thoughts on that history, but the history itself is a product of the public record. This too is consistent with the entire book.

I have attempted herein an honest read of history but one informed by my own experiences in the arena. The book's chapters provide an account of policy development in each era coupled with concluding thoughts and perspectives. For 2014 especially, I have further tried to anchor those concluding thoughts and perspectives in the longer history. Among my goals is to better understand that recent debate, as well as its current and future iterations, in terms of the long political and legislative history that underlies it. The lessons I learned about legislating, policy, and politics are foundational and inextricably connected to my read of this history.

Finally, the questions that I had about the two bills I worked on drove my research and writing of the fuller history. The fuller history, in turn, enriched my understanding of our system of government, posing new questions and opening new paths for exploration. Interested in government and policy enough to sacrifice my legal career, I became enamored with the system by which we govern ourselves while I served in it; writing this book has added further still. Chapter 9 seeks to tie these threads—the history of American farm policy and the politics and processes of governing—together. A theory developing from this effort is that the farm bills may serve as a window into self-governance, case studies in legislating, and policymaking.

At times, I still find it hard to believe that I had the opportunity to work in Congress and at USDA, that I was an active participant in history and government. I will be forever grateful for the experiences and for the advice and guidance I received along the way. It is not an exaggeration that the roots of this effort are to be found in those experiences. The book is indelibly linked to the questions and discussions among the many hardworking, dedicated participants with whom I served. Where there are mistakes in this history, however, they are my own, but they have been made in good faith.

ACKNOWLEDGMENTS

This book would not have happened without the combined experiences, support, guidance, and friendships during my eight years in Washington DC. Nor would it have been completed without the encouragement and support I have received here at the University of Illinois.

The single most important person through all of it has been my wonderful wife, Susan. Her patience and understanding, love and support have been critical and very much appreciated. She supported the high-risk career move that took us to Washington DC as newlyweds and put up with the long hours, travel, stress, and general craziness for eight years. She was also instrumental in our move to the University of Illinois. She is a wonderful wife, an incredible partner, and an amazing mother to our two children.

Professionally, I have been fortunate to have worked with many talented, dedicated people, many of whom I consider friends. There are, however, too many to properly thank and acknowledge individually. Instead, I will acknowledge the key teams and leaders that contributed to my education on farm policy and legislation. It begins with former senator Ben Nelson (D-NE), who took a chance on an unknown, unemployed lawyer in 2006. I was privileged that my learning on the job in the Senate was as part of a great team of Nelson staffers from 2006 to 2009.

Former Secretary of Agriculture Tom Vilsack also took a chance and appointed me to positions at USDA in the Obama administration—first as deputy administrator for farm programs at the Farm Service Agency and then as administrator of the agency. My two years at USDA were educational in many ways, and I gained much from Secretary Vilsack's leadership. I appreciate having had the opportunities and experiences, as well as the privilege of working with a great group of political appointees and career officials.

In early 2011 Senator Debbie Stabenow (D-MI) hired me to be chief counsel for her as chair of the U.S. Senate Committee on Agriculture, Nutrition, and Forestry. It was the most incredible professional experience during my time in Washington, from which I learned more than I ever expected. Senator Stabenow's leadership, tenacity, and legislative abilities during the difficult effort to write a farm bill provided lessons aplenty. I am honored to have been a part of the effort and proud to have been part of a great team of dedicated committee staff that worked together long hours, forged lasting friendships, and created great memories.

I want to acknowledge the contributions of my colleagues in the College of Agricultural, Consumer, and Environmental Sciences and on the *farmdoc* team. I appreciate the opportunity to work for this university and experience the world of higher education, research, and extension. My appreciation extends to the students who have taken my classes here and at the Vermont Law School. Thank you to all for being a part of my own learning process for teaching, researching, and communicating the results of that research.

Funding is no small issue in academia, and my work has benefitted from multiple sources. I have received funding from the USDA National Institute of Food and Agriculture, Hatch project #1002473, a portion of which can be considered to have contributed to the research and writing of this book. In addition, two generous endowments have contributed funding to my work: the C. Allen and Darren A. Bock Chair in Agricultural Law and the Gardner Agriculture Policy Program funded by the Leonard and Lila Gardner / Illinois Farm Bureau Family of Companies Endowed Chair / Professorship.

I want to thank Dr. David Orden at Virginia Tech and Dr. David Hamilton at the University of Kentucky for reviewing my manuscript. I appreciate the time and effort required, especially by Dr. Hamilton, who reviewed the revised manuscript. Thank you both for the great feedback and for helping direct me deeper into the history. The final product is much improved owing to their contributions. My work has also benefited substantially from many incredible conversations with Dr. Carl Zulauf.

I appreciate Bridget Barry and the team at University of Nebraska Press for helping guide me through this process. They generously answered many dumb questions and explained things to a first-time author. My appreciation for help in completing this book includes Julie Kimmel, the copyeditor. Finally, I want to thank Laurie Ristino at Vermont Law School for connecting me with Dr. Doug Hurt at Purdue University and Dr. Hurt for connecting me with Bridget.

THE FAULT LINES OF FARM POLICY

Introduction

Fault Lines and Farm Policy

Four wheels and steel cut through the prairie on one of Woody's ribbons of highway, and the traveler is struck by an ocean of green abundance unparalleled in human history.[1] Abundance may be great for the whole of humanity, but it can be a problem for the individual farmer working fields in competition with neighbors and farmers around the world.[2] This is the paradox of production that has haunted generations of American farmers with low crop prices. Low crop prices lead to depressed farm incomes. Depressed farm incomes jeopardize farmers and can cause problems in rural communities. American farm policy is rooted in this paradox. It has been carved by volatile crop prices. Seeking to address price problems brought together regional farming interests and formed the primordial political fault line; on this fault line, farm bills are written.

Contemporary commentators often consider eating a connection to nature and an agricultural act but lament disconnection between the eaters and the producers.[3] This is another paradox in farm policy: we all eat, but very few of us produce the food. A great chasm separates the consumer and the farmer. The same can be said for the taxpayer and the farmer receiving federal assistance. These chasms provide ample space for misinformation, misunderstanding, and political discord. These, in turn, often feed the criticism aimed at farm programs but fail to deliver more than limited impacts on the policy.[4]

It may well be that this is all in the process of changing. Recent years have witnessed a growing interest in food, farming, and the many varied strands connected to food and farming. Generically known as a food movement, this interest encompasses everything from local food production, organics, and farmers' markets to health and nutrition matters, the environment, and concerns about natural resources on an ever-more crowded, warming planet. Some of this movement is fueled by consumers becoming more connected to each other, online information, and interest groups, but not necessarily to the farmers. It is possible that the food movement will disrupt not only farming but also the politics and policies linked to agriculture.

Farmers generally grasp that the ultimate consumer, taxpayer, and voter is disconnected from farming and the production of food. Many farmers understand the pressures of the above paradoxes and the potential implications of the food movement. They are concerned that where disconnect leaves in its wake a lack of knowledge, it can cause misunderstandings that lead to misplaced policies with outcomes that do more harm than good. But they are also buffeted by cross pressures: calls to produce more to feed a growing and hungry world population; demands for sustainability and better environmental outcomes; concerns about the risks from markets and weather; and long-standing expectations for abundant, safe, and affordable food. And they receive assistance from policies that are themselves disconnected from much of farming and burdened by a long history.

The story of farm policy is the path cut by crop prices through history. The nearly hundred-year journey of farm policy began because crop prices were depressed by the weight of too many acres turned under the farmer's plow. America closed its frontier by planting it to wheat, right around the time that the world went to war in Europe. Farm policy was a by-product of the war. Prices peak and then fall back, time and time again. Each peak has produced substantial changes in policy; each fall has brought with it corrections to policies through periods of incremental change and slow evolution.

One theory underlying this book is that the farmer, consumer, and taxpayer cannot hope to have a meaningful conversation about food and farming if it is disconnected from history. Demands for reform divorced from history are blind and likely futile or counterproductive. Likewise, farmers and policymakers who support farm policies should proceed with a better sense of the burdens that history adds to the debate. This book is an attempt to tell the story of the farm bills, with the goal of providing a better, clearer understanding of farm policy and its development.

Reviewing the long path of development for American farm policy highlights a point worth remembering: history may not repeat, but it certainly recycles. Carved on the National Archives Building is the quote, "What's past is prologue."[5] One intention for the history herein is to provide perspective for future policy discussions, a perspective that may help development of more effective policy outcomes. A second famous quote serves as a warning: "Those who cannot remember the past are condemned to repeat it."[6] Farm policy history is full of mistakes, as any human undertaking would be. Reviewing it may help avoid repeating past policy failures and mistakes.

Policy is where politics, people, interests, and history come together to forge laws. History teaches many lessons, and some of them may have application for the future, especially as matters from history are recycled.[7] It can serve future debates and policy efforts through an understanding of where we have been, where we might be going (or should be going), and even what to avoid repeating.[8] Either way, if eating connects us to nature through farming, then farm policy connects us all to farming through politics and the expenditure of taxpayer dollars.

At its core, farm policy involves fundamental issues of risk. From its origins and through the first four decades, this was understood as the farm problem.[9] Commodities were oversupplied, and farmers were unable to align production and demand. They had no organized structure to do so, and more important, individual farmers were motivated to produce as much as they could on their farms, especially if other farmers cut production to raise prices. This free rider problem prevented farmers from operating as industry could, and they turned to the federal government for help.

It has never been just an overproduction issue. Basic farm commodities are produced around the world and have long been subject to restrictions, protectionism, and other market failures. Lost export markets can result from geopolitical problems far removed from actions on the farm, especially for commodities that rely heavily on exports. These two market challenges produce problems when oversupply builds up surplus and depresses prices. The challenge for farm interests, however, has always been wrapped in a basic reality: farmers are producing staple commodities used to produce food and clothing. Thus, the problem for the farmer can be a boon to consumers and industry if low commodity prices keep costs low as well. Food costs can be an inflationary problem in the larger economy.

Farm risk is made up of two fundamental matters that have always had consequences for farmers. The first are market risks; whether lost export demand or oversupplied markets, these risks return prices too low to cover cost and profit needs. Market risks can be long lasting, putting stress on farm management and economics across multiple growing seasons and with few good options in response. The second are weather risks, which arguably make farming unique in the policy arena. Good growing weather can result in massive crops that outstrip demand and lower prices. Bad weather can cause damage to crops that leave too little to cover costs and needs. This is also a systemic problem for society, albeit one that Americans have little familiarity with. We have not experienced a true food shortage but have seen its impacts and consequences around the globe. A nation that fails to feed its citizens will struggle to function and survive.

While generally systemic, these risks can be devastating at the individual farm level. The individual farmer occupies a very insecure position in the large commodities markets. That insecurity is magnified on a world scale. Farmers operate under the reality that these basic commodities are interchangeable. There is generally no difference between corn produced in one state or country as compared to another. The individual farmer has no control over the weather and no ability to affect the market price for the commodity being produced. To provide context for this discussion, consider a simple risk scenario.

A farmer with a 1,000-acre farm plants 500 acres in corn and 500 in soybeans. His is not a large farm; it is individually insignificant given that farmers in the United States planted over 90 million acres of corn and more than 90 million acres of soybeans in 2017.[10] At expected national average yields per acre of each crop, the farmer's 1,000 acres might produce 85,900 bushels of corn and 24,750 bushels of soybeans. Those totals are, however, only drops in the domestic bucket. According to USDA, U.S. farmers produced more than 14.6 billion bushels of corn and nearly 4.4 billion bushels of soybeans in 2017. Yet the 1,000-acre farmer likely paid nearly $300,000 to plant the crops and could be managing ground worth as much as $10,000 to $15,000 per acre or could have to pay $200 to $300 per acre in rent.[11] The costs add up quickly for a farmer, and fields are planted with seeds in the spring; revenue is unknown until after harvest months later. The farmer has no control over the weather and no ability to individually affect markets or prices. Farm policy was born of these realities; they exist regardless of what the policy outcome is and whether lawmakers get it right.

Policymaking and legislating are difficult, complex processes dominated by exercises in vote counting. The process and its rules channel policy concepts and disagreement about them through a transparent, equitable, and deliberate method of debate and decision. Farm policy, like all policies, is the product of negotiated political efforts in a large, diverse nation. Moreover, policymaking is an inescapably human and social endeavor. There is no purity in policy; outcomes are never perfect, and Congress will not produce a panacea. No one person, politician, or group gets everything that they want exactly as they want it. Policies and laws are the imperfect products of often-difficult negotiations among people with competing needs, views, interests, perspectives, and goals. Failure to compromise is failure to legislate; it is the death of policy and governance. Many great policy ideas reside in the dustbins of history or are detritus strewn alongside the rough roads through Congress.

The policymaking process favors coalition building as the method for accumulating sufficient votes. Building and maintaining a functional legislative coalition among different interests and demands can be difficult, requiring negotiation and compromise. The coalition is a source of both strength and weakness, located at the fault lines between the interests in the coalition. Thus, policies are built on fault lines, and the fault lines are the places where the forces of disparate interests, ideas, beliefs, and views come together to form a political coalition in order to produce legislation. Fault lines are subjected to intense pressures from inside and outside the coalition. Such pressures are largely the product of the issues and the times in which the legislation is written. What creates the coalition and what holds the competing forces together are the fundamental drivers of policy development.

Thus, coalitional fault lines are crucial to understanding the policies enacted into law. Fault lines are forged by the pressures threatening to tear the coalition apart and the adaptations that hold it together. Because they are the product of compromises, tracing a policy's fault lines and the pressures on them maps the history of the policy's developments and provides the best guess of what the future portends. Farm policy is no different. It is, arguably, an incredible example given that Congress has renewed it more than twenty times over the course of more than eighty years. It has failed, and it has been revised, reformed, and continued. It was controversial and faced strong criticism from its inception, but it has proved incredibly resilient. It now resides in omnibus legislation, which requires holding together numerous interests; it is consistently reconsidered, which requires maintaining the political coalition through changing times.

Throughout the history detailed herein watch for these basic policy development drivers. First, coalition building is necessary for policymaking and vote counting in Congress—necessary to overcome the status quo. Second, look for the external pressures that come to bear on the coalition and, especially, its fault lines. These external pressures can push the coalition to make changes to itself and to its policies. Third, the internal pressures that work on the coalitional fault lines have a large impact on policy development, for example, the path cut through history by crop prices and the way this path affects external and internal pressures, the coalitions, and the policy outcomes. These are the historical precedents worth keeping track of, and they will be discussed further in the concluding chapter.

A brief look at the Agricultural Act of 2014 begins this journey. The act was signed into law by President Barack Obama on February 7, 2014, and it expires on September 30, 2018.[12] Writing this farm bill, however, took nearly three years of debate and negotiation stretched over two different Congresses. It was stalled by partisan politics centered on the federal budget that nearly split the coalition apart. The 2014 debate thoroughly exposed the farm bill's two great fault lines. The first was the regional fight over the direction of farm policy that took place in the Senate. The other was an ideological, partisan battle over food assistance programs to low-income households (the Supplemental Nutrition Assistance Program [SNAP], formerly known as food stamps) that consumed the House of Representatives and nearly destroyed the bill.

The farm bill effort began in 2011 under intense pressure from deficit reduction efforts brought about by a new Republican majority in the House of Representatives that came to power in the 2010 midterm elections. Farm bill programs were a prime target for spending cuts leading the Senate and House Agriculture Committees to negotiate a compromise that laid much of the groundwork for the subsequent efforts.[13] The Senate passed a farm bill in 2012 over the objections of southern senators opposed to the changes in farm policy. House leadership refused to allow a farm bill to be debated on the floor over concerns about a difficult debate over SNAP spending in an election year. The Senate and House Ag Committees began again in 2013, with a southern power play in the Senate Ag Committee. The Senate managed to pass a revised farm bill, but the House proved difficult because controversial amendments to SNAP caused Democrats to drop their support. Combined with opposition from many Republicans affiliated with the Tea Party movement, the House initially defeated the farm bill. House leadership devised a two-bill strategy and muscled both bills through on party line votes over Democratic objections. Conference negotiations on the farm bill took place under the shadow of the federal government shutdown, itself a product of partisan fighting over health care and spending. Conferees finally produced compromise legislation in January 2014 that passed the House and Senate and was signed into law by President Obama.

The long, difficult process to get a farm bill in 2014 exposed the fault lines. It demonstrated the substantial pressures from budgets and partisanship on those fault lines. It raised significant questions about the policies being debated and the history behind them. Budget pressures resulted in elimination of billions in spending, which pitted the South against the Midwest in the Senate over farm policy and Republicans against Democrats in the House over SNAP. It took a major and embarrassing defeat in the House to force the bill through that chamber.

Disputes among the South, the Great Plains, and the Midwest over farm policy form the primordial fault line and have been the defining characteristic of the debate throughout its history. Because they formed the original farm coalition and have played the predominant roles throughout, this history concentrates on the three major bulk commodities: corn, cotton, and wheat. These three commodities also politically represent the three major production regions home to most of the influential policymakers.[14] They are not, however, the only commodities covered by farm policy, and as becomes evident, the farm bill has grown to encompass much more than assistance to farmers of a certain set of commodities. This history is unable to cover in detail all commodities and interests included in a farm bill.[15] These are limits of necessity given the scope and scale involved with nearly a hundred years of legislative and national history.

As a final matter, this book contains two appendixes for reference purposes. Appendix 1 is a compilation of charts that illustrate various relevant statistics and data provided publicly by the U.S. Department of Agriculture. The charts include data from the National Agricultural Statistics Service, the Economic Research Service, and other agencies within USDA. The information is available on USDA's website and the corresponding websites of each agency. Appendix 2 contains a chronological listing of all the bills covered in this history. It also contains a glossary of the terms used throughout the book, defining policy terminology and providing examples where appropriate.

1

The Origins of Farm Policy, 1909–1933

Introduction

History and policy do not offer much in the way of neat, precise beginnings; much is built out of what has come before. For American farm policy as we know it today—direct federal assistance for a select category of commodities—the clearest and most functional beginning point is May 12, 1933, when President Franklin Delano Roosevelt signed the Agricultural Adjustment Act (AAA) of 1933 into law. That bill, however, had roots throughout American history. The most prominent can be found tangled around World War I and policies formulated throughout the 1920s as farmers struggled in the postwar economy.[1] For the sake of an origin story, this history centers on 1921. That was the year two Illinois businessmen who ran the Moline Plow Company demanded policies that would provide "equality for agriculture" with direct federal intervention in the agricultural economy. Their argument distilled to the simple fact that as a business "you can't sell a plow to a busted customer."[2] Farmers had been busted by a confluence of events that caused overproduction and depressed prices.

Long-standing federal policies to encourage settlement had led to what was considered the closing of the American frontier. Much of that settlement took place in the Great Plains. Vast lands were opened to farming, and almost all of them went into wheat. World War I supplied strong demand and high crop prices that further drove expansion of farm land. Wartime demand, however, also accelerated a long trend toward industrialization and urbanization that drove up the cost of goods. When the war ended, farmers kept producing, causing a glut of commodities that depressed prices. The cost of goods continued to climb as did the rest of the economy in the Roaring Twenties. Farmers were caught with high costs, low prices, and depressed incomes.

What a twelve-year political struggle could not accomplish, the Great Depression and the election of Franklin Delano Roosevelt did. The new president and a compliant Congress were aided by a farm coalition consisting largely of the three major commodities: corn, cotton, and wheat. These three commodities represented the major crop production regions in the Midwest, the South, and the Great Plains (respectively). The long, difficult, and ultimately unsuccessful efforts in the twenties had forged a powerful political coalition that backed the new President's efforts to combat the Great Depression. Together, they produced the Agricultural Adjustment Act of 1933—the first farm bill.

The 1933 act gave the federal government a direct, centralized role in the affairs of American farming—a change largely demanded by farmers. Farmers had fought to create a primary federal role in limiting production to control supplies in exchange for increased prices for the fruits of farmers' labor. They asked the federal government to do for them what they were unable to do on their own. In part, the idea was to replicate what manufacturers and other businesses could accomplish by cutting production, laying off workers, or even closing facilities. Farming was seen as different. When the price of farm commodities fell, millions of individual farmers made individual decisions to try to produce more on their land, and these decisions when multiplied many times over made the problems that much worse. Individual farmers were reluctant to cut production for fear that their neighbors would keep producing and capture any price increases—a basic free rider problem. For farmers, the overriding issue was production that outpaced demand and depressed prices. This issue became known as the "farm problem." Farm policy was created to address it.

As will be seen, the development of federal farm policy fundamentally altered the relationship between traditionally independent farmers and the federal government they often viewed with contempt; in desperate times the farmers abandoned unprotected independence for policies that had once been the concept of radicals. The effort to write the first farm bill was difficult and produced questionable, controversial results. Policies born of an emergency quickly took on a different political life and character. The New Deal era produced no less than six different statutes to combat the Depression in agriculture, but these policies drew sharp distinctions among the interests in the coalition, making the fault lines more prominent and politically challenging. The statutes ignited an intense regional fight over farm policy that would be waged in full during the decades after World War II.

From Golden Era to Postwar Collapse (1909–21)

Before World War I, the American population grew faster than farm production.[3] This growth created strong demand for commodities, and that demand increased prices. Prices increased more than the cost of goods farmers purchased. As a result, farmers experienced strong incomes, and the prewar years from 1909 to 1914 were considered the golden era for agriculture. The golden era did not last long.

The year 1914 witnessed the end of federal homesteading policies and marked what is known as the closing of the American frontier.[4] The end of homesteading policies and the closing of the frontier also ended the availability of free or very cheap land. Land values quickly increased, some through speculation. Increased land prices caused many landowners, including farmers, to take on more debt. Importantly, America closed its frontier by planting it to wheat. Homesteaders were in the process of breaking up twenty-five million acres of native prairie and turning it over to produce wheat.[5] Located in a windy, semiarid region of the country, much of that land was only marginally suited for row-crop farming.

In addition, in 1914 Congress passed the Smith-Lever Act, which created an extension service to help farmers improve farming and increase production through new techniques and farming practices.[6] The extension service itself dated to the Morrill Act of 1862, creating the land-grant universities, and the Hatch Act of 1887, creating agricultural experiment stations.[7] The use of demonstration farms traced to the William McKinley administration and the work of Seaman A. Knapp in the late 1890s in response to demands from cotton farmers for help in combatting the spread of the boll weevil.[8] Counties began to provide funding for county agents in 1906, and the practice spread quickly as the scope of farm demonstration also expanded; Congress provided appropriations and the U.S. Department of Agriculture (USDA) followed via cooperative arrangements with state agricultural colleges for managing demonstration and education projects.[9] World War I spurred further action to help farmers increase production to meet wartime needs. Among other things, the Food Control Act increased the number of county farm advisors and extension agents throughout the country and helped solidify extension's place in American agriculture.[10]

The war effort also fueled efforts to organize farmers at the county level and through farm bureaus.[11] In many counties an important function of the farm bureau was to raise funds through memberships and provide governance via county boards of supervisors. The effort needed a more definite organization, and it expanded beyond educational and demonstration purposes to cover a broad array of issues facing farmers, including business and legislation. National meetings of the county farm bureaus in 1918 and 1919 led to the formation of the American Farm Bureau Federation, which quickly became the largest and most politically powerful farm organization, in Chicago in 1920.[12] By the time of the Great Depression and the New Deal's attempts to assist farmers in the 1930s, both the Farm Bureau and the extension service created by the Smith-Lever Act were considered "strongly entrenched" in farming, connecting the federal government with individual farmers.[13]

The farm problem, however, was predominantly the by-product of World War I. On June 18, 1914, the assassination of Archduke Franz Ferdinand of Austria sparked war in Europe. As part of its response, the American government encouraged farmers to expand production so that wheat could win the war—and the farmers did. Wheat acres increased from an average of just under fifty-one million during the years 1909 to 1913 to approximately seventy-seven million in 1919.[14] The demands of war created an artificial market for American farm commodities, but the war itself devastated U.S. agriculture's most important export markets in Europe. The war also transformed the U.S. from a debtor nation to the world's single biggest creditor, particularly as compared to the major European nations that borrowed to finance the war.[15] These changes affected trade—none more than the decision by the conservative U.S. government to cease lending to European nations in 1919 and demand repayment from the war-torn nations.[16] This move was an additional blow to the European market for U.S. commodities and damaged American farm exports; it coincided with the peak in wheat acreage.

Land that has been purchased, often on credit, and plowed under to produce a crop does not come out of production easily or quickly; farm production does not readily adjust to changes in the broader economy. Postwar events had little impact on farm production decisions; wheat in the ground had to be harvested. Farmers were left with no options other than to sell into already flooded markets. Prices were severely depressed, which damaged the farmer's purchasing power while magnifying the consequences of large debts on land purchased to expand production for the war effort. At the time, the 1920 crop was the most expensive ever planted, and it produced what was then the second largest crop in history.[17] Commodity prices collapsed at harvest, and heavily indebted wheat farmers were in trouble. According to Murray R. Benedict, by 1921 "American agriculture found itself in a more unfavorable position than it had experienced at any time in the memory of men then living, or possibly at any time since the nation's beginning."[18]

Forging the Coalition on Farm Policy (1921–29)

In their struggles with collapsed crop markets and depressed incomes, farmers demanded federal assistance. The demands for farm policy first arose in the region of the country that grows spring wheat, which runs from western Minnesota through Washington and Oregon. Pressure began to be felt in the capital. In the spring of 1921, Secretary of Agriculture Henry C. Wallace called a meeting of farm leaders in Washington DC.[19] At the same time, members of Congress from farm or rural districts got together to form what was known as the farm bloc to work on behalf of their farmer-constituents. These efforts quickly led to the development of federal policy concepts that would help farmers. Initially, these concepts revolved around fixing commodity prices or using federal purchases of surplus farm commodities that could be dumped in export markets.

The most influential farm policy proposal of the twenties came from George Peek and Hugh Johnson of the Moline Plow Company in 1922.[20] Titled "Equality for Agriculture," their proposal also sought federal intervention in the farm economy.[21] Peek and Johnson took aim at American tariff policy, alleging that it created inequality for farmers by protecting domestic industry from lower-priced imports. Tariff policy was unfair to farmers because it increased the cost of manufactured goods, many of which the farmer had to purchase to farm. Peek and Johnson also argued that, unlike industry, farmers could not regulate production to domestic demand; farmers too often produced surplus. In response, they proposed that federal policies reestablish a fair exchange value for farm products, which they defined as the cost to the farmer for producing the crop plus a profit, known as the "cost of production" measure. The Peek-Johnson concept also called for the federal government to purchase domestic commodity surpluses. The surplus commodities would then be sold in export markets with a tariff to protect the domestic market. The entire effort would be financed using an "equalization fee" that would be charged back on producers of the commodity at a rate to cover the loss on the export sales.

A regional, commodity-based coalition began to form around the Peek-Johnson proposal. The farm problem was spreading eastward from the wheat regions into the midwestern Corn Belt during 1923 and 1924. Secretary Wallace added his support to the proposal, apparently contradicting his boss, President Calvin Coolidge.[22] In 1924 the proposal was turned into legislation drafted by USDA officials and introduced in the House by Agriculture Committee chairman Gilbert N. Haugen (R-IA).[23] Senator Charles L. McNary (R-OR) introduced it in the Senate, and the bill became known as the McNary-Haugen legislation. The specifics would evolve over multiple bills from 1924 to 1928 but in general tracked the original Peek-Johnson concept. The bills sought to establish a government corporation that would purchase farm commodities in order to increase prices to an equitable level; the commodities could then be sold to domestic buyers at the equitable level. The corporation would sell any surplus in export markets at world prices, using tariffs to protect domestic markets. Farmers would be assessed an equalization fee to cover the costs.

The McNary-Haugen legislation began with little political support in Congress. It was defeated on the House floor in 1924 with strong opposition from southern and eastern representatives.[24] The South had yet to experience the farm problem because cotton and tobacco prices remained high thanks to reopened European markets. Support for the legislation built quickly, however. The American Farm Bureau Federation elected a president from Illinois in 1925, which changed that organization's position.[25] By 1926 farmers in the southwestern states had expanded cotton acres, especially irrigated lands in Texas and California. The expansion contributed to a collapse in cotton prices as farmers added nearly ten million cotton acres from 1922 to 1925.[26] Southern interests negotiated changes to the bill to add their support.[27] The legislation now covered the three main commodity and political regions: the South, Midwest, and Great Plains. While strengthened, the coalition was not yet strong enough to get a bill through Congress. Farm interests responded by applying pressure in congressional elections. Congress passed the bill in 1927, but President Coolidge vetoed it. This effort was repeated in 1928, when Congress again passed McNary-Haugen, but it was again vetoed by the president. The Senate failed to override the veto.

The Senate's failure to override President Coolidge's veto left the impression that McNary-Haugen was dead politically; the presidential election would soon confirm it. Farm policy supporters failed to derail the Republican nomination of Herbert Hoover, but they did force him to begin acknowledging the farm problem. He promised to push Congress on legislation, but he opposed McNary-Haugen. The Democratic nominee had been more responsive to the farm interests and more receptive to McNary-Haugen, but Hoover won in a landslide. Hoover carried every state in the Wheat and Corn Belts, meaning his opposition to McNary-Haugen did not hurt him with farmers at the ballot box.[28] Adding to the defeat, Republicans increased their House and Senate majorities. The dissonance between lobbying for farm relief but voting for candidates opposed to it was conclusive, and the election results were the end of McNary-Haugen policy. The long fight to get a bill enacted had failed, and farm interests had to reconsider their push for federal assistance.

Farm concerns had apparently registered with the new president, however. President Hoover called a special session of Congress to help farmers and pushed through a program to provide loans to agricultural marketing cooperatives in 1929.[29] The program also created "stabilization corporations" to help with surplus commodities. Hoover signed the bill into law in June. It had taken eight years and multiple bills to get a law on the books that was intended to help farmers. The final product, however, was one favored by business and industry instead of the farm organizations. In retrospect, this may have been somewhat of a blessing in disguise for farm interests and farm policy given what was on the horizon. The 1929 act barely had time to get started before the stock market crashed in late October. That crash took the national and world economies with it, igniting the Great Depression. The 1929 act completely failed to help farmers; the Great Depression overwhelmed its meager authorities. Arguably, the failure of the 1929 act coupled with the long fight for assistance throughout the twenties set the stage for the efforts that were to come next.[30]

The farm problem's persistence and its expanding national scope set in motion the political forces behind farm policy. Farm interests organized; they learned from failure, adapted and gained political sophistication. Most important, they built a coalition that spanned vast regions of the country and brought together three of the major commodity and farming interests. Dairy and livestock concerns were also involved. This geographical reach provided a broad political base that would help congressional vote counting. The creation of that coalition also created the primordial fault line of farm policy. From the beginning this fault line was subject to competing interests and political forces, most notably between cotton and corn (the South and Midwest).

These two commodities existed in very different market realities. Half of all cotton grown was exported; almost all the corn grown was used domestically and most of it to feed livestock. At the time livestock feeding generally took place on the farm where the corn was grown or on neighboring farms in local communities. This was an era before the large concentrated animal feeding operations that pull feed from far away. The McNary-Haugen concepts posed real difficulties for cotton interests. Dumping surplus commodities in overseas markets created major problems for an export commodity that would be compounded by the equalization fee to cover the costs of that dumping. Tariff policies also jeopardized export markets. Wheat further complicated the coalition because it was divided north to south in the plains, between spring and winter wheat production, and relied heavily on both exports and domestic uses. Desperate times finally brought these commodity interests together to fight for legislative solutions, but the scope of the problems also threatened to tear them apart.

The farm coalition faced formidable opposition from consumer, industrial, and political forces.[31] The national economy had been doing great throughout the Roaring Twenties, and policies to increase commodity prices could increase the cost of food to consumers. Farmers stood alone advocating for policies that might harm everyone else. Farmers were also running against the prevailing political winds of the time, which were dominated by a conservative ideology driven largely by business and industry interests. These interests opposed centralized governmental action and interference in the marketplace, except for tariff policies that protected their markets but harmed farmers. Business and industry were not particularly interested in increased commodity prices. Conservative business interests consistently defeated farm interests in the political arena. They were the force behind President Hoover's 1929 act, which became law over the McNary-Haugen legislation preferred by farm interests. The Great Depression, however, changed everything.

Great Depression Produces New Deal Farm Policy (1930–33)

The Great Depression tore a hole in the increasingly complex fabric of the American economy and sent waves of panic throughout the nation.[32] Benedict wrote, "Mere figures can give no adequate picture of the gloom and despair that gripped the country."[33] Farm policy provides a lens through which to understand the devastation wrought by the market crash and the onset of the Depression. For much of the twenties, farm interests had fought for policies once deemed radical but failed to overcome the insurmountable hurdles put up by industry and the reigning conservative ideology. The Depression eliminated those barriers and swiftly changed the political calculus, as well as the politicians in charge. What resulted were drastic policy changes as radical as anything defeated in the Roaring Twenties. Writing fifty years afterward, Harold Breimyer noted that the 1933 AAA "erased for all time the rural-agrarian heritage of a circumscribed role for government" and replaced it with "an urban-industrial commercial conceptualization and policy design."[34] His statement summarizes the magnitude of the tectonic force that was the Great Depression.

The Great Depression hit American agriculture hard. Farm income was cut by more than half, and farmland values collapsed; between 1930 and 1934, nearly a million farmers lost their farms. With the crash, "breadlines appeared in the cities and farmers were burning corn in place of coal because it was cheaper. . . . Radical movements of a kind scarcely known in earlier years began to spring up" in the Midwest.[35] The Depression produced political unrest throughout the American countryside the likes of which had never been witnessed. Congress, pushed by business interests, made it even worse with the Hawley-Smoot Tariff Act of 1930.[36] That act caused other nations to enact domestic protection measures that blocked U.S. exports, dried up trade, and took commodity prices to new lows. President Hoover was incapable of any adequate response, and in the 1930 midterm elections, American voters turned Congress over to the Democrats.

The 1929 act exemplified President Hoover's inability to respond to the Depression. It included authority for the Federal Farm Board to create stabilization corporations to purchase commodities, but in operation the stabilization corporations sold the purchased commodities back into the market and further depressed prices. The Federal Farm Board ran out of money and was defunct by 1932.[37]

Cotton had been the last of the major commodities to struggle with low prices in the twenties, but it was the hardest hit by the Great Depression. In 1931 American cotton farmers produced the third largest crop in history, but the market was already storing enough cotton to supply more than half of a year's demand.[38] President Hoover, shackled to his traditional, conservative ideology, could only beg cotton farmers to plow under a third of their crop.[39] The federal government's ineffectiveness left a vacuum; in desperation and despair, farmers grasped for once-radical solutions. The early years of the Great Depression were pivotal in forming a key but controversial element of farm policy. The policy focused on limiting or controlling the supply of commodities and dominated farm program debates for decades with ramifications far beyond the farm gate.

Beginning in the late 1920s, economists at universities and USDA developed an alternative policy for helping farmers combat depressed crop prices.[40] Their concept was based on centralized control over farm commodity production that made use of limits on the acres planted to the major crops and became known as the domestic allotment plan. This concept would influence the thinking of those who were forming a new vision for the broken nation, including the governor of New York, Franklin Delano Roosevelt. The policy was controversial. During the efforts of the 1920s, farmers and their interest groups often split over whether to restrict planting or output.[41] The concept combined with growing desperation in the Cotton South. Cotton farmers were offered few solutions. The president and others tried to convince them to voluntarily decrease acres planted to cotton, but that was adding insult to injury. Ineffective begging was soon followed by a more radical idea known as the Cotton Holiday or drop-a-crop.[42]

In 1931 the governor of Louisiana, Huey P. Long, picked up this heretofore radical idea and ran with it. He called a conference of the governors of each of the cotton-producing states and, using his force of personality to capitalize on the desperate times, pushed the delegates to overwhelmingly adopt his plan for a mandatory restriction on cotton plantings for 1932.[43] The plan faced opposition from the ginning industry and raised concerns that it would serve only to help foreign competitors; it never went into effect because Texas refused to go along.[44] Governor Long's efforts advanced in five southern states before Texas ended the effort. Although it failed in 1931, the concept altered the political landscape for farm policy.[45] For one, it demonstrated just how powerful the desperation for better prices could be in overcoming long-standing resistance to radical policies. It provides a powerful example of the Great Depression's impacts.[46] Southern farmers underwent a complete philosophical reversal and advocated for mandatory production controls.

Southerners assumed the lead role in farm policy and the farm coalition. In 1931 the American Farm Bureau Federation elected the Southern Democrat and cotton farmer Edward A. O'Neal as its president.[47] O'Neal steered farm policy away from the commodity purchase concepts of McNary-Haugen and toward the acreage-based production controls of the domestic allotment plan.[48] The southern ascension would quickly converge with an even more powerful political movement in Democratic presidential candidate Franklin D. Roosevelt.[49] Candidate Roosevelt and his political advisors worked with farm interests and academics to develop his farm policy positions, which evolved into a somewhat vague endorsement of the domestic allotment plan.[50] Roosevelt's comments were seen as crucial to helping swing farm support away from Hoover. FDR's campaign moved the Democratic Party away from "ethnocultural" matters toward economic matters, especially for farmers.[51] Doing so helped unite a powerful farm-labor coalition; FDR collected farm votes in the South and West to offset some weaknesses in the East with business and finance. Farmers helped Roosevelt, who, in turn, helped advance the policy efforts (and membership) of the farm organizations. FDR crushed President Hoover in the November 1932 elections, and his coattails expanded Democratic majorities in Congress—a "sweeping victory" that provided a political mandate for vast changes.[52] The New Deal dawned over the American political landscape.

Roosevelt's victory and his New Deal were decisive for farm policy but also for bringing about a shift in favor of southern cotton interests. The result would be heavily interventionist policy that put the federal government in charge of trying to control farm commodity production and planting decisions. Efforts to help farmers began in the lame duck session after the 1932 elections but carried over into the new Congress and administration.

First, the Ag Committees pushed through a bill to use remaining funds from the Reconstruction Finance Corporation to provide loans to farmers for the 1933 crop year.[53] Notably, the House added provisions to the legislation that provided authority for the secretary of agriculture to require acreage reductions. The reductions were an optional requirement to get the loan but exempted "the farmer, tenant, or share cropper who in 1932 planted not more than a minimum acreage of such crops."[54] The bill reflected the thinking that was evolving into the Roosevelt farm relief proposal, which the House was also working through at the time.[55]

Led by the Farm Bureau, agricultural interests pushed the House Ag Committee and Chairman Marvin Jones (D-TX) to produce a farm relief bill based on the domestic allotment plan. A version passed the House in the lame duck session after the 1932 election.[56] The Republican-controlled Senate, however, let the bill die, and the issue was handed off to the new president and Congress.

The provisions that would be included in the 1933 AAA had not been formed anew at the time of the 1933 debate. The basic proposal was developed with farm leaders and the agricultural advisors to president-elect Roosevelt, many of whom would go on to top positions at USDA. The concepts were crafted into legislative text by experienced policy, legislative, and legal experts from USDA, Congress, and the American Farm Bureau Federation.[57] The overriding focus was on providing USDA the power to reduce supply by leasing farmland out of production. This was the concept that had been developed and advocated by economists as the 1929 act's Farm Board collapsed under the weight of surplus and the Depression.[58]

The domestic allotment plan was different than the McNary-Haugen concepts, and it represented different farm coalitional dynamics. Corn was grown nationwide and was almost completely consumed domestically as livestock feed. Acreage-based production controls (allotments) would have an entirely different impact for corn farmers and livestock feeders. Cotton, the industrial commodity largely exported, possessed a smaller acreage footprint that was limited regionally to the South. Democrats dominated in the South and in Congress, especially on the Ag Committees. Influence in the new administration, the Farm Bureau, and Congress allowed cotton interests to dictate farm policy despite their having far fewer acres and a different market situation than corn. This influence added stress on the fault lines of the farm coalition, but the emergencies and urgencies of the Great Depression initially kept potential problems in check.

On January 25, 1933, Farm Bureau president O'Neal warned the Senate Ag Committee, "Unless something is done for the American farmer we will have revolution in the countryside within less than 12 months."[59] Farm groups mostly united, promised political protection, and produced results. With their support, President Roosevelt appointed Henry A. Wallace of Iowa (son of Henry C. Wallace) to be his secretary of agriculture. Secretary Wallace immediately went to work with farm groups on relief legislation.

All attention turned to the new Democratically controlled Congress and its Southern Ag Committee chairs: Marvin Jones (D-TX) in the House and Ellison D. "Cotton Ed" Smith (D-SC) in the Senate. The two chairmen were very different. Chairman Jones had grown up on his father's small subsistence farm in the Texas panhandle. He had wanted to escape the drudgery of "chopping cotton beneath a blistering Texas sun."[60] He went to law school and eventually won a seat in Congress in 1916, along with future Speaker of the House Sam Rayburn. He would rise to chair the House Ag Committee and become a key player in New Deal farm policy, working with the Roosevelt administration and leading efforts to help farmers and tenants. He was considered "one of the best managers of a bill on the floor of the House" because "he did his homework."[61] In 1940 Jones accepted an appointment to the U.S. Court of Claims and retired from the House.[62]

Chairman Smith was known more as a political showman than as a legislator. He had been raised on his family's plantation near Lynchburg, South Carolina.[63] Chairman Smith had helped organize the Southern Cotton Association and campaigned riding on a bale of cotton, but he had voted against McNary-Haugen because he "felt it discriminated against cotton."[64] He also held unfortunately conventional southern views on race, segregation, and white supremacy, views that grew more strident and demagogic during the thirties.[65] One historian described him as a "senator of ordinary ability who remained in office too long" with the "modest accomplishments of his earlier years" having been "obliterated" by his racist image in his later years; he died in office after losing in the 1944 primary.[66]

Secretary Wallace and his team at USDA worked hand in hand with the Farm Bureau to draft a bill that President Roosevelt sent to Congress on March 16, 1933. Chairman Jones and the House Ag Committee worked through the weekend to report the bill on Monday, March 20, 1933.[67] The House overwhelmingly passed the farm bill on March 22, 1933, six days after the president's message to Congress.[68] With passage, the House had quickly embarked on Roosevelt's "new and untrod path" seeking to address "an unprecedented condition" through "the trial of new means to rescue agriculture."[69]

The Senate was not as easy. The president's "new and untrod path" encountered a philosophical fight between the southern-dominated Farm Bureau and the populist, plains-dominated National Farmers Union (NFU).[70] The fight began in the Senate Ag Committee and carried to the Senate floor. Senators backing the two sides largely disagreed over how to calculate price-support levels and whether the federal government should have the authority to control planting and production decisions.[71]

For price supports NFU and its allies demanded the use of a cost-of-production calculation.[72] Senator George W. Norris (R-NE) led the fight for NFU's cost of production and won an amendment in the committee to provide the secretary an option to use it. Secretary Wallace, the Farm Bureau, and southern senators combined to oppose him.[73] They all wanted to peg supports to the 1909 to 1914 base period, using a ratio of commodity prices and the cost of goods.[74] Doing so provided a fixed reference point for the policy; assistance to farmers going forward would be clearly measured on a particular historic scenario.

Secretary Wallace, the Farm Bureau, and the southern senators were concerned that using cost-of-production calculations would not provide that fixed reference point because it was tied to unknown (and unknowable) future outcomes. In short, these calculations were oriented toward the unknown risks instead of toward the known past. This was an important matter because without a fixed level for supporting prices, the secretary, the Farm Bureau, and southerners were concerned that lower prices would continue. But this concern was more significant for cotton and its southern supporters.[75] For cotton, the cost-of-production calculation was applicable only to that portion of the crop consumed domestically. Because much of cotton was exported, this calculation would produce a lower support level.[76]

The strongest point of disagreement, however, was over production controls. The cost of production was very similar to the fair exchange value. Calculating price-support levels occupied a central role in the debate, but it arguably should be seen as more of a proxy battle; the real fight was over reducing acres to boost prices. This would become the common battleground for farm policy in future debates. In retrospect the 1933 Senate debate has all the indicators of a harbinger for farm policy. Even in the midst of the Depression's urgency, farm interests battled with each other over these intertwined but fundamental matters. Central control over planted acreage and the methods for calculating price-support mechanisms constituted the defining aspects of the fault line from the beginning.

For cotton interests such as Senator John H. Bankhead (D-AL), the point of the 1933 bill was to increase crop prices, which meant reducing acres.[77] The NFU perspective represented by Senator Norris strongly opposed acreage reductions, the centerpiece of the cotton-led policy. These two perspectives were also caught up in disagreements over which farmers to help: the planter class in the South or the small, family farmer of the populist plains.[78] While the Senate debate dove into technical policy specifics, the larger disagreement about what the policy should achieve loomed. Southerners wanted policy that would reduce acres to increase crop prices but that would also put a fixed support floor under prices.

The National Farmers Union represented the first stage of opposition to this viewpoint. It opposed acreage reductions and the complex fair exchange price-support calculation.[79]

Senators were unable to settle the debate, and President Roosevelt personally intervened.[80] His efforts were insufficient to bridge the fault line but were sufficient to move the bill; the issue was kicked to the conference committee.[81] Conference negotiators were also unable to reach agreement, and they sent cost of production back to the House separately for a vote. The House defeated it, which signaled to recalcitrant senators that a farm bill would not pass Congress with that provision included. The next day the Senate voted to pass the farm bill conference report without it.[82] President Roosevelt signed the Agricultural Adjustment Act of 1933 into law on May 12, 1933.[83]

The intensity of the congressional debate in some ways reflected the exploding unrest in the countryside. Radical agitation organizing efforts were prevalent throughout farm country, mostly under the banner of the Farmers Holiday Association. The Farmers Holiday Association focused mostly on farm mortgages and pushed the idea that farmers should withhold all farm production from markets until they got a price equal to the cost of production. In places the movement used threats of violence to prevent foreclosures and "penny sales" to retrieve farms from public auction.[84] The latter was collective action in response to farm mortgage foreclosures. Neighbors would take over bidding at a foreclosure sale, purchase everything for a penny, and leave it all for the bankrupt farmer; the banker left almost empty-handed. The penny sale served as a powerful anecdote regardless of how widespread its usage.

The message had to have gotten through to Washington DC. President Roosevelt signed the 1933 AAA into law one day ahead of the proposed strike by the Farmers Holiday Association, and the bill helped calm the political unrest.[85] It had been a twelve-year effort to get a direct farm relief law on the books; it culminated in broad authority for the secretary of agriculture to intervene in the farm economy.

This first farm bill was foremost emergency legislation, designed with the goal to "reestablish prices to farmers."[86] The directive was to "give agricultural commodities a purchasing power with respect to articles that farmers buy."[87] The vast, flexible powers included the authority to reduce acreage or production in exchange for rental or benefit payments. These provisions were meant to achieve the goal of improving prices by balancing production and consumption. They applied, however, to only the select group of basic, bulk commodities. The bill went further still. It also provided the secretary of agriculture the authority to levy a processing tax to cover the costs of the payments and mortgage relief for farmers— extraordinary powers delegated to an unelected official within the executive branch.

Secretary Wallace and USDA had to move fast to put the 1933 act into action. Among other things this required setting up a new bureaucratic infrastructure to operate the programs and deliver assistance to farmers.[88] The first efforts were drastic and controversial. The Department of Agriculture paid cotton farmers to destroy ten million acres of growing cotton to avoid further depressed prices and also instituted a plan to slaughter millions of small pigs and pregnant sows to try and avoid a glut in the pork market. Given the Depression, USDA also decided to convert them to salt pork for distribution as food aid.

The 1933 AAA's unprecedented authority marked "a major turning point in the philosophy of American government."[89] Gilbert Fite wrote that "there was nothing like it in all of American history" because "millions of individualistic farmers were welded into a great co-operative effort through the vehicle of government benefit payments."[90] The extreme actions taken immediately by USDA emphasized that point, as did Secretary Wallace: "To have to destroy a growing crop is a shocking commentary on our civilization. . . . Plowing under of 10 million acres of cotton in August, 1933, and the slaughter of 6 million little pigs in September, 1933, were not acts of idealism in any sane society. They were emergency acts made necessary by the almost insane lack of world statesmanship during the period 1920 to 1932."[91] Wallace's statement was justification for drastic measures loaded with controversy. The secretary was confronted with negative reaction from a public suffering the Great Depression. There was, however, more to the story of the 1933 AAA, and it drastically complicates matters.

New Deal Collides with Reality

The Great Depression was unparalleled in the nation's history; it stands to reason that the responses to it would be as well. An emergency situation, the Great Depression required urgent measures that were bound to produce unintended consequences and problems. Substantial concerns both at the time and in retrospect have focused on the particular consequences of acreage reduction policy on poor, and often black, tenant farmers and sharecroppers. As noted at the time, "That the Administration should have chosen the cotton South, with all its deep-seated antagonism of a decaying feudalism, for its first experiment in crop control seems surprising."[92]

Importantly, the bill and the agency created to implement it highlighted that the central focus of the policy was on adjustment of agriculture. Cotton had a particularly difficult problem known as "surplus labor"; too many small, inefficient farmers were stuck in outdated production.[93] The South was considered home to a "great majority of low-income, underemployed farmers," most of whom often existed "in an environment of stark poverty."[94] Southern agriculture in general and cotton in particular were considered "woefully backward" in terms of mechanization and equipment.[95]

Tractors were available but were in limited use, and the mechanical cotton picker would not be introduced until 1942.[96] Cotton was mostly planted, tended, and harvested by hand, leading to a conclusion that the system limited productivity and efficiency, but this production system for cotton also contributed to the slowness with which the South adopted mechanization and technology.[97] The bottom line was that adjustment, modernization, and mechanization necessarily meant that a lot of farmers would no longer be involved in farming. There was not room for hand labor in a modernized production system. This conclusion would be controversial in any setting, but it was magnified in cotton because it had drastic impacts for small, poor, and mostly black-operated farms.[98]

When the AAA called for the plow-up and acreage reductions in cotton country, planters and landlords often rented the land farmed by poor farmers to the federal government and kept the federal payments for themselves.[99] The attempts to adjust cotton farming through modernization, mechanization, and fewer acres concentrated the problems. A "great wave of evictions" swept up tens of thousands of poor, mostly black, farmers and was arguably the thorniest of the 1933 AAA's problems.[100] Researchers found that many of the evicted families were left homeless, trying to survive "on the rivers in flatboats, in the coves and swamps, on barren hillsides, and on the roads," or they were driven into cities and towns, falling further into destitution and onto relief rolls.[101]

At the heart of the matter was the sharecropper system in the South that was used mostly in cotton and tobacco.[102] Sharecropping was not a traditional or common landlord-tenant relationship; it was a specially designed wage labor system.[103] Under the sharecropping system, "a landlord or credit merchant furnished housing, land, seed, fertilizer, and work stock."[104] The sharecropper generally provided only the "labor incident to the production and harvesting of cotton" and was typically considered a liability to the landlord until the harvest was completed.[105] The sharecropper and his household were "paid with a share of the crop value at the end of the season," which was partially intended to guarantee "the availability of the household's labor for the downstream harvest."[106] The landlord typically handled all sales of the harvested cotton, returning "only what may be due [the sharecropper] after deductions for advances in the form of 'furnish,' interest and any other indebtedness of the cropper to the landlord has been taken out."[107]

Functionally and legally, this arrangement made the sharecropper a wage laborer and not a tenant; wages were paid from the harvested crop after all debts to the landlord were repaid. By comparison, cash tenants paid a specific amount to rent the land and were entitled to the crops they grew on it. Share tenants also rented the land but paid rent through a sharing arrangement with the landlord, such as paying rent in kind from the crop produced. Unlike share and cash tenants, sharecroppers, day laborers, and wage hands possessed no legal rights in the land or the crops they helped produce on it.[108] The nature of this relationship placed the landlord firmly in control; the crop and the land on which it was grown belonged to him.

The sharecropper was not only an "employee of the landlord" but one who had borrowed to produce the crop that would pay the wages.[109] Sharecroppers could remain in debt year over year under often-excessive interest rates that could range from 40 to 110 percent with little recourse against the landlord.[110] The share-cropping system dominated the cotton-producing region of the southeastern U.S. where plantations were strongest, as were problems of discrimination and unequal treatment in the Jim Crow era.[111] Comparing the idealistic ambitions of the New Deal with these real-world challenges has raised substantial consternation about how federal policies helped white landlords consolidate cotton farms at the expense of black tenants and sharecroppers.

The 1933 AAA was designed to work through agreements with producers to rent acreage out of production.[112] Congress, however, failed to define the term "producer" and did not provide any guidance to USDA as to who could enter an agreement and receive the benefits of the bill or the programs initiated under its broad authorities.[113] Neither landlords nor state and local laws considered sharecroppers to be actual producers or farmers; they were laborers, not tenants.[114] In light of the sharecropper system in the South and the extraordinary measures taken for cotton, this would turn out to be a consequential omission.

In the final legislative text for the 1933 AAA, Congress left the matter to USDA as Secretary Wallace had suggested was the policy decision when he answered questions in a hearing with the Senate Ag Committee.[115] When the bill went to USDA for implementation, however, the "cotton program was put into effect with such speed that little thought was given to the special problems raised by farm tenancy in the South."[116] USDA designed a contract for the 1933 plow-up that permitted only landowners to sign the contracts, although they were instructed to divide the payments appropriately.[117] It was later discovered that before Congress passed the 1933 AAA, a Washington attorney had warned President Roosevelt and Secretary Wallace that the cotton program could cause severe harm to sharecroppers and tenants unless the cotton con-tracts protected them.[118] Much attention has focused on the work of the Cotton Section of the AAA, which was run by men who knew the southern system and believed it would be disturbing if payments were made to tenants and sharecroppers directly.[119]

The Cotton Section officials had a difficult job and were faced with practical challenges. They had to convince cotton farmers to plow under ten million acres of growing cotton and had very little time to accomplish it. They needed large landowners to comply if there were to be any hope of making the program work. They also had to deal with processing large numbers of contracts and checks—an effort that would have been magnified if they included tenants and sharecroppers. It was more convenient and efficient to deal only with landlords. Moreover, doing so would be less dis-ruptive or disturbing to the system and the landlords.[120]

Of all the USDA officials involved in the cotton program, the most attention has been directed at Oscar Johnston.[121] He caught the attention of many because he was finance director at USDA, AAA comptroller, assistant director of the Commodity Credit Cor-poration (CCC), and director of the Cotton Pool. In addition to these high-level positions at USDA, Johnston was the manager of the largest cotton plantation in Mississippi (and likely one of the largest in the world), Delta and Pine Land Company (D&PL).[122] He gained further notoriety when it was reported that the planta-tion was the largest recipient of federal payments under the AAA from 1933 to 1935.[123]

One historian found that in the 1933 effort to convince cotton farmers to participate in the program, Johnston informed the plant-ers "that if a landlord signed a contract at a time when he had no agreements with his tenants, he would not have to share rentals" paid by the government. If, however, "the landlord already had agreements with managing share-tenants, they would be entitled to half of the rental money." The historian concluded that the point "must have been clear to all landlords present that they should sign the 1934–35 contract before making arrangements with their ten-ants."[124] Others researching the issue concluded that landlords at the time were purposefully "delaying the renewal of their tenant contracts until the 1934 reduction plan [was] made clear."[125] John-ston's D&PL, however, was reported to have been "the fairest plan-tation studied in dealing with its tenants."[126] Reportedly, Johnston fully complied with the requirements on distributing benefits to tenants and was held to high standards.[127]

From the perspectives of the sharecroppers and those who have researched the issue, Oscar Johnston's obvious conflicts of inter-est were the most troubling. He was in a direct position to benefit D&PL and, through it, himself owing to the decisions he made in his positions at USDA. He reportedly did not think that the AAA had the authority or capability to make reforms in the southern system and that attempts to do so would destroy the program.[128] His conflicts were seen as more troubling because D&PL rented to the government the worst acres it owned and intensified pro-duction on the remaining acres, producing a larger crop in 1933 than it had produced in 1932.[129] Research further uncovered that he used his position in D&PL and the cotton industry to influence decisions at USDA and in Congress.[130] He also helped reinforce the opinion within the Roosevelt administration that pushing reforms on the planters would wreck the program.[131]

The problems continued when USDA developed a second con-tract soon after the 1933 plow-up, one designed for the 1934 and 1935 crop years. It was not the rushed effort of the 1933 contract, and more thought was put into the sharecropper and tenant issues. But researchers concluded that the "cropper d[id] not fare so well" under it either.[132] These conclusions were largely based on provi-sions in the contract that were considered unenforceable, more aspirational than legally binding.[133] The contracts also contained specific provisions that were interpreted to exclude sharecroppers and any tenants who did not furnish their own equipment and materials or manage the farm.[134] USDA had no practical ability to enforce any of the contract provisions against the landlords.[135] In fact, interpreting them was the scene of an intense bureaucratic battle between reform-minded lawyers and the landlord-minded Cotton Section. The Cotton Section ultimately won that battle, and one result was that many of the liberals were subsequently purged from USDA.[136]

This historical record has led many to conclude that the AAA authorities were designed and used intentionally to help southern cotton planters push poor black sharecroppers off the land and consolidate their holdings thereby easing the transition to mechanized and modern cotton farming. Acreage reductions, in particular, are seen as providing the method for potentially increasing prices while cutting back on labor needs, problems, and costs; additional funds covered sharecropper debts and allowed planters to buy tractors and other modern equipment to replace them.[137] The planters would have had strong self-interested reasons, as well as the means, the sophistication, and the connections, to make certain that whatever came out of DC would work in their favor.

Adding to conclusions about the consequences of the cotton program is the knowledge that the New Deal was stocked with many of the young, urban liberals who were idealistic reformers.[138] They were long on idealism and far-reaching plans for social reform, including centralized land use planning, but were short on farm experience and knowledge. Moreover, Secretary Wallace and USDA were under intense time and political pressure to get the cotton program up and running. They were seeking to push out assistance to a severely depressed farm economy with few good options. The president and the secretary had committed to the concept of production controls through acreage reduction as the method to increase prices for the farmer.[139] In hindsight, the cotton planters appeared to have held the upper hand in this battle, and the most difficult consequences fell hardest on the poor tenants and sharecroppers.[140]

Concluding Thoughts on the Origins of Farm Policy

It is difficult to imagine a more abrupt change in federal policy for farmers, as well as in the relationship between the two, than the Agricultural Adjustment Act of 1933. It provides a good example of the tectonic forces of the Great Depression and New Deal. An emergency response, it was the culmination of a dozen years of effort to overcome an entrenched policy and ideological status quo. The farm depression had preceded the Great Depression, and its longevity helped drive policy development, including formation of a coalition to work through the process. The McNary-Haugen bills were ultimately too heavy of a lift in a conservative government during the Roaring Twenties. To help farmers was to hurt consumers and industry, creating problems for other sectors of the economy.

That the Great Depression completely changed the equation is evident in the drastic policy that became law in 1933. Where McNary-Haugen was built largely on the idea of moving surplus commodities into foreign markets to help prices, the 1933 AAA was built on the premise of paying farmers to plant less in a system of centralized planning. It was cotton in the driver's seat instead of corn or wheat, and the policy solution was built on a more industrial design fitting the industrial commodity rather than those used for feed and food.

The most drastic measures were initially restricted to cotton and pork. A severe drought in 1933 prevented a plow-up of the wheat crop, but USDA required steep acreage reductions under the allotment program.[141] This was likely fortuitous given the potential for severe controversy if USDA had required plowing under a food crop as people were starving and standing in food lines. Additionally, domestically fed corn was spared the worst of the acreage responses, receiving price-supporting loans instead.

The New Deal's emergency responses provided different treatment for the three key players in the farm coalition. Over time this would set in motion further problems for farm policy, but the Depression held much of it in check. The policy was untested, questionable, and controversial. It was a far-reaching regulatory scheme that had been previously unthinkable in the countryside and, in fact, had long been opposed.[142] At the core of its conflicts and controversies was the authority to require farmers to cut production and acres. It was counterintuitive and loaded with problems.

Looking back, the consequences of the AAA in cotton country stand out with ten million acres of a growing crop plowed under and the strictest reductions in acreage. More difficult were the consequences for tenants and sharecroppers; farming small acres was inefficient and a roadblock to larger, more efficient and mechanized farming in the South.[143] In action, the New Deal policies resulted in sharecroppers being driven off the land and out of farming with dire consequences for a particularly disadvantaged cohort of citizens.[144] Many were caught between institutionalized discrimination and the forces of modernizing agriculture fueled by emergency federal policies.[145]

Consolidating farm holdings eased modernization, and federal payments provided capital to purchase tractors and equipment but also left the displaced, poor tenants and sharecroppers to fend for themselves or seek help from the government or other sources—a surplus labor problem made worse by the Great Depression.[146] The damaged economy limited available off-farm employment opportunities; thus, more workers were available for wage labor. But wage labor had been cheapened by the squeeze of too many workers in a system switching hand labor out for mechanization.[147] Many of these workers did not move out of southern farming areas and were victims of the way in which the old system was brought to an end.[148]

Sharecroppers presented the most troubling of problems for the 1933 AAA, but not the only one. With the benefit of hindsight, it is difficult to comprehend how reducing acres would help farmers. Acreage reductions were based on the idea that the farmer should act more like business and industry, yet he needed federal help to do so. Farming presented unique challenges for controlling supplies because millions of independent farmers made business decisions according to the vagaries of nature and weather. Production controls could not control the weather. A great growing season could still produce bumper crops on reduced acres while a bad season could cause shortages. Reducing planted acres was also a limited method for supply control because the government could not at the same time prevent increasingly intense production practices on reduced acres that would boost yields and supplies.

Farm income and purchasing power depended on growing and selling crops, but Congress passed a bill that would result in farmers doing less of what produced an income. Controlling production was intended to improve prices over time, but whether prices would improve enough to offset acreage reductions was unknown and required direct assistance to the farmer. Acres taken out of one crop could be planted to others, causing competition with, and problems for, other commodities that would increase tension within the farm coalition. The entire program was also voluntary, riddled with free-rider and other problems, such as the pressure on individual farmers to increase production when prices improved.

History provides a safe distance from the emergencies of the Great Depression and the urgencies of the 1933 AAA. From that vantage point, it is easy to conclude that the policy was destined to fail and generate myriad problems in the wake of its failure. The story of this failure unfolds in the following chapters. For the conclusion of this era of farm policy history, the fact that federal production control made it into the 1933 AAA and was implemented in such a drastic manner clearly illustrates the larger transformation in American society and self-government wrought by the Great Depression and New Deal. These events helped acreage reduction policy clear the barriers of the existing status quo in farm policy. Beginning in May 1933, the AAA became the status quo that future changes, reforms, and improvements would have to overcome.

2

Adjusting to the New Deal and War, 1933-1945

Introduction

The 1933 cotton plow-up problems give the impression that farm policy was off to an inauspicious start. A more complete picture would include evidence that the heavy federal intervention was delivering some relief. Crop prices improved slightly, and farmers experienced some improvement in cash receipts and net income from farming for 1933 and 1934.[1] With all that was going on at the time, it is difficult to distinguish whether improvements were due to the AAA, other New Deal programs, speculators, drought, or other factors in the economy. Certainly the extensive mortgage relief delivered by Title II of the 1933 AAA was also a key factor.[2] As FDR, Wallace, and their teams settled in and the initial smoke of emergency response in the first hundred days cleared, the longer policy and political game of the New Deal began to unfold.

Secretary Wallace and USDA had been forced to move fast to put the 1933 act into action for depression-wracked farmers. Less noticed than slaughtering hogs and plowing up cotton were the consequential developments in DC. For one the AAA required a new bureaucratic infrastructure to operate programs and deliver assistance across the country. This involved standing up the Agricultural Adjustment Administration within USDA. Among those brought on to the effort was George Peek, who was appointed to be the first administrator. Bureaucratic battles broke out almost immediately, and Peek's tenure did not make it to the end of the year.[3] On the ground USDA's footprint had to grow quickly to implement cotton plow-ups, program sign-ups, and more. The department enlisted the cooperative extension service and county farm bureaus and created governing committees at the county and state levels. New program benefits developed a constituency, as did creating a large bureaucratic system. Both added to the substantially changed circumstances in the agriculture economy and in the relationship between farmers and the federal government.

The desperation of the Depression had produced action but had also magnified existing conflicts and challenges that ran through the farm coalition. For one, actions under the 1933 act set cotton apart from the other commodities. Steep acreage reductions were implemented for wheat but drought allowed it to avoid a politically fraught plow-up. Corn, meanwhile, avoided any drastic measures in the initial operation of the AAA. In the center of the political storms was this untested policy for controlling supplies by reducing the acres planted. The collision in the southern sharecropper system was only one issue.

From the perspective of policy and development, the big question immediately looming was how to respond once the emergency situation began to change. Emergency response measures managed to stave off a populist revolt in the countryside. The new federal policies intruded significantly on farm life and management. This intrusion was bound to set off reactions on the ground and cause all actors to respond. Further responses to these measures were also inevitable as policies and politics adapted to interaction with the blunt forces of constantly changing circumstances; these responses were also unknown.

The AAA was a major policy and political achievement. Roosevelt's first hundred days remain legendary, establishing a standard that would challenge every president who followed. Swift achievements spawned further planning and more far-reaching goals within the growing New Deal apparatus, some of which bordered on utopian. For example, there were reportedly hopes that "the emergency efforts would evolve into a system of complete control that would restrict commercial agriculture to the most efficient farmers operating the best of our lands, convert the other lands to other uses, and move the other farmers into other occupations."[4]

This chapter of farm policy history is very much a story of how grand plans and ambitious policy intentions can falter, stumbling when confronted with reality. Agricultural adjustment in the New Deal involved multiple adjustments: in Washington and Congress, at USDA, on the farms, and among the farmers and those ostensibly working on their behalf. The saga of cotton sharecroppers is the most troubling and conspicuous. The ensuing twelve years would also witness an unprecedented man-made environmental disaster; defeat before the Supreme Court; more bureaucratic problems; attempts at long-range, centralized planning; and the brutal force of another world war. Through it all, the nation and the federal bureaucracy were skillfully guided by President Roosevelt, who would win four straight presidential elections and maintain strong majorities in Congress—a political achievement that will not be repeated.

As events intrude on policy, they affect its development and force adjustments. This process creates a "continuous thread" that "runs through the evolution of an agricultural policy" such that the most recent programs "become the foundations for programs of the future."[5] In that way, the turn to a "collectivist type of capitalism" formed out of the Depression becomes foundational.[6] The 1933 AAA constituted a clean break in history. It had to overcome a long-standing status quo in opposition to centralized federal intervention. As a result, the act became the status quo against which each subsequent policy development would have to prevail. A large swath of farm policy history and development is tied to the fateful decisions codified in 1933 and the federal-industrial model it created; there would be no turning back to a pre–New Deal status quo. America's experiment in self-government was forever altered. The ensuing iterations of farm policy provide clear examples and lessons in policymaking.

Adjusting to the New Deal

Despite the extraordinary measures taken for farmers, problems persisted, especially in the South. That is not to say that issues were limited to the South; nationwide, farmers and their interest groups also continued to raise concerns about purchasing power, cost of production, and other policies.[7] The events in cotton country, however, continued to dominate the discussion. Within the year southerners in Congress were pushing additional measures for cotton. They were concerned about the effectiveness of the voluntary acreage reduction program. Their efforts led to the Bankhead Cotton Control Act of 1934, which instituted compulsory controls on cotton production and taxing noncompliance.[8]

Although obscure, the Bankhead Act adds pieces to the overall policy puzzle. First and foremost, it amounted to an acknowledgment that acreage reduction was not effectively controlling supplies. In a very real sense, the bill was a sign that the policy had failed. Supporters of the Bankhead legislation defended the bill based on the "tremendous carry-over of American cotton that is overhanging the market" and claimed that cotton farmers were demanding compulsory controls for their unique crop.[9] It was, they admitted, a "new departure in the policy of our Government."[10] More accurately, it should have been an admission that supplies could not be controlled on the basis of reductions to the acres planted.

Representative William Bankhead (D-AL) and his brother, Senator John H. Bankhead II (D-AL), were the lead authors of the policy. Representative Bankhead was in his ninth term at the time and would eventually go on to be Speaker of the House. Senator Bankhead, the elder brother, was a freshman senator. Both were part of an Alabama political dynasty, their father had represented the state in both the House and the Senate; John occupied the seat once held by their father.[11] While from Alabama, they were not from cotton; speculation at the time was that John's ambitions included replacing Chairman "Cotton Ed" Smith at the helm of the crop's political leadership.[12] Speculating further, the senator's political ambitions might well have combined with the standard prejudices of the region and era to result in both problematic policy positions and the zealous pursuit of them.

The bill was not intended to replace the acreage reduction program but rather to supplement it with compulsory reduction of harvested bales—"governmental restraint as to the amount that can be produced and put into commerce."[13] Comparing the two, Representative Wall Doxey (D-MS) explained that "the principle of this Bankhead bill is compulsion" whereas "the acreage reduction program that is now being employed . . . is voluntary."[14] The problem, however, was not one of low participation because about 85 percent of cotton farmers had signed the cotton contracts. It was more the concern that they were being harmed by "that 10 percent of greedy and avaricious men" who did not participate.[15] Of course, if the 10 percent of farmers not complying were large enough, they could affect supplies disproportionately. Concern about participation from large plantations had weighed heavily on USDA officials in the implementation of the program, as discussed in the previous chapter.

More to the point, the bill was punitive. It was aimed at the "slacker" or the "chiseler" farmer who would not "come in and help us get rid of this tremendous and disastrous surplus" of cotton.[16] The real problem appears to have been that cotton farmers were violating the spirit of the controls in the 1933 AAA by increasing intensity and the use of fertilizers to get more bales on the reduced acres. Many had rented the least productive acres to the government and boosted production on good acres. Representative Bankhead informed his colleagues, "Evidence shows that all over the Cotton Belt fertilizer sales as compared with last year have increased more than 100 percent. . . . In some sections of the country they have increased as high as 300 percent."[17] Unpredictable weather was certainly a factor as well, helping deliver bumper crops despite reduced acres. These should have been strong signals that the centralized system for controlling supplies by reducing acres was unlikely to succeed. Instead of seeking out more effective policies, they doubled down on the concept of supply control.

To supporters of the Bankhead bill, a prime example of the problem was working at the department. Oscar Johnston, manager of the largest cotton plantation and a high-ranking official at USDA, was found to have "broke the spirit and letter of the 1933 contract" when D&PL rented to the government the worst acres it owned and intensified production on the remaining acres, resulting in a larger crop in 1933 than it had produced in 1932.[18] Johnston and similar "scientific farmers" knew all too well that acreage reductions "rewarded intensive cultivation."[19] Johnston focused on ten thousand acres of the best land and fertilized it intensively even in the face of the Bankhead allotment.[20] Delta and Pine used better seed, and Johnston knew that acreage records for 1928–32 used by AAA included a lot of land that didn't produce cotton (ditches, gardens, etc.). Reduction in acreage was easy for those with an inflated base.[21] D&PL's size allowed it to split allotments across county lines, and its resources enabled it to store excess bales to avoid the tax.

Oscar Johnston was a notable opponent of the Bankhead bill, and his conflicts of interest were quickly attacked by Representative Bankhead and his allies. Johnston argued that the Bankhead bill would hurt tenants.[22] He was reportedly concerned about the potential for bale limits to replace the acreage controls he preferred, policy that worked best for large operations that could manage around them. As Johnston saw it, under the Bankhead policy "rental payments might have evaporated" and the tax "chiefly threatened efficient, scientific agricultural management" such as his on behalf of D&PL.[23] But Johnston was fighting against a tide that apparently included a large portion of cotton farmers. Congressional supporters went over his head and secured President Roosevelt's support for the bill.[24]

Reopening the farm bill within its first year to demand more federal intervention for cotton has to be considered a risky move. It put significant pressure on the regional and commodity alliance, exposing the fault lines that would later threaten the coalition.[25] Opponents of the bill were concerned that compulsory controls would spread to all commodities and regions.[26] They were also concerned that the cotton industry did not fully appreciate what it was reported to have demanded.[27]

Senator Thomas Gore (D-OK) said that "I fear that the farmers are winding a boa constrictor about themselves that one day may break every bone in their bodies" and that he could not support it.[28] Additionally, if American cotton farmers cut their production, they were likely surrendering market share to foreign competition.[29] Representative Harold McGugin (R-KS) argued that the cotton industry was hurting itself, "taking away the world market for cotton from the South and giving it to other countries of the world."[30] Others questioned the severity of the cotton situation and whether the bill was simply a bonus. Representative Clifford R. Hope (R-KS) noted that the price of cotton had doubled and "more nearly [approached] the parity price than the price of any other agricultural commodity" while the surplus had begun to decline.[31] There were also concerns about the expanding USDA bureaucracy and pushback against the Roosevelt administration and the Democratic leadership in Congress over this new path in federal policy.[32]

Keeping it in perspective, the Bankhead bill came on the heels of a ten-million-acre plow-up in 1933 and steep acreage reductions for 1934. These reductions cut back the acres planted to a crop that produced income in a time of severe depression. To the extent that doing so actually reduced cotton supplies, this had further implications for an exported commodity in a world market with foreign competitors. American acres could easily be replaced by acres in other countries, severely diminishing the impact on prices. Congress was adding further to these problems with bale-based allotments and reductions. Combined, the 1933 AAA and the Bankhead controls would require cotton farmers to submit to a reduction in acres planted to cotton and to a reduction in the amount of cotton they were allowed to produce or market. Failure to comply with the latter risked potentially ruinous taxation. If acreage controls were misguided, adding bale controls was more so. Altogether the bills created a large and intrusive federal role in farming in a region known for animosity toward federal interventions.[33]

Some justification for the Bankhead bill can be found in a dispute internal to cotton about the shift in acreage from the Southeast to the irrigated acres in Texas and California.[34] The extra layer of federal controls could have been an attempt to slow or halt that shift. It was also intended to push southern farmers to diversify production and become less reliant on cotton. On the House floor, Representative Bankhead raised concerns that "the only cash price that our people have is the cotton crop, because we have never yet, unfortunately, diversified our agriculture in the South."[35] Opinion at the time was that the South lacked a diverse mixture of cash crops. In response, legislators crafted these policies as part of an effort to get southern farmers to be more like farmers in the Midwest and Great Plains and to grow some of the same crops.[36] The effort seemingly ignored the impact this would have on the rest of the farmers across the nation.

It was also understood that cotton production was too dependent on hand labor, which impeded the adoption of crop diversification, mechanization, and technology by many farmers.[37] Many economists and experts at the time diagnosed a problem of excess or surplus labor in the southern farm economy. They concluded that agriculture "simply could not absorb so many people and provide an adequate standard of living for them."[38] A significant goal of these policies was to drive more efficient cotton production, but the issue of labor was problematic.[39] Many cotton farmers were also interested in reducing labor costs so they could stay competitive with foreign and synthetic fiber production. Clearly the policies were designed to address many issues rooted in price and production concerns. Seeking to increase prices and control acres, however, had consequences for the many actual humans considered surplus farmers. It was clear that members of Congress were aware of these problems.[40]

On the House floor, as members debated the Bankhead bill, Representative George B. Terrell (D-TX) noted, "More than half of all farmers in the South are tenants, or share-croppers."[41] More directly, Representative McGugin pushed the point in debate: "When you take away 40 percent of the cotton production, what are you going to do with 40 percent of your cotton producers?" He added that cotton farmers "are not going to be able to drive from the plantations of the South 40 percent of your share croppers, black and white, and cast them upon the highways without bringing social despair to your fair land; and yet that is exactly what the Agricultural Adjustment Act, as applied to cotton, is doing today."[42]

As an example of how Congress wrestled with this difficult issue, the House debated whether to permit tenants and sharecroppers the ability to vote in the referendum that would determine whether allotments and taxes would apply.[43] This debate highlighted the challenges because a version of the bill might have excluded many sharecroppers; they were entitled only to wages based on the harvested crop (after debts were covered) but not the crop itself.[44] Permitting them to vote added administrative complexities but might have also worked against landowner interests. Officials at USDA had sought a revision that would've excluded tenants and sharecroppers in order to ease administration.[45] Representative McGugin accused the committee of "basing the right of suffrage" on land ownership, which he called "a feudal system" and an indication that the committee was too concerned about "the welfare of the largest planter," not the poor sharecropper.[46] Rather than fight the issue, Chairman Marvin Jones (D-TX) withdrew the amendment.

Debate of the issue pushed Congress to take steps to address the problems that had resulted. Representative Gerald Boileau (R-WI) claimed that the House Ag Committee had "gone . . . as far as is humanly possible to protect the interest of the share-cropper and the tenant farmer . . . and there will be no danger of their being discriminated against."[47] Representative Bankhead added that effects on sharecroppers and tenants had been "taken into consideration" and that "every reasonable precaution [had] been taken in the preparation of this bill to see that all interests are properly, fairly, and humanely protected."[48] The final 1934 Bankhead text included provisions to encourage the secretary to protect the interests of tenants and sharecroppers.[49]

The AAA issued regulations on behalf of tenants, but subsequent evaluation concluded that these regulations were favorable to the landlords, lacking any real protections for tenants or sharecroppers.[50] Many sharecroppers and tenants had significantly misunderstood the Bankhead system and signed away their shares or simply failed to apply for certificates. Concerns focused on the method for implementing the punitive tax for compliance and the impact on the tenant's or sharecropper's portion of the crop used for rent. Lawyers at USDA concluded that they were not able to revise any contractual relationship between the landlord and tenant or sharecropper.[51] Further concerns involved the allotment and sign-up system, viewed as effectively excluding sharecroppers because they could not complete the detailed signed statements required.[52]

In the longer run of history, the Bankhead Cotton Control Act was more of a minor diversion than any significant policy shift. It was in operation for only two crop years before it was repealed,[53] and it was quickly buried by more dramatic events. The combined effects of the 1933 AAA and the Bankhead Act were substantial, however, and far more consequential than their brief existence would tend to indicate. Together, these major interventions in cotton farming to reduce supplies would ripple through not only cotton production and the rural South, but eventually throughout the nation and Congress.

Fallout from sharecropper treatment under the reduction programs would set off complaints to DC and in the press, spark lawsuits, and result in the formation of the Southern Tenant Farmers Union (STFU).[54] The STFU was controversial in no small part because it was a union that did not segregate black and white tenants in its membership, meetings, or actions. It also included socialist and communist elements that helped raise alarms about the organization and blur perspectives on its animating issue. In eastern Arkansas, planters responded to the STFU with violence and threats of violence, much of it led by a group called the "nightriders" in what was known as the "reign of terror."[55] Fallout quickly spread to the capital from the South. In DC, fights within USDA over tenant and sharecropper issues would lead to a purge of New Deal liberals in 1935.[56] In both locations the core of cotton power held firm, prevailing over the liberal idealists and reformers as they scrambled to react. In the words of one historian, "It was an old story. The South's land problem forever confounded well-intentioned liberals in and out of government."[57]

The overriding focus on controlling cotton production to reduce surplus required the cooperation of large landowners; without reducing acres from large operations, the program would have little chance of working.[58] Any evaluation of the outcomes or benefits of the early New Deal policies for cotton struggles with this issue. Production controls converged with the sharecropping system to produce drastic consequences for poor and often black farmers. The landowner could plant the entire cotton allotment and force the sharecropper out, while also excluding the sharecropper from any share of the assistance and checks. Each step under the reduction policies managed to harm sharecroppers.[59]

This single policy choice—seeking to increase prices by controlling production and acres—has determined much of the direction and development of farm policy. It sharpened the edges and deepened the divides around the primordial fault lines. It even set the stage for the second great fault line over food assistance to low-income individuals. Cotton policy proved vast and consequential, but a collision with nature intruded first. In the longer term, fallout from the New Deal cotton policy contributed to major demographic changes that came to affect Congress and national politics (matters explored in subsequent chapters).

Adding in the Dust Bowl

As if the Great Depression were not challenge enough, farmers were soon hit by a natural disaster the likes of which the nation had never experienced. The Dust Bowl was the catastrophic product of a multiyear drought across all those broken acres in the windswept high plains. The winds lifted dry dirt into massive dust storms that drove people off the land, caused property and health damage, and affected flights; western soils dumped as far away as Chicago, Boston, and New York. The most notable event took place on April 14, 1935, and was known as "Black Sunday."[60] Black Sunday was instrumental to the development of natural resource conservation policies that focused specifically on farming.[61]

Despite overwhelming evidence of soil erosion problems, USDA and Congress had been slow to address the issue. Hugh Hammond Bennett, known as the father of soil conservation and a political force to reckon with, had been championing soil conservation efforts for decades. Black Sunday provided a powerful incentive for congressional action. In testimony before the Senate Agriculture Committee, Bennett used the dust storm to his benefit: "A senator who had been gazing out the window interrupted Bennett. 'It's getting dark outside.' The senators went to the window. . . . The sun over the Senate Office Building vanished. . . . Light filtered through the flurry of dust. For the second time in two years, soil from the southern plains fell on the capital. . . . 'This, gentlemen, is what I'm talking about,' said Bennett. 'There goes Oklahoma.'"[62]

Bennett's comments were on Friday, April 19, 1935, five days after Black Sunday hit the plains. President Roosevelt signed soil conservation policy into law just over a week later. The act was designed to "provide for the protection of land resources against soil erosion" and established the Soil Conservation Service within the Department of Agriculture.[63] Bennett served as the service's first chief until his retirement in 1951. Soil conservation had pushed its way into a conversation that had been dominated by low prices, farm foreclosures, and the Great Depression. It became a significant issue and the "Soil Conservation Service gained popularity with many farmers and congressmen because it was striving to curb erosion in the midst of overwhelming dust storms."[64]

The Great Depression helped President Roosevelt achieve success for his New Deal, which included farm policy. The Dust Bowl delivered soil conservation policy. Together, they created dual federal roles in farming to improve prices, alter planting decisions, and change farm practices to reduce soil erosion. Within a year the Supreme Court responded to this expanded federal role; the response forced Congress to merge the two.

In 1936 the Supreme Court concluded that the 1933 act was an unconstitutional exercise of congressional power and struck it down. Specifically, the court concluded that the AAA invaded "the reserved rights of the states" by trying to "regulate and control agricultural production, a matter beyond the powers delegated to the federal government."[65] The Supreme Court decision staggered the farm coalition, placing in jeopardy the far flung efforts that Congress and USDA had scrambled to piece together for adjusting American agriculture.[66] With the Depression continuing and plains dust blowing, it was politically untenable for Congress and the FDR administration to accept the decision without response. The immediate reaction was to further combine their efforts and use soil conservation to resurrect production controls. Within a few months, Congress passed and the president signed into law the Soil Conservation and Domestic Allotment Act of 1936.[67]

The basic premise of the 1936 act was to institute acreage allotments based on normal production estimates and estimated domestic consumption. The payments to farmers, however, were to reduce planted acreage of "soil-depleting crops" in favor of soil-conserving crops. The soil-depleting crops were the major cash crops facing surplus. The soil-conserving crops were grasses, legumes, and other forage crops seen as building and protecting soil. Once again, the policy mechanism consisted of federal rental payments to take acreage out of production to relieve oversupplied commodities. This was a continuation of the policy of reducing supply to improve prices, with all of its attendant problems. Ostensibly, the federal funds were renting surplus crop acreage for conserving uses.

Conserving soil from erosion to help address the massive multi-state devastation of the Dust Bowl helped avoid the constitutional conundrum presented by the Supreme Court. The 1936 act was a bit of legislative sleight of hand, putting conservation in service of price supports and production controls. The soil conservation objectives appeared to be more popular, especially outside farming, and certainly crossed state boundaries. The policy also provided more generalized national benefits.[68] The program design, however, left little doubt that the purpose was farm prices and incomes, not conservation. Murray Benedict memorably pointed out that soil conservation programs "did not contemplate a gentle shower of federal checks evenly and undiscriminatingly distributed over eroding and noneroding areas alike."[69] Notably, the payments were to be made through the Agricultural Adjustment Administration and not the Soil Conservation Service.

The bottom line for the 1936 act was yet another emergency response. Congress responded to the Dust Bowl, the continuing economic depression, and the Supreme Court ruling. The act was unlikely to be a long-range policy or an opportunity for rethinking and debating the appropriate direction for farm programs. It was largely status quo policy under the guise of conservation and until something else could be worked out for the long term. That status quo, however, remained problematic, and the stresses of it were beginning to show in the House.

The impact of these policies on sharecroppers was becoming a bigger issue for Congress, and USDA's implementation decisions were the focus of blame. Representative James Wadsworth (R-NY) attacked cotton acreage reduction policy for "the throwing out of occupation of thousands and thousands of share croppers, an unexpected result; but now, as we look back upon it, an inevitable result."[70] Representative Malcolm Tarver (D-GA) accused the bill's drafters of "not doing anything for the sharecropper and the tenant farmer. . . . No reference is made to the payment of any benefits except to the landowner." He added that leaving it to the secretary of agriculture had proved unacceptable because the secretary had "permitted the tenant, and the sharecropper, and the small farmer, in many instances to be the victims of rank and unjust discrimination."[71] He pointed to USDA testimony that acknowledged only landowners were considered farmers and that neither tenants nor sharecroppers were considered farmers in the eyes of USDA officials. Representative Maury Maverick (D-TX) pointed out that numerous complaints and charges had been filed "about conditions existing of sharecroppers, tenants and agricultural workers all over the South and in many other parts of the Nation" because "the sharecroppers, the workers and the tenants were not adequately protected, and I believe we should do something to protect them."[72] Representative Luther A. Johnson (D-TX) noted the criticism aimed at USDA's operation of the 1933 AAA and the 1934 Bankhead Act, under which "the small farmers and the tenants did not receive the benefits to which they were entitled," and said he hoped "that the Secretary of Agriculture will see to it that under this new law the small farmers and the tenants are given their fair and equitable treatment."[73]

There was also recognition that while a shift to a conservation focus might protect the policy from the Supreme Court decision, it would not help sharecroppers and might make matters worse. Representative Hope noted a serious objection to the bill because it "may further increase the problems of the tenant and sharecropper."[74] The perception that conservation could further damage sharecroppers was likely based on the presumption that sharecroppers farmed the least productive acres, those most likely to be rented to the government by the landowner for conservation purposes. For example, Senator Gore noted that "the pending bill, since it addresses itself to soil conservation, may be difficult of application in such a way as to help the tenant farmer who owns no land at all."[75]

The House Ag Committee was on the defensive, trying to rescue the policy from the Supreme Court's decision. Political blowback over sharecroppers and tenants presented a significant threat to these efforts, and Representative Doxey argued that "the language of the majority decision of the Supreme Court makes it exceedingly difficult for Congress now to enact practical and beneficial legislation in behalf of the farmers, especially the small farmer—the tenant and the sharecropper." Doxey did, however, acknowledge the "complaints against the A.A.A.," and that he had worked to draft the 1936 bill "to incorporate provisions that would directly help the small farmer—the tenant and the sharecropper. However, my efforts in this regard met with little success."[76] He indicated that the need to develop a bill quickly forced the committee to write general authorities and again leave many of the specific decisions to USDA.

This was clearly unsatisfactory to many members given USDA's track record. In response, both Representative Maverick and Representative Tarver, as well as Representative B. Frank Whelchel (D-GA), pushed amendments to force USDA to distribute assistance equally to landowners, tenants, and sharecroppers.[77] Representative Tarver was seeking to "demonstrate clearly to the Secretary of Agriculture the purpose and intent of Congress that in carrying out the provisions of this legislation the tenant and the sharecropper shall receive a fair share of the benefits that are to be paid."[78] The pressure on Chairman Jones to address the issue led to an agreement to effectively include these amendments in the House bill and avoid a further damaging fight.[79] The debate and amendments serve as a measure of the level of political pressure that was coming to bear on the policy from this issue.

In conference the House amendment was altered to provide USDA flexibility and protect contractual relationships.[80] The Senate had not included the same protections in the bill it passed. Senate Ag Chairman Smith noted that the revision was a Senate demand agreed to by the House conferees and that it "very considerably modifies the provision as adopted by the House."[81] Chairman Jones acknowledged the modification but claimed it had "substantially the same purpose as outlined in the original bill."[82]

When pressed on the matter, he responded that "I think that is the best that we can have, and I think it will assure that they will be fairly treated."[83] Within the context of the previous three years of problems, administrative complexities and the challenges posed by legal relationships on the ground added further layers of confusion and concern to a troubling component of New Deal farm policy. Such matters offered little satisfaction in light of the considerable harm to a vulnerable group of American farmers.

The 1936 act was short-lived; it was another temporary, emergency measure for farmers overwhelmed by depression and dust. The act may have been a reflection of Secretary Wallace's ambitious goals for the long-term direction of farm policy, albeit muted by congressional goals and diminished by continuing the dire consequences in cotton country.[84] Including conservation goals and concepts in a farm program could have been a significant policy development but was in reality limited by the overriding focus on supply controls to increase prices. In three years' time, farm policy had arrived, was adjusted, defeated, and adjusted again, but Congress continued forging ahead with programs to deliver assistance to the countryside. Whether these efforts and programs had any future once the emergencies of the Great Depression and Dust Bowl had passed was a matter of doubt.

Attempts at Adjusting to a Permanent Policy Footing

The various New Deal policies may have helped some farmers and clearly hurt others, but they hadn't come close to curing commodity surplus problems, especially for cotton. In 1937 cotton farmers were set to produce another record crop (over eighteen million bales) despite planting ten million fewer acres.[85] Cotton's move to mandatory production controls could not get it out from under surpluses. Increased foreign competition and declining export markets received most of the blame.[86] That cotton was going in a direction that was uncomfortable for wheat and opposed by corn created substantial pressures on the farm coalition's main fault line. When Congress tried to further adjust New Deal emergency policies to place them on a permanent footing, the effort exposed the growing divide and worsened it.

This next round began when farm interests, led by the American Farm Bureau, pushed Secretary Wallace and Congress for additional assistance. The one thing they all agreed on was the non-recourse loan program because it put a floor under commodity prices.[87] The loan program necessitated checks on the production incentive it created, but the ability to control supplies depended more on the weather and farm practices. The demands from the farm coalition coalesced with Secretary Wallace's long-term planning efforts at USDA. Wallace had been calling for an "ever-normal granary" that would balance supplies by holding grain from good years to cover needs in bad crop years.[88] He saw that the loan program could be an effective component of his concept. The political problem was that he did not have an equivalent policy proposal for cotton. The American Farm Bureau pushed Congress to merge price-supporting loans and Wallace's ever-normal granary concept, but leadership of both Ag Committees remained reluctant.

The American Farm Bureau responded by turning up the pressure at the White House and in a series of field hearings.[89] Much of the pressure was coming from cotton, which stepped up demands for help with the record 1937 crop.[90] President Roosevelt, likely battered from the four previous years of policy efforts but comfortably reelected, responded with a demand of his own. He called on Congress to come up with "some plan, some machinery, some law under which the Government's financial interests could be protected."[91] Congress passed a resolution promising to produce permanent farm legislation and added $65 million to an appropriations bill to make payments to cotton farmers if they agreed to further reduce plantings in 1938.[92] President Roosevelt authorized loans for corn and cotton, and he called Congress back for a special session in November 1937.

The final provisions of the 1938 act display all the markings of the tough debates in both the Senate and House. What continued to bind the coalition together was the goal of achieving "justice for agriculture" at a time when farm interests fought to receive equal treatment by the law—defined largely as the "right to an offset to the tariff."[93] Some members also remained concerned about soil erosion in the wake of the Dust Bowl, but the ultimate priority was to increase crop prices. For example, Senator George S. McGill (D-KS) explained that the bill was intended for "conserving our national soil resources" by preventing surplus production that hurt farmers and the public, but that ultimately "the philosophy of the bill [is] to attain parity prices."[94] Across the Hill, House Ag Committee Chairman Jones advocated for continuing to use the conservation system of payments to farmers, which he considered "restitution" made "in the best possible form . . . for the conservation of our soil in the national public interest."[95]

The Supreme Court's 1936 decision weighed on members as they debated permanent farm policy. Representative John Gwynne (R-IA) offered, "We are apparently on the threshold of a great expansion of the interstate commerce clause. . . . We and the Court between us are going to determine what things done inside the State directly affect interstate commerce."[96] After years of providing assistance with mixed results, Congress was also facing budget pressures that some thought limited what they could achieve. Senator John Bankhead noted the "clamor all over the country of late, and it has reached down into the common walks of life, that we should quit carrying this Government into larger and larger indebtedness."[97] Likewise, Senate Ag Committee Chairman Smith lamented the "considerable agitation in the press as to the limitations on the amount of money that can be used for the purposes of the pending bill."[98] These issues did not, however, constitute the major points of conflict in the debate. The real conflicts were over production controls and price supports.[99]

The issue of production controls remained the biggest conflict, and it was splitting the South and Midwest. By the time Congress was debating permanent policy, farmers and members had more than three years of experience with control policies. Some southern Senators continued to fight for strong controls on cotton.[100] But there were indications of possible cracks in, or weakening of, that support. Chairman Smith made it clear that he "came here under the order of a majority of the cotton growers of America" but was reluctant to support the bill; it was not the bill he preferred but "what they [the cotton growers of America] demanded. . . . I am going to say to them, 'If it is a success, I congratulate you; if it is a failure, shake not thy gory locks at me.'"[101] Controversy continued to build around cotton policy. It did not help that controversial changes were made to the Senate bill's cotton provisions in a Sunday evening session at the request of Senator Theodore Bilbo (D-MS).[102]

The 1937 debate demonstrated the continued challenges for programs stretched across cotton, corn, and wheat.[103] Wheat and corn were used to feed humans and livestock, so limiting the acres planted to them came with significantly more risk. For example, if the weather turned unfavorable in the growing season and too few acres had been planted, then the policy could be blamed for food shortages and price spikes. This is what led to Secretary Wallace's proposal for an ever-normal granary. While senators used acreage allotments for corn and wheat, they were not reductions, nor were they compulsory.[104] Instead, if favorable growing conditions produced an excessive crop on the allotted acres, USDA could introduce limits on how much of the crop could be marketed—known as marketing quotas—but farmers had to accept them in a referendum. This was the ever-normal granary concept: if the crop came out too big, then the federal role was to help the farmer store that crop until prices could recover and to protect against a potential short crop in the future.[105] In other words, "so far as wheat and corn are concerned, the bill . . . does not seek to control production or marketing until a certain stage of overflow in the granary has been reached."[106]

Cotton considered itself to be in a unique situation that required different policy treatment. It was used for industrial purposes and not for food or feed, plus it was mostly exported; a short crop would not create the same type of problems but permitting too much cotton to be planted would continue to depress prices. Cotton interests adhered to the compulsory program, their priority was trying to keep production aligned with consumption by reducing acres.[107] Making this point, Senator Bankhead explained to Senator McNary, "We have a different method of approach. . . . We try to avoid what is, as we see it, the waste of producing more than is needed and thereby reducing the price . . . avoiding . . . in advance producing a crop which must be impounded on the farm."[108] They wanted to avoid having surplus stocks in storage, which they viewed as weighing on the market and depressing prices. There were, however, increasing concerns about the consequences of reducing cotton acres even among southerners who supported it.[109]

Outside the South, concern was turning into outright opposition to a policy that was failing in large part because of the farmers themselves. Many of them were responding with more intensive practices to improve yields on reduced acres. Senator William McAdoo (D-CA) pointed out, "There is no authority in the bill to control or regulate the fertilization of that soil."[110] Add in the unpredictable impacts from weather and reducing acres was also looking more and more counterproductive. Cutting back the acres American farmers planted to cotton gave away export markets to foreign producers at cheaper prices. The policy would likely exacerbate the economic problems of U.S. cotton farmers.[111] Southerners were forced to acknowledge that if acreage reductions failed to control supplies, the logic behind them would require steeper and steeper reductions that would harm farmers and not be politically sustainable.[112] These issues did not, however, deter the demand for strict acreage limits.

Politically, the demand set cotton apart from corn and wheat, creating problems for the coalition. What made the debate over acreage reductions more difficult was that the policy pitted cotton's interests against the "very delicately balanced machine" of American agriculture.[113] Acreage reduction policy applied intense pressure on the fault line.

Cotton could not be grown in the Midwest or much of the Great Plains regions. Corn and wheat, however, could be grown in the South, as could other feed grains that competed with corn such as sorghum. Under their preferred policy, southern cotton farmers would receive payments to reduce acres. Those acres could go into direct competition with other regions and commodities, especially feed grains that competed with corn.[114] Cotton acres could also go into producing forage for livestock and dairy, which raised opposition from those interests as well.[115] At the same time, cotton farmers were not limited on how much they could produce on their reduced acres. Thus, more intensive farming practices could increase yields on fewer cotton acres while acres were diverted to subsidized competition with wheat and corn. The policy left the impression that it was partially designed to force diversification in the South at the expense of farmers in the other regions.

In the House the fight over acreage reductions for cotton centered on an amendment by Representative Gerald Boileau (R-WI). His amendment was designed to protect the dairy industry from expanded production on former cotton acres, which USDA data indicated was taking place.[116] Representative Boileau's attempts to limit production on the reduced cotton acres, however, produced strong objections from southerners that their farmers were being treated unfairly.[117] It is clear from the House debate that cotton's demands created direct conflict with the interests of its coalition partners and had real consequences on the ground for farmers. Cotton was creating, or at least adding to, the political problems for farm policy. This was further magnified by the outright rejection of acreage reduction policies for wheat and corn.[118] Acreage reductions created the most distinct policy difference among the major commodities.

Acreage reduction policy was one part of a dual conflict among agricultural interests about the proper direction for long-range or permanent policy. The House and Senate both struggled with the design of price supports. The Senate established a scaled concept under which price-supporting loan rates would depend on the level of supply.[119] The more a commodity was oversupplied, the lower its loan rate. Given the very different market situations for corn, cotton, and wheat, lawmakers should have anticipated that this policy would have the effect of producing different loan rates for each. Specifically, chronically oversupplied cotton would come out with consistently lower loan rates. Loan rates could also be lower at the same time and for the same reasons that the market prices would be lower: surplus commodities in the market. Added to that, the Senate wanted to make direct payments to farmers that would supplement the loan rate and were called parity payments.[120] The way it was designed, the more a commodity was oversupplied the lower its loan rate, and the lower the loan rate the higher the parity payment. Where a flexible loan rate would likely hurt cotton, the payments would be expected to benefit cotton farmers because of the larger spread between prices and the loan rate. If the concern was subsidizing the diversion of cotton acres into competing production, these payments would worsen the problem.

Corn Belt representatives, led by Representative Scott W. Lucas (D-IL), picked a fight in the House. They demanded mandatory loans for corn, arguing that 85 percent of corn went to feed livestock on the farm or on neighboring farms.[121] They also took aim at speculators in the commodities markets and argued that the mandatory loan would allow farmers to store their corn after harvest rather than sell it at the mercy of "nefarious schemes" and the lowest prices.[122] They faced opposition from southerners but also representatives from wheat-growing districts who thought it was a dangerous policy of congressional price fixing.[123] Representative Hope called the Lucas amendment "unwise" because it would "change the entire theory of this bill and make it a price-fixing bill."[124] Southerners were especially opposed to it if the loans were not linked to production controls, considering it too favorable to midwestern corn farmers.[125] The regional commodity dispute blended disagreement over acreage reductions with an apparent competition among members to secure the spoils for their farmers. Representative Frederick Biermann (D-IA) made the point that "again and again we have given special treatment to cotton" while special treatment for corn was justified because "corn is not exported, it is consumed in this country" and thus a "loan on corn is different than a loan on a farm commodity with a large exportable surplus."[126] Down this path awaited further complications.

The regional disputes over policy were difficult but not enough to block the bills. The House passed its farm bill on December 10, 1937, and the Senate passed its version on December 17, 1937.[127] Passage in both chambers moved the fight over acreage reductions and price supports to conference, where negotiators struggled to find resolution. The two matters were tied together. Production-control policies were considered necessary to make price supports work because the federal government could not put a floor under commodities that paid above-market prices and also permit unlimited production at those prices. To support commodities and serve as the buyer of last resort, it was argued, USDA had to place some restrictions on how much could be planted or marketed. It was a necessary evil, resorted to in a world that was unfair to the farmer.

The House had wanted to continue use of soil conservation to pay farmers directly, but the Senate demanded parity payments.[128] Conferees compromised by codifying the parity concept and continuing the conservation payments, which were popular with farmers.[129] Southerners also continued to push for additional benefits for cotton farmers. This added to the challenges for conference negotiators and created leverage against cotton, further feeding concerns that helping cotton while reducing acres created competition for farmers of the other major commodities, including dairy.[130] Dragging dairy into the fight put cotton in an even tougher spot, forcing it to defend its programs against both corn and dairy.[131] Corn interests remained frustrated that the price-supporting loans depended on farmers' accepting some form of control and that different corn-producing regions were given different treatment.[132] The commodity and regional disputes highlighted the growing complications with farm programs.[133] Deal cutting in conference produced different programmatic treatment for each of the main commodities and resulted in regional discrepancies.[134]

The Agricultural Adjustment Act of 1938 created very prescriptive farm policy, beginning with the production-control features.[135] The 1938 act was the first to explicitly use the term "parity" and define it. Parity prices were those that gave the commodity the "purchasing power with respect to articles that farmers buy equivalent to the purchasing power of the commodity in the base period" (1909–14).[136] The bill provided payments to farmers for diverting acres from the major commodities to achieve soil-conserving purposes, plus parity payments if Congress separately appropriated funds for them. The bill supported commodity prices through nonrecourse loans for farmers who accepted acreage allotments and, if necessary, marketing quotas. As part of the conference compromise, the loans were more favorable for corn than for cotton or wheat.[137] For example, corn loans were available at higher market prices (75 percent of parity) than cotton and wheat loans were (52 percent of parity) and at more favorable loan rates. The loans favored corn, but cotton received the acreage reduction policy it demanded, a negotiated settlement due to the intense fight by midwestern House members and the demand for extra benefits to cotton farmers.[138]

The 1938 AAA also created the federal crop insurance program. Initially, the program was only for yield losses in wheat production, and farmers could pay premiums either in cash or in kind with wheat. The Federal Crop Insurance Corporation (FCIC) was required to "fix adequate premiums" for insurance policies.[139] Midwestern and northeastern farm interests had requested crop insurance in the Senate's version of the farm bill, and it was accepted by the conferees. The insurance program was viewed as an experimental method to insure against yield losses for a crop that often struggled in the challenging climate of the Great Plains.[140] It constituted further evidence of the deal making in conference, which provided acreage reductions for cotton (with payments), mandatory loans for corn, and crop insurance for wheat.

Congress passed the conference report in February 1938, and President Roosevelt quickly signed it into law.[141] The prescriptive, complicated program details contained in the final legislation revealed the increasing divisions among the commodity and regional interests. They further exposed the fault lines and demonstrated their impact on the development of farm policy—harbingers for disputes to come. In fact, the final compromises in the 1938 negotiations set the stage for the major corn-cotton dispute that followed World War II, none more so than acreage reductions.

The bill had also witnessed increasing partisanship. Republicans were critical of the production-control policy and considered it a bad deal for farmers that would have political repercussions.[142] They considered acreage reductions to be impractical and unworkable, increasing foreign competition and only benefitting large farmers.[143] Production controls had created "a real battle in conference" but survived as a key part of what some claimed to be the last chance at rational policy to help farmers; failure would result in more radical policies.[144] Acreage-reduction policy for cotton continued to cause the most problems, affecting farmers and damaging the cotton industry. Congress pushed ahead in spite of the problems, acceding to the demands of their farmer-constituents.

If farmers supported acreage reductions, it was likely because they permitted some farmers to consolidate land holdings and reduce labor costs—efforts to remain competitive with increasing production by foreign farmers and of synthetic fibers.[145] In addition to cutting labor costs and consolidating lands, the federal payments contributed capital to purchase tractors and better inputs, which further reduced labor needs and costs. Under the programs, mechanization was spreading eastward from Texas.[146] Criticism of the policy focused on the consequences for farmers who were unable to remain in farming; some critics were troubled by the inescapable racial element of the policy.[147] Their concerns were magnified by the fact that the displaced sharecroppers had few options in an economy still trying to recover from the Great Depression. Many flooded the labor market for harvest, further depressing wages and limiting options. They also increasingly looked to federal policymakers for assistance.[148] From the first cotton plow-up to the Bankhead controls to the STFU and the resulting "reign of terror" in Arkansas, these consequences were obvious and becoming increasingly difficult to ignore politically.

Thus, the 1938 AAA marked a turning point on the issue. For the first time, Congress took significant steps in the legislation to address the issue and provide protection to tenants and sharecroppers. The bill required that payments be divided with tenants and sharecroppers. It precluded a landlord from increasing federal payments by reducing tenants and sharecroppers. Most important, the bill defined producers to explicitly include tenants and sharecroppers and required USDA to protect their interests.[149] This was the result of growing political pressure on Congress that caused intense fights on the House and Senate floors.

A clear sign of trouble arose in the Senate. The special session called by President Roosevelt became entangled with anti-lynching legislation and the growing political unrest surrounding issues of race and civil rights.[150] How much the two were linked is unclear, but senators expressed increasing concern about the treatment of the "small cotton farmer or sharecropper" by the adjustment legislation.[151] There was little opposition to an amendment by Senator Joshua Lee (D-OK) that required federal payments to "be divided among the landowners, tenants, and sharecroppers of any farm" proportionally.[152] Senator Elmer Thomas (D-OK) added support for changes to "make it possible for the sharecropper and the tenant farmer . . . to make some money" so that they could advance and become landowners.[153]

The more intense debate again took place in the House. Representative John Robsion (R-KY) claimed that the 1933 AAA had failed. As proof, he cited that a "great commission established the fact that the A.A.A. took about a million farmers, farm tenants, and sharecroppers off the farms on account of the cotton cut-out" and most were "forced onto relief."[154] Representative Tarver again spoke out on the issue. He proclaimed that the 1934 Bankhead Act had "operated very harshly on the little man, the tenant, the sharecropper, and the small producer, and we all professed much sympathy for them."[155]

The House Ag Committee bill exempted from the allotment requirement any farmer who produced three bales of cotton or less, but this was not considered satisfactory as long as the system operated through the landlords.[156] Pressed on the issue, Chairman Jones acknowledged that the sharecropper's rights were "dependent on his contract with the landlord" but argued that it was a matter with which Congress could not interfere.[157] Representative Tarver pushed the issue, calling out members who have "advised their constituents of their sympathy" over treatment under the Bankhead Act to demonstrate sincerity and "take steps to prevent a recurrence of the hardships and trials" from it.[158] He complained that "the Department of Agriculture by some sort of administrative legerdemain which I have never been able to understand, has construed the word producer to not include a sharecropper."[159] To help remedy this problem, Representative Tarver was able to get the House to accept his amendment that would require fair representation on local and county committees for tenants and sharecroppers.[160] This did not, however, go far enough to address "one of the principal evils of previous farm legislation" at the local level.[161]

Representative George Mahon (D-TX) sought to limit the landlord's ability to reduce tenants or sharecroppers by precluding an increase in the landlord's benefits if he had previously reduced their numbers. Mahon argued that "I know of too many instances where large operators have discharged their tenants, bought tractors, and are working from 500 to several thousand acres of cotton with hired labor" and "tenants have been unable to rent other farms and have been forced on the relief rolls." He added that the "government should not pay a man benefit payments for putting his neighbors on the relief rolls," but the committee's provision was "a case of locking the stable after the horse has been stolen for perhaps 2 or 3 years."[162] In defending the committee provision, Jones noted that "I wish there were some way to go back" but it was impossible.[163] Republicans joined the fight against the New Deal policies, pointing out that these large planters did not "need a subsidy from the Government."[164]

Representative Tarver faced significant resistance to his attempt to revise the definition of the term "producer" to include tenants and sharecroppers.[165] Chairman Jones explained that the sharecropper did not have control of the farm unit and thus was not a producer. Representative William Whittington (D-MS) registered opposition to the "discrimination" against large landowners and argued it could raise constitutional takings concerns.[166] Representative Richard Kleberg (D-TX) wanted to know "whether we are drafting a bill for the benefit of agriculture or whether we are finally going to have a bill which is solely for the benefit of tenant farmers and sharecroppers."[167] Representative John E. Rankin (D-MS) argued, "The man who owns his land . . . is worth more to his country . . . than is the Negro tenant or the Mexican tenant [or] . . . the large absentee landlord who controls thousands of acres of rich land, works it with Negro or Mexican tenants, and raises nothing but cotton."[168] Ultimately, the Tarver amendment was defeated by the House, but the debate had exposed significant tensions.[169]

The final version of the 1938 AAA that was worked out in conference demonstrated the most significant progress to date on the issue. It required that payments be made directly to landlords, tenants, or sharecroppers and divided among them on a proportional basis. Conservation payments were distributed according to "the extent [to] which such landlords, tenants, and sharecroppers contribute to the carrying out of such practices."[170] The bill also addressed reductions in the number, or changes in the status, of tenants and sharecroppers on a farm. It precluded such changes from being used to increase payments to the landlord but permitted local committees to approve the changes.[171] Congress also permitted assignment of payments for sharecroppers and tenants to the landlord.[172]

The 1938 act carried larger political consequences. President Roosevelt had won an overwhelming reelection in 1936, carrying nearly all the votes in the South and helping Democrats to again expand their majorities in Congress.[173] The 1938 act did cause a few political problems for the president in Congress and his party in the midterm elections. The most notable defeat was of Senator McGill, one of the main authors of the 1938 act.[174] But Senator McGill was not alone. In the 1938 election, Republicans attacked the production-control features of the act and won seats in farm country.[175] Many Democrats at the time blamed their losses on the 1938 AAA and low commodity prices. Roosevelt, however, went on to win an unprecedented third term in 1940.[176]

Adjusting to World War II with High, Fixed Loan Rates

The 1938 farm bill likely failed, but an accurate conclusion was lost to the onset of yet another world war. World War II damaged export markets for U.S. cotton and wheat, but the domestic feed market continued to drive relatively strong demand for corn. Under the 1938 act, corn farmers were getting loans at 70 percent of parity while wheat and cotton were receiving lower loan rates at 52 and 57 percent of parity respectively.[177] On top of that, cotton farmers were required to make substantial reductions in planted acres.[178] These issues combined to drive demand for appropriations to make parity payments, which benefitted cotton the most.

More consequentially, the flexible loan rate feature of the 1938 farm bill gave way to a demand for high, fixed loan rates across all commodities during World War II.[179] Supporters of the 1938 bill had warned that its failure would likely lead to more radical policies. Congress proved them correct. In 1941 Congress began fixing the loan rates for all commodities at a high percentage of parity.[180] The move was part of the continued regional fights among corn, wheat, and cotton. Because corn had lower levels of carry-over supplies, corn farmers were receiving higher loan rates than either cotton or wheat farmers.[181] This was, however, exactly how Congress had designed the program in 1938.

The Department of Agriculture was unable to keep supplies under control, and Secretary Claude Wickard requested that Congress increase penalties on farmers if they violated corn and wheat marketing quotas; the Senate had agreed.[182] The House Agriculture Committee decided that any increase in penalties should be coupled with an increase in the loan rates. They provided for mandatory loans at 75 percent of parity—a fateful step.[183] Members were well aware of the potential consequences. They knew that high, mandatory loan rates would lead to increased production and more surplus commodities. Providing farmers with a fixed price floor could contradict market signals and lead them to continue producing oversupplied commodities. For House members local concerns about equalizing treatment among commodities prevailed.[184]

The Senate responded by amending the House provisions to increase the loan rate to 100 percent of parity. Conference split the difference and fixed all loan rates at 85 percent of parity.[185] The legislation not only broke with the 1938 act's flexible schedule; it did so without requiring any additional production controls or reductions in planted acres.[186] American Farm Bureau president O'Neal declared victory.[187] Victory or not, the new loan rates crossed a threshold Congress had previously avoided and established a new status quo that would be difficult to change.

The attack on Pearl Harbor in December 1941 pushed the U.S. fully into World War II, and another round of wartime demand strengthened commodity prices but also caused concerns about inflation. Congress and President Roosevelt fought over the proper policies during wartime: Roosevelt was concerned about controlling inflation with price controls; farm interests in Congress demanded price supports.[188] They settled on loan rates fixed at 90 percent of parity for two years after the war ended. This was subsequently extended to all commodities that were asked to increase production for the war effort. In short order and under pressures of war, Congress had followed the 1941 Joint Resolution with the Steagall Amendment of July 1, 1941, which fixed price supports at 85 percent of parity for all commodities that were asked to increase production for the war effort, and then at 90 percent of parity for two years after the war. Notably, the situation during the war became desperate enough to require rationing foods, while farm commodity prices and farm incomes increased more than they had in World War I.[189]

While Congress was pushing loan rates to 90 percent of parity, the Supreme Court upheld the 1938 act in the landmark case *Wickard v. Filburn*.[190] Filburn, an Ohio farmer, challenged the penalties he was charged for producing more wheat than he was permitted. The farmer argued that the wheat was going to be consumed on the farm, either as animal feed or as flour. Therefore, he argued, it was not a matter of interstate commerce under the commerce clause of the Constitution, and his production was outside the reach of Congress.[191] The Supreme Court disagreed. It concluded that Congress's powers under the commerce clause extended to "those activities intrastate which so affect interstate commerce . . . as to make regulation of them appropriate means to attainment of a legitimate end."[192] The court added that it was permissible "for the Government to regulate that which it subsidizes."[193] The decision was a complete change in the court's view on New Deal farm policies. After 1942 farm price-support and production-control policy bore the imprimatur of the U.S. Supreme Court as a permissible subject for congressional action and regulation.[194]

Concluding Thoughts on the New Deal Adjustments

Clearing the Supreme Court was the final achievement for New Deal farm policy in the American political system. The court's imprimatur established a lasting precedent for congressional authority to intervene in the economy. Whether it was for conservation, support, or regulation of production, the federal reach could permissibly extend to every farm in the country. As the court made clear, this reach was a by-product of the receipt of federal assistance; subsidization and regulation went hand in hand. This cleared the way for a long-running political process through which policy could be developed over the course of many legislative efforts. Within the farm coalition, however, achieving the court's stamp of approval was mere prelude.

The roughly twelve-year reign of Franklin Roosevelt and his New Deal was unprecedented. It permanently altered the landscape of American government; few areas of the economy were as clearly affected as agriculture. The 1933 AAA had been a part of the first hundred-day response of a new president. It had failed one Supreme Court test but passed another. With all the act's problems and controversies, its New Deal programs delivered substantial relief to farmers and through them many parts of rural America. The bill provided billions of federal dollars in payments and loans but also helped with everything from mortgage relief to rural electrification.[195] As just one data point, net income from farming went from $2.285 billion in 1932 to $7.723 billion in 1941, increasing steadily almost every year that some version of the AAA was in operation.[196] The political benefits had also been immense, as demonstrated by Roosevelt's four electoral victories and the strong Democratic control in Congress throughout the era. By some measurements then, New Deal farm policy was incredibly successful, and it established a very high bar for federal policy. Going forward, however, the challenges would change substantially; development of farm policy was about to enter a very turbulent and difficult period.

The forces of history worked quickly, moving from Depression to war and then to postwar.[197] President Franklin Roosevelt died on April 12, 1945, and Vice President Harry Truman became president.[198] Hitler committed suicide a few weeks later, and the U.S. and its allies declared victory in the European theater of war. Japan surrendered in August 1945, bringing World War II to an end. President Truman declared the official end to hostilities on December 31, 1946. The war's official end started the two-year clock on the 90 percent of parity price-support policy.

President Truman struggled in the shoes of his predecessor.[199] In the 1946 midterm elections, Democrats lost control of Congress for the first time since 1930. Farmers noticeably broke from the Democratic coalition, especially in the Midwest. Historian Dr. Allen J. Matusow has written that "by election day 1946, the delicate fabric that was F.D.R.'s New Deal coalition lay in tatters" as wartime prosperity changed political calculations and farmers "joined businessmen in protesting the restraining hand of government." He added, "Peace inaugurated a rancorous and destructive struggle among competing groups for consolidation of their uncertain wartime gains. . . . Farmers, for instance, had repudiated the New Deal coalition in a fury that could not soon be assuaged."[200]

As mentioned in the introduction to this chapter, the New Deal years and World War II were a period of adjustment—adjustments in agriculture, for farmers, and in policy. All this adjusting occurred in the framework of emergency legislation to combat the Great Depression and, to some degree, the Dust Bowl. World War II ended that process. It effectively ended the Great Depression and thus the emergency legislative responses to the Great Depression.

For all intents and purposes, New Deal farm policy ended in 1941, not when Japan attacked Pearl Harbor and America entered in the war, but when the flexible system of 1938 gave way to the high, fixed parity loan rate. This shift constituted a fundamental change in the commodity programs and in the philosophy of the policy. The farm bill was no longer an emergency response designed to assuage some of the pains inherent in farm economy adjustments. Instead, it became a blatant effort to maximize and solidify previous gains—a grab for policy spoils based on speculation about postwar uncertainties negotiated in the fog of war.

During the war the commodity programs supported the war effort by encouraging production through 90 percent of parity price supports. The wartime demand improved prices and should have loosened acreage restrictions, although cotton acres continued to decline steadily (see appendix 1, figures 1 and 2). Farmers prospered during the war, as did the rest of the economy. For example, net income from farming topped $13.7 billion in 1945.[201] For farmers this prosperity was clouded by concerns that prices would collapse again after the war ended. Their supporters continued to push for and defend the high, fixed loan rate policy.

During the New Deal era and World War II, farming changed nationwide. New Deal policies have been criticized for benefitting larger, wealthier, and commercial farmers, while forcing small farmers out of farming.[202] Nationwide, people left farming in large numbers and the size of farms increased—a trend that didn't start with the Great Depression but was certainly accelerated by it.[203] These changes were mostly due to technological and economic factors, as well as other issues like the Dust Bowl; farm policy's impacts, however, cannot be ignored.[204] Understandably, farm policy had a far bigger impact on the politics of agriculture. These political impacts would, in turn, drive farm policy in what became a difficult feedback loop. At the close of this era, the political situation was quickly coming to a head.

The legacy of cotton policy remains the most difficult aspect of the New Deal's adjustment programs. Under New Deal acreage policies and price supports, the southern farm population shrank dramatically, and cotton fell from its place as the South's main cash crop.[205] The mechanical cotton picker, introduced in 1942, provided another key method for southern farmers to reduce farm labor.[206] It was adopted slowly, however, and by as late as 1950, "only about 5 percent of the nation's cotton was picked by machine"—mostly in the Mississippi Delta, Arizona, and California.[207] One additional challenge was weed control, and further mechanization depended on advances in chemicals.

Extensive research has concluded that the combination of acreage reduction policies and mechanization drove sharecroppers in the traditional cotton-producing states of the South off the farm in large numbers in the 1930s and throughout the war years.[208] Federal payments and acreage reductions helped farmers consolidate land holdings and mechanize production. A sign of problems to come, those acres were not going out of farming.

Adjustments in cotton policy in 1934 and 1936 did little to ease the problems created by the 1933 plow-up and advancing mechanization partially underwritten by federal support.[209] On this issue World War II also altered matters substantially. The war and defense production helped ease the surplus farm labor issue, pulling many into the armed forces or into urban areas to secure good jobs in manufacturing.[210] With opportunities outside the region, many former sharecroppers migrated north; the postwar economy continued that pull for labor. The greatest contribution to the plight of the sharecropper, then, was not federal policy changes but rather the prospect of work in northern factories.[211]

From the perspective of 1945, New Deal farm policy was bookended by the cotton plow-up and 90 percent of parity. The policies and programs, unfolding at great speed and under enormous challenges, helped many farm families and rural communities. They hurt a great many people as well. The New Deal forever changed the agricultural situation in this country but failed to reform certain of its most troubling components. Neither the effort nor policy development ended with war. Going forward, however, status quo was established at 90 percent of parity with acreage controls.

Vastly different experiences under depression compared to war helped to develop a new perspective on policy. The nation had survived the Great Depression. New Deal policies attempted to control production as an emergency response with questionable success. Full employment and strong demand due to World War II had ended the Depression and returned strong prices and incomes for farmers.[212] The emergency had passed. Linkage of these matters—full employment, strong demand, and strong prices—had been absorbed in some corners of the federal policy apparatus. From those corners emerged an opposing policy perspective that came to be known as the abundance view.[213] This view emphasized improving employment and consumer income and came to include opposition to price supports, production controls, and the concept of parity. The view's adherents considered the 1938 act policy for a depression economy and an ill fit for the productive, booming postwar economy. They wanted to transition farmers out of the policy with direct payments—to subsidize income rather than support prices.

The abundance view also included assistance to subsidize or improve demand, such as by assisting low-income households and the unemployed in purchasing food. One example can be found in the debate surrounding a proposal known as the National Food Allotment Plan, designed to allow low-income families to purchase stamps that could then be used to purchase surplus commodities and improve their diet. The plan was first introduced by Senator George Aiken (R-VT) in 1943 and was based on the Food Stamp Plan that the Roosevelt administration had created and operated during the later years of the Great Depression. The first version was defeated on the Senate floor as an amendment to a bill extending the life of the CCC in February 1944. Senator Aiken revised the plan and reintroduced it in 1945. He would continue to push for the policy but struggle with concerns that it would jeopardize the price-support program.[214]

The opposing views that developed out of the Depression and the war quickly produced a fundamental policy disagreement that tested the farm coalition. This dispute was rooted in the different lessons taken from these two seminal events and boiled down to an argument about surplus and abundance: Could great productive capacity be a force for good, or was it another pending disaster for farmers? The difference proved a catalyst for the difficult farm policy fights that were on the horizon. In 1946 Secretary Clinton Anderson organized long-range policy planning at USDA.[215] He put Assistant Secretary Charles Brannan in charge of the effort, and the next round of history began.

3

Transition and Turbulence after War, 1945–1949

Introduction

The shift from New Deal emergency legislation to the parity era in farm bill history began with the 1938 attempt at permanent legislation, which codified the parity concept. The parity era continued through the fortieth anniversary of farm policy in 1973. As will be explored in this chapter and the next, the farm coalition experienced incredible difficulty as it developed policies during the majority of these years—beginning after World War II and continuing through 1969. Congress produced multiple bills, each seemingly more contentious than the one before it, and the effort suffered a few significant setbacks. Throughout, Congress made minor, incremental progress toward revising the policy status quo that emerged from World War II; legislators were challenged to fit policy to changing times in farming.

Commodities began piling up in federal storage, and these surpluses provided clear, politically powerful evidence that the policies were failing. The debates in Congress indicated a frustration with the policies' apparent inability to deliver improved crop prices and farm incomes. The farm coalition that had been forged in crisis broke apart over its fault lines in the turbulent transitions of this era; the coalition itself was the source of much of its own undoing. One analysis explained that after the desperation of the Depression had passed, farm interests concentrated on "how individual crop, variety or area interests would be treated in subsequent legislation," which degenerated into a "luxury of internal conflict."[1] The roots of this problem were visible in the 1937–38 debate and would grow into a complete breakdown of the policy. Division within the coalition began in earnest in the years immediately following World War II and with the two farm bills produced during the Truman administration explored in this chapter.

The internal conflicts were certainly driven by self-interested actors in competition for the program benefits they were debating. But these conflicts were also the product of challenges inherent in reconciling interests with extraordinarily changing circumstances. Programs designed in response to one set of problems struggled to adjust when circumstances changed, especially when those changes resulted from the policies implemented. For example, USDA assistant secretary Brannan was put in charge of the postwar planning efforts at the department. Those efforts struggled to design recommendations that were relevant to the unknowns of the postwar economy; foresight was extremely limited. To many in USDA, World War II had delivered a lesson in the benefits of having abundant commodity supplies to feed a growing economy experiencing strong employment and incomes.[2] Adherents of this view came to see the problems of the Depression as too little demand, not oversupply; unemployed (or underemployed) consumers lacked the ability and purchasing power to buy what the farmer supplied.

This abundance view conflicted with the traditional view rooted in concerns that farmers would inevitably overproduce and that the resulting surplus would do great harm to farm incomes and, ultimately, the larger economy. The abundance view emphasized support for production through flexibility and less federal intervention, coupled with assistance to the consumer to help bolster purchasing and consumption. The traditional view adhered to the New Deal policies as modified during the war: production controls in return for high, fixed price supports. These views collided with conflicting self-interests from within the commodity coalition. The result was a dispute over acreage policy and loan rates that became increasingly intense and irresolvable. Neither side accurately predicted the immense changes taking place in agriculture and neither meaningfully responded to them in terms of policy development.

Initially, the postwar farm economy remained strong. Unlike after the First World War, the United States helped Europe recover and rebuild. Known as the Marshall Plan, America's European Recovery Program included efforts to combat severe food shortages caused by the war's devastation.[3] These efforts fueled demand, which, in turn, continued to deliver strong prices and incomes to American farmers. Because of the Marshall Plan, farmers were able to avoid another price collapse in the war's aftermath.[4] But farm policy faced a new challenge with 90 percent of parity price supports scheduled to end on December 31, 1948. Without new legislation, farm policy would revert to the 1938 act's flexible loan rate provisions. Pressures new and old came to bear on the farm coalition. The immense stress opened a major schism and a bitter, internecine fight over the way forward.

The traditional concern was that the farm economy would return to surplus commodities and depressed prices. The familiar responses were Depression-era adjustment policies that combined flexible price supports with production controls.[5] Farmers were comfortable with high, fixed price supports that were the same for all commodities. Returning to the 1938 system would result in both lower loan rates (and potentially prices) and different loan rates. The efficacy of these policies in the new postwar economy was unknown, but wartime price-support experiences provided reasons for concern. The 90 percent of parity price support had been designed during wartime demand, but it carried a great risk of overproduction if it continued without that demand. Price supports could create a powerful production incentive that insulated farmers from the adjustments that the market signaled were necessary. Simply put, if the market signaled a need for adjustment through lower prices, high loan rates interfered. This discrepancy was not a surprise, nor was it something that policymakers did not or could not understand. The problem was that any postwar market signal for adjustment conflicted with political signals and the general desire to deliver to an important constituency.

That was not the only problem, however. The real surprise was a new production reality. After the war American agriculture witnessed a production revolution.[6] Fewer farmers were producing more commodities than farmers had in the thirties, and on less land. Much of this was due to the adoption of mechanized farming practices combined with the introduction of hybrid seeds, synthetic fertilizers, and chemicals to improve yields. The full impacts of this production revolution were yet to be felt in the early postwar years, but they provided enough support for traditional farm interest concerns about overproduction and low prices. In a nation where the consumer did not spend much on food, farmers remained insecure because real risks remained that demand could not keep up with supply.[7]

By 1947 farmers and their supporters in Washington had fourteen years of experience with farm policy. They were experiencing strong prices and had the security of a high floor price for their crops. Prices and supports were underwriting mechanization and the adoption of more intensive production practices. But there were signs of problems ahead. The regional, commodity fault line had been widened by the 1938 effort and the different loan rates it produced. Partisan pressures were also growing as the FDR coalition broke apart and Republicans gained political advantages. The Great Depression faded further into history, increasingly obscured by a booming economy. The epicenter of this partisan pressure could be found in the Midwest, where New Deal farm policies had been least intrusive. This contrasted sharply with the South, where cotton acreage reductions had been incredibly intrusive. These two regions and their commodities had always occupied opposing poles on the policy spectrum, but the postwar years drove them further apart.

The first round of this dispute between the Midwest and South took place within the American Farm Bureau in 1947.[8] Midwestern Farm Bureau members, dominated by corn and hog interests, opposed high price supports and, especially, production controls. Southern members, dominated by cotton interests, were strong supporters of the 90 percent of parity and production control system. The policy differences were rooted in both market realities but also in a sort of philosophical difference that had been prominent in the 1937–38 debate. To corn the price-support system was merely a backstop against the risk that prices could collapse.[9] Corn farmers had to be cautious about inflating prices with high loan rates because it might damage their feed customers. Corn was fed by many of the same farmers who raised it or by their neighbors; it relied on domestic demand, much of which existed in the Midwest. Controlling production by reducing acres planted to individual crops was counterproductive. It would hurt both corn farmers and feed customers. Moreover, controls for other commodities diverted acres that created competition in the feed market.

For cotton the high, fixed loan rate was protection against the vagaries of the world market and cheap foreign competition. Production controls, like reducing acreage, were necessary to avoid surpluses. They were also needed to move southern farmers to diversify their operations and decrease reliance on cotton; decreasing reliance necessarily involved more than adding new crops. As discussed in previous chapters, cotton acreage reductions were also a method for adjusting southern farm labor to advance mechanization, removing sharecroppers from cotton farms, and advancing consolidation. These were strong parochial interests for cotton; cotton farmers were less concerned with creating competition problems for corn and wheat farmers in the rest of the country. The elevation of self-interests over the interests of the entire coalition invited trouble.

The Midwest won the early round in Chicago. The American Farm Bureau elected as its president Allan Kline, an Iowa corn and hog producer. He replaced the southern cotton farmer Ed O'Neal. The Midwest also won on policy. The Farm Bureau convention endorsed "mandatory variable price supports—with or without quotas."[10] These victories escalated the regional fight over farm policy. That escalation, however, coincided with the unprecedented and largely unforeseen growth in the farmer's crop-producing capabilities.[11]

The Postwar Stalemate and the 1948 Farm Bill

Republicans captured the House and Senate in the 1946 midterms, and the new Congress began working on a bill to replace the 1938 act.[12] The bill was being crafted in the House under the leadership of Ag Committee chairman Clifford R. Hope (R-KS). In the Senate Arthur Capper (R-KS) was the Ag Committee chairman beginning in 1946, but the long-serving senator decided to retire rather than seek reelection in 1948.[13] The record indicates that much of the leadership on the Senate bill was provided by Senator George Aiken (R-VT). Hope and Aiken were from very different parts of the country, and the effort to pass a compromise bill dragged into the 1948 election year.[14]

Chairman Hope, born into a successful Iowa farm family that later established itself in Garden City, Kansas, has been described as an "agrarian idealist" who believed in the fundamental importance of farming to society; he held Jeffersonian ideals "tempered by the special interests of the wheat country" he represented.[15] He had earned a law degree and served in World War I before he was elected to the Kansas legislature and then to the U.S. Congress. On the other side, Senator Aiken was a Vermont fruit farmer who had held local and state offices before he served thirty-four years in the U.S. Senate, where he was often known as a progressive, maverick Republican.[16] These backgrounds add perspective to the different routes taken by these two policymakers.

Warning signs of the pending trouble for farm policy generally, and the 1948 bill specifically, were apparent early. Farm organizations, the Truman administration, and congressional Republicans initially agreed that farm policy should return to flexible price supports.[17] The South, however, strongly opposed the flexible loan rate policy; southern farmers were concerned that support would be allowed to decline and "be of little, if any, benefit to the growers at the time the price support is a necessity."[18] The House and Senate moved opposing policies. Chairman Hope sponsored a one-year extension of 90 percent parity supports, contending that supports kept the promise made to farmers and would allow them to make readjustments during the unsettled times after World War II.[19] Supporters argued that another year was needed to "taper off wartime price support and build a bridge to a comprehensive long-range program" after the election.[20] To supporters, including some from the Midwest, the parity system was "the keystone" of farm policy and "the attainment of economic equality for agricultural producers."[21] The intervening war years had not altered traditional farm interest concerns about depressed prices and incomes formed by the years after World War I and during the Great Depression.

Opponents countered that the "first problem before the Nation today is the high cost of living" and that Congress should focus on that instead of continuing supports "when prices are the highest in the history of the Nation" with a bill that "will continue the high prices to consumers."[22] To opponents, continuing policy designed during wartime would damage the entire national economy, especially the consumers they represented. Their position would have aligned with the abundance viewpoint that wages needed to be increased to spur demand. This view was complicated by a desire to control prices and inflation. The challenge for these members, however, was that they were advocating on behalf of consumers in a farm bill debate. While the debate exposed partisan undercurrents in an election year, farm interests held great strength in the House.[23] With the upper hand in the lower chamber, they won easily.[24]

Unlike the House bill, the Senate bill presented a break from both the 1938 act and wartime high, fixed price supports.[25] Led by Senator Aiken, the Senate Ag Committee had unanimously agreed to revise farm policy to make it more flexible and better aligned with the market conditions of the time. The Senate bill floated price-support loan rates between 60 and 90 percent of parity (depending on the relationship of total supply to normal supply) and sought to update the old parity formula to use more recent prices. Senator Aiken argued that a "flexible price system that reflects supply and demand conditions" provided a "better guide to production needs" whereas "high price supports . . . have the tendency to freeze production or acreages in undesirable patterns rather than to encourage adjustment in production to meet demands."[26]

Flexible price supports, in this view, allowed farmers to adjust production decisions according to market demands. The Senate bill was more oriented to the market and provided less insulation for the farmer from the market's signals. By comparison, high levels of price support had significant drawbacks. They locked the farmer into producing for the government loan program rather than for the market, and they were expensive. High, fixed supports provided too much federal support too frequently. In general, the senators saw a need for a long-term program instead of emergency response policies. They argued that with no surplus on hand, it was the ideal time to move farm policy away from the wartime production incentives that caused absurd results in the postwar economy. They were quick to point out that continuing the wartime policy was likely to provoke a backlash from the American public and taxpayers. They had some support from farm state senators, including those from the South.[27]

The Aiken bill drew swift, strong opposition from southerners who viewed it as unfair to cotton. Senator Richard Russell (D-GA), a powerful senator with substantial influence in the chamber, countered Senator Aiken, arguing that the bill was "short-sighted" given the uncertainties of the postwar world.[28] Senator Russell wanted to substitute the House-passed bill that extended 90 percent of parity, but the Senate soundly rejected his amendment.[29] He also sought to revise the parity formula in a manner that would likely have benefitted cotton more than the other commodities. He argued that policy changes "should be based on the free market, not upon the restraints and controls of a war period . . . unfair to a great many commodities which were held down by controls."[30] Senator Russell and his allies saw the Aiken bill as an attempt to "reduce the parity on cotton, after the cotton farmer has been robbed of millions of dollars. . . . It is an outrage and a disgrace that cotton should be reduced from 90 to 75 percent in the face of what has happened to cotton."[31] The election year offered Democratic opponents more partisan attacks against past Republican policies.[32] In the Senate southern opposition was not strong enough to block or change the Aiken bill, and it passed overwhelmingly.[33]

The two competing visions for farm policy met in conference committee, which was under a tight time line. Congressional leadership needed to adjourn for the Republican presidential nominating convention and the campaign to keep control of Congress while ousting President Truman.[34] As a result, conference was brief but bruising. The hastily arranged compromise combined the one-year extension passed by the House with a switch to the Senate's flexible price-support system beginning on January 1, 1950.[35] Democrats in the House were particularly furious and predicted that they would have to rewrite the bill in the next Congress.[36] Congress swiftly completed its work and left town for the campaign.[37] President Truman quickly signed the bill into law, ignoring the questionable marriage of competing visions for farm policy.[38]

The bill provided ominous indications of a growing partisan element to the regional disagreement over price supports. The flexible loan rate system was being labeled as the Republican and Midwestern vision. The fact that it had been supported by President Truman and some Senate Democrats did little to diminish this line of attack. On the other side, the 90 percent of parity belonged almost exclusively to the South and Democrats but included Chairman Hope and some Republicans in farm districts. Adding partisan pressures on the commodity fault line was a dangerous development and promised further complications.

President Truman won an improbable, come-from-behind victory in November 1948, and Democrats recaptured Congress, returning southern control to the Ag Committees.[39] The Democratic sweep was a surprise given Truman's unpopularity and the rapidly unravelling Democratic Party coalition.[40] As examples of the unravelling Roosevelt coalition, President Truman had been forced to navigate attacks from his southern flank over the growing influence of the civil rights movement and from his left led by Roosevelt's secretary of agriculture (and vice president before he was replaced by Truman) Henry Wallace.[41]

The 1948 act became law just ahead of a bin-busting harvest that decreased crop prices going into the fall. Buried in a previous bill from the Republican Congress in 1948 was an obscure farm issue that the president put to use in the campaign. Charles Brannan, appointed agriculture secretary after Clinton Anderson had resigned to run for the Senate seat in New Mexico, helped make it a campaign issue for Truman. In short, the issue involved the CCC, first created by Roosevelt and chartered by Congress in June 1948. The Charter Act had included a provision pushed by commercial grain dealers that limited the CCC's authority to purchase or lease storage for forfeited commodities. The record harvest and falling crop prices raised concerns that the CCC would have to default on its obligations.[42]

Standing in the shadow of Roosevelt's ghost, President Truman conjured the Great Depression and attacked the Republicans in Congress. He accused them of having "stuck a pitchfork in the farmer's back."[43] Farmers helped push the president and Democrats to victory, especially in the crucial midwestern farm states.[44] The farm vote results refreshed the political potency of price supports in the congressional mind.[45] Truman and the Democrats won, but playing partisan politics on the farm policy battle lines over price supports would continue for the next two decades. The 1948 election "politicized" farm policy "as much or more than ever before, and collegiality gave way to partisanship."[46]

President Truman reappointed Brannan to be his secretary of agriculture. Secretary Brannan inherited hundreds of millions of dollars' worth of perishable commodities accumulating "in a cave in Kansas" thanks to high, fixed price-support policies.[47] Even with high loan rates, farm incomes were falling.[48] What is clear in hindsight was unknowable at the time; farming had entered a period of technological revolution that was drastically increasing the yields from America's farmlands.[49] Much of this revolution involved the final stages of trading animal for mechanical power, along with expanded adoption of hybrid seeds, fertilizers, and chemicals to improve yields.[50] This intense pace of technological advancement, however, did not result in production adjustments at the farm level; farmers stuck to their existing production systems despite the fact that technology was increasing their market risks.[51] There were also reasons to lay blame at the feet of farm policy, given its rigid system of acreage allotments and loan rates.

The outlook darkened. A stalemate in Congress was followed by signals from voters that implied a preference for Truman and Democratic policies. Slipping commodity prices supported a political push to return to the comfort of high, fixed price supports. That wartime policy, however, when coupled with the technological revolution, caused a complete breakdown in the commodity support system and the politics of the farm coalition.

Fighting the Brannan Plan with the Agricultural Act of 1949

Holding true to their words from the year before, Democrats returned to power in Congress and went to work rewriting the farm bill. Representative Harold D. Cooley (D-NC) and Senator Elmer Thomas (D-OK) were handed the Ag Committee gavels.[52] Chairman Cooley was born into farming in Nash County, North Carolina, and followed his father into both farming and law before he was elected to the House in 1934. He served for thirty-two years until his defeat in 1966, with sixteen years as chair of the House Ag Committee.[53] Senator Thomas grew up on a family farm in Indiana, earned his law degree, and moved to Oklahoma in 1900. He settled there, built a successful law practice, and entered state politics in 1907. He was subsequently elected to the U.S. House in 1922, with support from labor and farm interests during the initial farm depression that preceded the Great Depression. In 1926 he was elected to the U.S. Senate, where he joined the Senate Ag Committee and was part of the New Deal efforts to help farmers and rural communities. A series of issues piled up during his Senate career, and he was defeated in the 1950 primary; among those issues was the Brannan Plan debated in 1949.[54]

Secretary of Agriculture Charles Brannan was controversial before he sent to Congress the farm bill proposal that would bear his name and carve for him a unique spot in the history of American farm policy.[55] Brannan, a lawyer from Colorado associated with the New Deal liberals of the Roosevelt administration, had taken a high-profile role for President Truman in 1948. His plan quickly consumed the farm bill debate. Opponents claimed the secretary sought to replace price-supporting loans with direct, compensatory payments that would subsidize cheap food for the consumer and leave the farmer dependent on a government check. In fact, Brannan did not propose compensatory payments for the bulk, storable, or basic commodities, nor did he propose to eliminate price supports or production controls. Compensatory payments were an alternative for perishable commodities that he saw as a better method than the food stamp plan proposed by Senator Aiken to help prevent a depression.[56] Regardless, opponents attacked the plan extensively. They claimed it would create an extremely costly and intrusive federal role in farming.[57]

In retrospect, the policy concepts in the Brannan Plan do not appear to have been all that new or revolutionary. The only arguably new component was his proposal to place limits on the amount of program benefits a farmer could receive, which he based on a calculation of units linked to the average-size family farm.[58] The details of the Brannan Plan do not provide satisfactory reasons for its becoming so politically toxic in Congress and with some of the farm groups. In fact, the opposition to the plan was more political than substantive. Brannan appeared to be choosing sides in the long-running disputes among farm interests and against the American Farm Bureau.[59] Brannan was a liberal Democrat proposing changes to farm policy that were easy to mischaracterize as motivated by partisanship or worse. His plan made for an easy target, especially in the House, where southerners raced to replace the hated Aiken provisions of the 1948 act.[60] Brannan's proposals were caught up in the collision of competing directions for farm policy. The 1948 elections fueled an uncharacteristically partisan House debate, the implications of which were recognized by, and clearly concerned, farm district members.[61]

Representative Stephen Pace (D-GA) sponsored the bill reported by the committee, which embodied some aspects of the Brannan Plan for perishable commodities only.[62] The Pace bill continued nonrecourse loans for the basic commodities (i.e., corn, cotton, and wheat) at 100 percent of a new parity formula, with continued authority to limit production via acreage allotments and, if farmers voted on them, marketing quotas.[63] The bill's Brannan-like changes to the nonbasic crops served as a clear acknowledgment that 90 percent of parity support—"a wartime measure in order to bring about greater production of food"—as applied to perishable commodities had caused indefensible consequences, especially for potatoes and eggs.[64]

The painfully obvious problems with perishable commodities had produced recommendations from farm interests and Secretary Brannan to replace price-supporting loans with direct compensatory payments for perishable crops. The Pace bill did so on a trial or experimental basis only. The hope was that it would provide a solution without further jeopardizing all of farm policy.[65] The bill ran into problems on the House floor largely because it was associated with Secretary Brannan.[66]

Republicans attacked the Pace bill as the "camel's nose under the tent" for the Brannan Plan, the "brain child of Henry Wallace, Rexford Tugwell, and Alger Hiss" that had been "resurrected" by Brannan in order to "sell it to the American farmers as a 1950 election issue."[67] Republicans argued that the Brannan Plan was designed by labor interests as a backdoor method to provide low-cost food to the consumer that would leave the farmer with subsidy payments to offset low market prices.[68] House Ag Committee ranking member Clifford Hope said, "We have before us today a bill which I believe goes a good deal further in its implications than most bills which come before this Congress."[69] He pushed for another one-year extension of 90 percent parity price supports and conservation payments.

Representative Albert A. Gore Sr. (D-TN) offered a substitute that extended for another year the 90 percent of parity price supports.[70] He argued that farmers did not support the "cluster of delusionary promises of food both cheap and expensive at the same time" that was the Brannan Plan. He was opposed to making direct payments to farmers and "putting the hands of both the farmers and consumers into the pockets of poor old Uncle Sam for a livelihood."[71]

Representative Gore's efforts drew the ire of committee leadership. Chairman Cooley commented that the committee members "regret very much that the gentleman from Tennessee did not find it possible to attend at least one of the very great many hearings which we have held" on the farm bill. He added that Gore's substitute would not terminate the Aiken provisions and that "our beloved agricultural expert from Tennessee" had not agreed to allow the House to decide whether "to kill this snake" or "breathe a new breath of life into it and keep it alive."[72] Representative W. R. Poage (D-TX) contributed his view that there must be a "mental darkness that undoubtedly befogs" Gore because he would continue the Aiken policies and lower cotton price supports.[73]

Representative Gore outmaneuvered Chairman Cooley and the committee on the floor, however. He agreed to three amendments that brought support to his amendment, including one to repeal the Aiken provisions outright.[74] A coalition of Republicans and Southern Democrats rejected the Pace provisions and passed the Gore substitute farm bill.[75] The House of Representatives had once again backed 90 percent of parity price supports for the major commodities.

The Senate did not move until October. By then USDA had informed Congress that corn was oversupplied and in need of marketing quotas but that the secretary was unable to proclaim them because of the 1948 act.[76] A power struggle had broken out in the Senate Ag Committee over the Brannan Plan, which was supported by Chairman Thomas. Senator Clinton Anderson, the previous secretary of agriculture, and others opposed the chairman and the Brannan Plan. Senator Anderson won, and his bill was reported by the committee.[77] As a result, the Brannan Plan did not come up for direct consideration or a vote in the Senate. Instead the committee bill returned to the flexible price-support policy contained in the Aiken bill that lowered loan rates according to supply levels.[78]

The committee bill rested on two related concerns. The first was about continued expansion of intensive row-crop farming. Senator Anderson said, "Some day we shall have to go back to a type of agriculture which is easy on the soil and will conserve the soil."[79] The second, and larger, concern was with surplus production induced by price-supporting loans. Senate Majority Leader Scott W. Lucas (D-IL) argued that a "workable system of price supports requires that they be at a lower level during periods of abundance" and that farm income "will not suffer since the farmers will have a greater abundance of a particular commodity."[80] He added that high price supports and surplus would be so expensive as to be politically unsustainable. The argument was that the wartime policy of 90 percent of parity would lead to further surplus because that price-support level encouraged farmers to increase production even if markets were oversupplied.[81] More surpluses would require additional controls on farming and federal costs; ultimately, political opposition to both would bring down the entire farm program.[82] The battle lines in the Senate were drawn. From the Midwest it was "whether we should have high, rigid supports or whether there should be flexibility in our agricultural system."[83]

On the other side was the northern plains wheat region led by Senator Milton Young (R-ND) and aligned with the southern cotton region, led by Senator Russell. Their position was to continue the fight against flexible loan rates and for 90 percent of parity, bolstered by concerns that the committee bill would lower supports to wheat and cotton.[84] They opposed lower price supports because the farmer was not treated equally in the economy or by federal policies.[85] Their concerns were rooted in the farmer's costs of producing the crops and the special treatment typically afforded the industries that sold goods to farmers.

Those concerns could not be divorced from the damage done to a farmer's bottom line by acreage reductions. The circularity of the argument was clear, if not addressed. American farmers were producing surplus crops, which required acreage limits and production controls. In turn, those limits required a "rigid" price-support system or incomes would be hurt and farmers would opt out of the program, adding further to the surplus problems. Of course, high price supports also fueled surplus production, especially for crops losing export markets in part owing to the inflated prices and production controls.[86]

The political liabilities inherent in flexible price-support policy were not lost on these senators either. The loan rates served as the price floor with the CCC acting as the buyer of last resort when prices fell. A flexible loan rate system meant a lower loan rate given surplus conditions. A lower loan rate meant a lower price to the farmer.[87] The senators' conclusion was that the flexible support policy would not provide "a fair deal to the American farmer . . . [but] fail the American farmer when he stands in greatest need of assistance at the hands of his Government."[88] The senators also made sure to note that Congress had raised the minimum wage and provided benefits to other sectors of the economy. Failing the farmer would lead to punishment at the ballot box for those who pushed it.[89]

The dispute between the Midwest-Northeast alliance and the Northwest-South alliance over high, fixed price supports was made more intense by the challenges of acreage reductions. Reducing acres of surplus wheat and cotton spread the surplus problem to other production areas, especially corn. This policy created federally supported competition that pitted the major commodities against each other in a manner that transcended abstract policy or politics.[90] It also created issues about equal treatment for all commodities and regions. Senators fought to secure the best deal for their favored commodities and accused others of getting a better deal.[91] The Midwest-Northeast allies accused cotton and wheat of having "ganged up against the other farmers of the country," and the Northwest-South allies responded with accusations that opponents were trying "to stir up sectionalism" at the same time as corn farmers were getting a better deal.[92]

Partisan political bruises from the 1948 campaign did not help.[93] Senators debated farm policy for multiple legislative days. They initially defeated 90 percent of parity only to have threats from the House to revive it. Senator Anderson and Majority Leader Lucas were forced to return the bill to the Ag Committee.[94] They brought a revised farm bill back to the Senate floor two days later and confronted another intense effort to save 90 percent of parity that was ultimately unsuccessful.[95] These maneuvers demonstrated that there was no path forward in the Senate for a farm bill with 90 percent of parity, and the Senate finally passed Senator Anderson's farm bill containing flexible price supports.[96] The Senate had preserved flexible price support policy, but barely.

Senate passage merely transferred the fight over loan rates to the conference committee.[97] Conferees crafted a compromise that continued the loan rate at 90 percent of parity in 1950 and then permitted flexibility between 90 and 80 percent in 1951. In 1952 the range would increase, from 75 percent to 90 percent, but the actual loan rate was left completely to the discretion of the secretary with the schedule serving only as the legal minimum loan rate he could provide.[98] Thus, the secretary could maintain loan rates at 90 percent, which was important to supporters of that policy and part of the victory for cotton and wheat in conference.[99] The final deal included concessions to corn farmers by raising the supply levels that would trigger marketing quotas and help prevent any actual need to use them.[100]

Increased scrutiny from outside the farm coalition also appears to have added pressure for additional changes in food and agricultural policy. For example, the 1949 act provided authority for the CCC to donate commodities to the school nutrition program, to the Bureau of Indian Affairs for the relief of Native Americans, and to federal, state, and local public welfare organizations to assist needy persons.[101] It also expanded support policy to include other commodities.[102] The expanding set of political pressures helped forge these necessary deals and bring about minor adjustments, if begrudgingly.[103] Combined, these compromises were enough to allow the Senate and House to pass the bill by large margins.[104]

Reviewing the Fight over Price Supports and the Brannan Plan

President Truman signed the Agricultural Act of 1949 into law on Halloween 1949.[105] This may be fitting because the 1949 act haunts farm bill debates to this day. The Agricultural Act of 1949 stands as permanent farm support law. It is suspended for the operation of current farm programs. If Congress does not act to timely reauthorize them, farm supports revert to the 1949 act provisions.[106]

That the 1949 act continues to play this role is notable in part because it was not settled policy, nor did it represent any form of consensus on direction.[107] The 1949 act was part of a stalemate on farm policy that included the 1948 act. The coalition was split mostly with midwestern and Republican interests seeking flexible loan rates and a more market-oriented policy against the southern and Democratic interests wanting strong price supports and acreage reductions. The divide was philosophical or ideological: Should Depression-era federal involvement be scaled back to allow farmers to compete in the booming economic expansion, or should the system built up in depression and war be maintained to protect the farmer in a booming market that raised costs and depressed prices?

Disagreements over the postwar direction for farm policy were intense. They "engulfed the House Agriculture Committee in the most heated partisanship it had ever known."[108] President Truman drove the partisan wedge deeper in the 1948 campaign. Secretary Brannan had his proposal rejected by a Congress that could barely patch together a compromise. At the time members may have agreed on the need to move away from high, fixed price supports used in the war. They knew that such policies encouraged production, but they could not agree on when and how to abandon them. Chairman Cooley's perspective on a "period of transition" is instructive because it also involved "the necessity of taking out of production 28 to 30 million acres of land . . . [and] the necessity of making drastic changes in the pattern of American agriculture."[109] The political liabilities were immense. Lowering commodity prices for farmers would be difficult to manage—more so if it involved taking land out of production. This called into question congressional ability to undertake any transition.[110]

To put this in perspective, the 1948 and 1949 farm bills were the first efforts to write farm policy without the larger threats of economic emergency or war. Neither bill offered much in the way of policy development; they were less transition than stalemate. This stalemate established the terms of the contest going forward. Loan rates and acreage reduction "would dominate agricultural debate for years to come."[111] Most of the debate centered on loan rates, but the real fight was over acreage reduction policy. More than loan rates, acreage reduction is what split the coalition apart because it created federally induced competition among farmers with real-world implications. Even though outsized attention was given to loan rates, acreage reduction went to the heart of the fault line in the farm coalition.

Flexible price-support policy was framed in terms of making farm policy more market-oriented. Opponents of 90 percent of parity wanted to decrease the likelihood that farmers would plant for the loan rate when prices were down. They argued that the lower loan rate would, in turn, force farmers to adjust planting decisions to better reflect market signals. The very nature of price-support policy calls that argument into question. A loan rate of 90 percent or 75 percent of parity constituted a high price floor. Either rate would be significant to farmer decision making given that the CCC was the buyer of last resort at the loan rate. Flexibility between these rates, therefore, was less likely to force farmers to adjust their planting decisions than it was to create political problems.[112] The most prominent of these was different loan rates for the major commodities and the appearance of lowering prices for farmers.

If there were a political justification for flexible loan rates, it may have been that such a system served as some protection against southern competition for the feed market. Southern farmers were considered too reliant on cotton as their main, if not only, cash crop. Southerners in Congress left a long record on this point; conceding that the acreage reduction policy was an attempt to force diversification of southern agriculture so that it was not completely beholden to cotton as the single cash crop.[113] Reducing cotton acres would push southern farmers to grow other crops. These other crops were mostly feed grains, such as sorghum, but also increasingly soybeans. They were feed sources that competed with corn in that important domestic market.[114] More problematic, some form of controls would also be needed on corn to allow for the former cotton acres to switch to feed grains and compete.

Lower prices and supports could be expected to have a more significant impact on those farmers trying to divert cotton acres than on the farmers in the Midwest. Different crop values and market realities add weight to this point, along with the production advantages in the Midwest. It stands to reason that farmers in the I states (Iowa, Illinois, and Indiana), for example, could better manage relatively moderate price declines for feed grains. With good soils, favorable weather, and their most important markets local, the difference between 90 percent of parity and something less was likely diminished. Potential southern competitors would presumably struggle with the same relatively moderate price declines, especially if they had to divert acres, buy equipment, and find markets. Flexible loan rates may have been a midwestern counter to the subsidized competition from acreage controls in cotton, but they appear to have been a poor choice politically. For one, this would be a tough sell in a Congress where southerners held the gavels and outsized power. For another, the policy change from the status quo (90 percent) could be blamed for lower prices. It is therefore difficult to accept that fighting for these minor variations in the loan rate were worth the political capital expended.

Another likely reason for the attention paid to loan rates was the recognition that high, fixed price supports were necessary to the primary southern goal of reducing cotton acreage. In fact, high loan rates were vital to that policy. Farmers could maximize production on reduced acres and capture the higher supported prices or loan benefits, which, in turn, helped underwrite the diversion of cotton acres to other crops. This reduced the damage to farmer incomes and decreased the risk that cotton farmers would abandon the programs entirely. Cotton therefore needed high loan rates to protect income, subsidize diversion to feed grains, and avoid being squeezed out of that diversion by more productive, established midwestern producers because of relatively moderate price declines. These benefits also explain why southerners so fiercely defended 90 percent of parity. Viewed from that perspective, market-oriented arguments and flexible loan rates can be seen as part of a proxy attack on subsidized acreage diversion.

Acreage reduction policy was the real battleground. Acreage reductions were unworkable to, and unacceptable for, midwestern farmers because corn was predominantly fed either on the same farm it was grown or on a farm nearby. Cutting back on corn production could only bring about more problems than it solved. This held especially true because a growing demand for protein in the booming postwar economy created plenty of demand for meat and thus corn in the Midwest.[115]

By comparison, cotton was chronically oversupplied in a world market with increasing competition from foreign producers and synthetic fibers—competition partially the by-product of acreage reductions and high price supports. From the Cotton Holiday to the 1933 AAA plow-up and the compulsory 1934 Bankhead Act, acreage reduction was foremost cotton policy. Acres planted to cotton fell dramatically in the thirties, which should have been disastrous for southern farmers and the textile industry.[116] High, fixed loan rates helped soften the blow from acreage reductions while subsidizing the diversification of crop practices on southern farms. Effective diversion of acres out of cotton required more, however, and had to be coupled with acreage limits on those competing crops to make room for southern acres in the market. Thus, acreage restrictions (or allotments) could not be just for cotton; midwestern corn acres had to be held in check or reduced so that they could be replaced with diverted cotton acres. Therein lay the true heart of the problem and the dispute.

These policies were loaded with complexities and problems, often working at cross-purposes and, at times, toward competing goals. For example, acreage controls may have also been designed to address concerns with acres shifting out of the traditional Southeast and toward Texas, Oklahoma, and California. If so, then the policy was meant (at least to some degree) as a method for protecting the traditional production areas from newer, irrigated western acres but at the same time pushing traditional cotton areas to diversify to other crops.

Set adrift from war and depression, policy development appears more confused and contradictory. The only clear goal left was making sure farmers had good prices for the crops, but prices had remained strong postwar and inflating them further added more problems. Thus, a fundamental question lingers at the end of the 1940s: Why continue such a controversial and problematic policy? Diverting acres to other crops carried great political risks, especially given that it created conflicts with the other members of the coalition and caused disruptions on the ground and for farmers.

There are not clear, satisfactory answers eighty years after the fact. These bills were probably what they appear to have been: a stalemate in the long process of adjusting or reforming the status quo (acreage controls and 90 percent of parity). They were a confusing amalgamation of conflicting goals and policies because this early in the postwar process there was no clear direction or strategy. The 1938 system had failed, and it was politically difficult to return to that flexible schedule after receiving 90 percent of parity. The transitional concepts embodied in the 1948 act—some of which might have been inflated by a misreading of the 1948 election results—were unacceptable to powerful southerners and many farm constituencies.

Cotton was the only crop with compulsory acreage reductions, and southern members were the most adamant about continuing the policy. Self-interest and overwhelming political power probably allowed them to ignore the impacts on midwestern farmers. But they continued the policy in spite of the damage being done to cotton farmers and to the cotton industry, such as the lost market share to foreign producers and to synthetic fiber production. The conclusion remains that the goal was to push diversification of southern cropping practices and, especially, to advance labor adjustment, modernizing away from sharecropping and hand labor.[117] The controversial nature of this policy would be further magnified with an increasing focus on civil rights and the significant role played by federal policies and tax dollars.[118]

By 1949 employment demand during the war years had pulled many sharecroppers and former sharecroppers into northern cities and regions.[119] The controversies surrounding the policy persisted, further complicating farm bill politics; supporters, having settled on this policy, remained defiant in its defense. Few compelling arguments existed for this policy, and disputes focused outsized attention on loan rates. The two were inextricably linked. Acreage reductions addressed crop diversification and sharecroppers while high price supports eased the burden of those reductions on the farmers seeking to consolidate and mechanize. These realities complicated any efforts to make course corrections despite abundant evidence that acreage policies were a problem for the policy and the politics of the commodity coalition. The loan rates effectively subsidized both direct competition for the feed market and politically fraught farm labor adjustments.

This also sheds some final light on the Brannan Plan's demise. It was simply crushed between the corn and cotton forces fighting over much bigger issues. Brannan stumbled when he sought to expand acreage controls to more crops; he made it worse with compensatory payments and limits on benefits.[120] History provides a much longer view, however, and Brannan may have simply been ahead of his time.[121] How farm policy eventually caught up with the Brannan concepts, and how the farm coalition's dynamics pushed it there, occupies the next twenty-five years. Loan rates and acreage reductions define the commodity dispute until then. The debate created extraordinary pressure on the coalition's fault line. The 1949 act capped two volatile years of farm policy debate, but the real feuds were just getting started.

4

A Surplus of Problems and Disagreement, 1950–1969

Introduction

The problems and disagreements that plagued farm policy in the fifties and sixties grew out of the primordial fault line that had featured ever-more prominently in previous debates. The driving force for policy in this era was a cost-price squeeze on farmer incomes; commodity surplus weighed down prices as costs rose in a booming economy. The fifties were a particularly problematic decade for agriculture but lacked the animating forces of Depression and dust, leaving much room for partisan politics and changing national demographics.[1] In other words, the debates lacked the caliber of catastrophic events that could force abandonment of narrow self-interest in favor of abrupt, drastic changes. The sixties went further into dysfunction and problem; farm policy reached rock bottom in the Kennedy administration. The decades-long buildup of pressures at the fault line culminated in a complete breakdown that required a salvage effort and forced détente among the feuding interests. Finally, it was of great consequence that this era began in the wake of war but ended in a very different political and demographic landscape. As Congress and the coalition sought to figure out directions and programs, farm policy underwent a slow and arduous evolution that unfolded over the course of seven major legislative efforts.

The prevailing concern for farmers was the diagnosis of a cost-price squeeze that damaged farm incomes in a national economy otherwise prospering in the postwar era.[2] The cost-price squeeze was arguably more of a cost issue for farmers because they were spending more to produce their crops. They adopted technology to improve yields at a pace previously unknown, paying more for equipment, fertilizers, hybrid seeds, and other chemical inputs. For example, the Federal Reserve Bulletin from August 1956 indicated that gross farm income increased from 1950 to 1955 but that net farm income decreased as production costs, especially fertilizer and seed costs, increased.[3] The booming economy would also have increased costs of many manufactured goods. The squeeze was on: as they paid more to produce larger crops those large crops were holding down prices in terms of surplus commodities.

Problems began with USDA's push for expanded production just before the Korean War ended. But it was the post–World War II technological revolution that made the most difference; farmers produced larger crops on fewer acres. Mechanization provided a key example. By reducing the number of work animals that required feed, mechanization permitted nearly sixty million acres to become available for food production.[4] Over these two decades, it caused surplus supplies "so excessive that public and private storage facilities overflowed" while decreasing the number of farmers and farm workers substantially.[5] At the same time and for some of the same reasons, production controls, still based on reducing planted acres, failed.

Surplus supplies were blamed for depressed prices, but this may be misleading. The average prices received by farmers during the marketing years of the 1950s were at levels above where they had been before World War II and during the Great Depression. Prices had not collapsed, but the surplus may have held them back from keeping pace with the booming national economy; that is, prices were lower than they would have been without the surplus, and this became more noticeable as costs increased. The surplus was also a big problem politically because it provided an obvious demonstration of failure, reinforcing the view that the farm programs were relics of wartime.

On June 24, 1950, the North Koreans launched a surprise attack and invaded South Korea, triggering war on the peninsula.[6] Another war, of course, had geopolitical implications around the world but also consequences for domestic policy.[7] Under the 1949 act, price supports were scheduled to transition toward flexible loan rates in 1951.[8] By April 1952, and with the Korean War approaching its second year, the Truman administration called for a repeal of the flexible loan rate provisions. Congress responded by extending 90 percent of parity loan rates for an additional two years.[9] Doing so locked the high rates through 1954, ensuring that a high floor on crop prices would remain in place for more than a decade. Secretary Brannan had also called for an increase in production for the 1952 crop year. Together, high loan rates and production increases proved fateful when hostilities in the Korean War ended abruptly in July 1953.[10] Crops were already in the ground or being harvested when hostilities ceased.[11] Truman's USDA had gambled on continued wartime demand, but without it massive surpluses developed quickly, weighing down prices.[12] President Truman would pass the surplus problem to his successor, President Dwight D. Eisenhower, and so would Secretary Brannan to his successor, Ezra Taft Benson.

Eisenhower and Benson also inherited a farm policy debate in Congress that had reached a stalemate.[13] Farmers began with loan rates pegged at 90 percent of parity; too high for the market, these rates encouraged overproduction and added to the surplus. Making additional reductions in planted acres under the circumstances would only further damage farmer incomes.[14] At the time, the political environment for the farm coalition was growing significantly more perilous.

Fears of communism consumed the fifties and produced strong suspicions about all federal governmental involvement in the nation's economy, including the agriculture sector.[15] Democrats were on the defensive in the early fifties. In the 1950 midterm election, they lost twenty-eight seats in the House and five in the Senate, including some key members, such as Senate Majority Leader Scott Lucas (D-IL).[16] Then they lost the White House for the first time in twenty years (five presidential elections) when President Truman decided against running for another term. Dwight D. Eisenhower, former Supreme Allied Commander in Europe during World War II, president of Columbia University, and (briefly) NATO Supreme Commander, defeated Illinois governor Adlai Stevenson.[17] National and congressional politics were also realigning as voters continued to move from rural areas to urban and suburban areas and as African Americans left the South for northern cities.[18]

Combating Surplus and Benson in the Fifties

Eisenhower's successful presidential campaign had included only a vague promise about achieving full parity in the marketplace. He and Republican candidates that year were more clearly against production controls, the Brannan Plan, and the Truman administration's policies in general. Election analysis indicated that farmers had helped Eisenhower and Republicans sweep the 1952 elections, but that the Korean War was the far more significant reason.[19] Republicans had regained full control of the federal government for the first time in twenty years.[20]

After he had been elected president, farm district Republicans warned Eisenhower against repeating the mistakes of 1948 on farm policy. President Eisenhower instead followed the recommendation of the American Farm Bureau president from Iowa and appointed the conservative Ezra Taft Benson to be secretary of agriculture. Benson was considered a "zealous economic libertarian" and "enemy of the New Deal farm program" who pushed "freedom to farm" concepts that would remove the federal government from farming, beginning with flexible price supports.[21] For his first speech as secretary, Benson outlined a vision for achieving parity prices in the market but added that "it is doubtful if any man can be politically free who depends on the state for sustenance" because a "completely planned and subsidized economy weakens initiative, discourages industry, destroys character, and demoralizes people."[22] These were not the words of someone friendly to the existing parity system.

Before returning to farm program debates, Congress passed a law to use the surplus for geopolitical purposes.[23] The bill moved in the wake of the truce that ended hostilities in Korea, indicating the focus on combating communism around the world.[24] Assisting friendly nations and developing foreign trade for (and exports of) American agricultural commodities that were in surplus was noncontroversial, and Congress easily passed the bill.[25] The law may also be notable as the first turn toward Republican policies that prioritized market-based and export solutions.[26] It also differed from the Marshall Plan's efforts to help Europe redevelop, moving in the direction of export subsidies and foreign food aid—designed for both strategic purposes in the global struggle with communism and to dispose of surplus commodities.[27] In fact, it recycled elements of the McNary-Haugen bills of the 1920s but updated them to make strategic use of excessive commodity supplies.[28] The surplus problem continued unabated, which forced a reckoning on the controversial price-support and production-control elements of farm policy.

President Eisenhower and Secretary Benson represented a philosophical change on farm policy, and in 1954, they decided it was time for reform. Aligned with midwestern Republican corn farmers and the American Farm Bureau, they faced a Congress in which southern cotton and western wheat interests held significant power. For Eisenhower and Benson, reform began with breaking the twelve-year run of 90 percent of parity loan rates, policy they blamed for causing the growing surplus problem.[29] They believed that lower loan rates would discourage surplus production, forcing farmers to adjust to market signals, which would, in turn, help balance supply and demand, lift prices in the market, and wean farmers off federal assistance.[30] The administration's proposal carried substantial burdens and political risk. Flexible calculations would produce lower loan rates in years when crops were over-supplied, thus lowering prices. Politically, it was asking members of Congress to reduce the prices for their farmer-constituents at a time when incomes were being squeezed by costs.[31]

The 1954 debate produced the first, minor shifts in the politics for farm policy but not without real resistance from some farm interests. Both Ag Committees moved to their respective floors with a bill to continue the status quo commodity programs. It was clear that any reforms in policy would have to happen on the floor, with a coalition that could override the committees.

Returned to the chair, Representative Clifford Hope (R-KS) reprised his 1948 position by reporting a bill maintaining high price supports. He was clear that the committee was not going to adopt flexible price supports with decreasing farm incomes and increasing surplus.[32] The administration's reforms were a heavy lift for farm district Republicans. Democrats, especially southerners such as Representative W. R. Poage (D-TX), argued that reducing the price supports caused farmers to increase production to make up the difference on price, "a lesson Secretary Benson has yet to learn."[33] His views represented the concerns from cotton country that to "reduce prices just aggravates the production situation in a one cash-crop area; at least until you have bankrupted a large part of the farmers."[34] A flexible system, Poage argued, was "a sliding scale which will slide right from under you every time you need it."[35]

Senator George Aiken (R-VT) became chair of the Senate Ag Committee and reprised his role from 1948 by pushing for flexible loan rates. In 1954, however, he was defeated in his committee by Senator Milton Young (R-ND) and a faction that wanted to continue 90 percent of parity for another year.[36] Senator Young's argument was of a piece with Representative Poage's—that lowering crop prices only led farmers to overproduce in a self-defeating attempt to overcome the lower prices. Young added that the consumer didn't get any benefit out of lower prices and that given droughts in wheat country, carryover stocks could be beneficial.[37]

Commodity prerogatives held less influence on the House and Senate floors, where President Eisenhower had important allies in the American Farm Bureau and midwestern Republicans. During floor consideration the House narrowly agreed to an amendment by Representative Robert Harrison (R-NE) that was backed by Farm Bureau. It established "a little flexibility" by allowing loan rates to move between 82.5 percent and 90 percent of parity.[38] Chairman Aiken also found a more favorable environment on the Senate floor when he attempted to reverse his committee, and the Senate agreed to accept the same flexible loan rate provisions that had passed the House.[39] Before passing its bill, the Senate also added a controversial provision that penalized farmers with loss of conservation payments if they did not comply with acreage allotments.[40]

The compromise on loan rates held through conference and the final bill provided for price support from 82.5 percent of parity to 90 percent of parity on the 1955 crop. It also contained the

Senate's language for requiring compliance with acreage allotments to receive conservation payments.[41] This modicum of flexibility left a floor high enough to alleviate low price concerns. It was intended to be a "temporary bridge to the permanent provisions of the Agricultural Act of 1949" (i.e., the sliding 75 to 90 percent of parity scale) to "gradually shift from high, rigid supports to flexible supports."[42] It also represented a victory for the administration and Farm Bureau, won in the House with urban votes and Republicans from the old farm bloc.[43] The 1954 bill was also significant for being the first time the House had agreed to any flexibility in loan rates since passing the 1938 AAA.

The victory can also be attributed to an erosion in the wheat-cotton alliance evidenced by wheat's effort to craft a different policy. In desperation, wheat was turning back to concepts based on the McNary-Haugen legislation from the twenties. A wheat marketing certificate program and the price-support loans provided two levels of price support: one for wheat consumed domestically, which required millers to buy wheat at a fair (higher) price, and one for wheat exported to be sold at the world price.[44] While Congress did not agree to this policy in the final bill, wheat's efforts further demonstrated the shortcomings of the existing system and weakened the 90 percent of parity alliance.[45] That wheat sought a different policy—and almost received it—was at least tacit acknowledgment that acreage reduction and supply controls had failed.[46] Wheat's move combined with pressure from the Eisenhower administration to realign the balance of power between corn and cotton, at least enough to contribute to the step toward flexible loan rates.

The Agricultural Act of 1954 passed the House and Senate on votes that did not fully represent the political challenges that had surfaced in the process.[47] Opponents of 90 percent of parity loan rates were backed by evidence of its problems in the form of surplus—obvious to rural and urban members alike. They attacked that wartime measure as ill-suited for the postwar economy.[48] With history and experience on their side, the opposition gained intensity from the diverting of upward of thirty million acres into competing crops.[49] Moreover, American commodities were losing export markets, including to cotton production in Brazil and wheat from Canada, which cut into support for the status quo policy.[50] Compounding the political battles, senators from wheat states also attacked the baking industry for increasing prices while wheat farmers suffered from lower prices.[51] This could have driven away potential support from non-farming regions but could also have been a reaction to those regions walking back their support.

Defenders of 90 percent of parity loan rates blamed USDA mismanagement for the surplus situation because Secretary Brannan had mistakenly pushed expanded production in 1952–53.[52] They also blamed the technological revolution that was increasing yields. Arguably, this should have been a liability for them because it demonstrated that policies to control supplies were not effective. For example, with twenty million fewer acres of wheat in 1953 than in 1919, production still surpassed that record crop from thirty-five years earlier, but consumption had not kept pace.[53] Less debated but likely central to their opposition, flexibility could result in each commodity receiving a different loan rate as had happened initially under the 1938 AAA; a fixed price-support level for all commodities was equitable. The flexible system tended to provide corn a higher loan rate because it had typically not been in surplus.[54] But with the minor step toward a flexible system, President Eisenhower had established his proverbial beachhead against the wartime policy of high, fixed loan rates.

The 1954 farm bill victory for Farm Bureau and the administration would quickly be tested at the voting booth. Opponents of the bill predicted that it would be another Republican policy that a Democratic Congress would have to fix.[55] Democrats campaigned against Republicans for lowering prices to the farmer, and voters returned Democrats to majorities in the House and Senate in the 1954 midterms—a House majority they would hold for forty years.[56] These results added to the prevailing view in congressional politics that high price supports were winning campaigns for Democrats and that farmers were not happy with Benson, Eisenhower, and the Republican push for flexible loan rates.[57] The election results therefore also furthered partisan erosion of the farm coalition.

Commodity prices declined and did not help the 1954 act either. Secretary Benson lowered the wheat loan rate as southern Democrats regained control of the congressional farm bill process in 1955.[58] Senator Allen Ellender (D-LA) returned as chair of the Senate Ag Committee. Senator Ellender had been born on a sugar plantation in Louisiana and had pursued a career in law before he went into politics during the reign of Huey Long.[59] He replaced Senator Aiken, who had been a leading supporter of flexible loan rates. Senator Ellender was appointed to the Senate after then-senator Long had been assassinated in 1935. Ellender went on to be one of the longest-serving senators in history. He died running for reelection in 1972 but was in the middle of his long career when he assumed the gavel and a lead role in addressing the problems of farm policy. Representative Harold Cooley (D-NC) resumed the chair on House Ag.

The farm bill effort got off to a rocky start in 1955 and then deteriorated further. As predicted, Chairman Cooley moved quickly to return to 90 percent of parity, attacking Secretary Benson and "every blustering, blundering bureaucrat in the Department of Agriculture [who] has turned against the tillers of the soil."[60] The House Ag Committee voted 23 to 12 in favor of returning to 90 percent of parity.[61] On the floor Chairman Cooley blamed the Korean War, defended the farmer from an orchestrated attack, and argued that the surplus problems could not be blamed on 90 percent of parity loan rates.

Southerners held a strong hand in the House on loan rates, but playing that hand escalated the regional fight over acreage reductions; cutting cotton acres grew competition for corn in the feed market.[62] When combined with partisanship, competition among the commodities caused the farm coalition to turn on itself in a flurry of charges, attacks, and counterattacks. Midwestern Republicans linked farm policy to communism and labor, while singling out peanuts in a divide-and-conquer strategy that opened an all-out war between southern Democrats and midwestern Republicans.[63] Southerners were barely able to rescue their bill on the House floor, which exposed the weakness of their position.[64]

The southern Democrats were also embarking on a direct collision course with the popular president. President Eisenhower had informed Congress that the country "was headed in the right direction," and he "urgently recommend[ed] . . . we continue resolutely on this road."[65] He backed up his position with a veto threat on the House bill. Senate chairman Ellender declined to enter the fight, moving instead to hold hearings and get feedback from farmers.[66]

In January 1956 President Eisenhower made the next move. He proposed a new farm program linked to conservation that he called the Soil Bank.[67] He blamed 90 percent of parity for the growing surplus, depressed prices, lost export markets, increased foreign production, and declining exports and imports. He also blamed acreage reduction policies for having harmed efficient farm management while spreading surpluses. The president argued that the Soil Bank would help address both surplus and diverted acreage problems by reducing production acres through single-year and multiyear contracts.[68] The Acreage Reserve Program would reduce surplus crop acres by paying farmers to put some acres into conserving uses in any crop year, while the Conservation Reserve Program (CRP) would retire additional lands over multiple years; the latter program would concentrate on acres that arguably should not have been in production. In effect, the president was seeking to recycle some of the 1936 Dust Bowl policy; instead of trying to hold down blowing topsoil, the effort would hold down growing surplus. Eisenhower's proposed programs also amounted to a $1 billion federal investment to prevent diverted acres from spreading the surplus problem. The president further warned that the only way to continue price supports was with flexible loan rates.[69] The Soil Bank was caught in the fault line from the start, however. Southern Democrats viewed it as midwestern Republican policy cooked up by the Farm Bureau to be more favorable to corn.[70]

In the Senate Chairman Ellender proclaimed "an urgent need for immediate and effective action to restore to the American farmer a reasonable share of the consumer dollar, and to increase his income" because farmers were "caught in a . . . price squeeze, and were helpless, on their own to combat it." He added that this squeeze was the result of increased costs on items purchased to produce a crop while prices for the crop had fallen. One result was that farmers were spending less, an "impact being felt right on down the line amongst the merchants, the laboring people, the service trades and other segments of our population."[71] The farm groups, however, "stand divided, and seem to be vying with each other as to the best approach to meet the challenge."[72]

Senators were divided too. Ranking Member Aiken acknowledged that the technological revolution was making it "very difficult for many farmers to keep up . . . and some farmers are getting trampled" but that it was a "farm squeeze" and not a "farm crisis at this time." He continued, "American agriculture is undergoing a revolution which makes the industrial revolution of almost a century ago look like a slow-motion picture. . . . Despite his efforts to conduct his operations more efficiently and to convert to modern methods of farming, the farmer today is not getting his fair share of the unprecedented national prosperity." Too much attention, however, had been paid "to make certain that the farmer gets a greater share of the national income" and too little had been paid to policies to conserve natural resources and expand export markets for American commodities.[73]

The desire "to put more dollars into the farmers' pockets without delay" was struggling to bridge a regional divide that was growing wider under pressures from an election year that featured President Eisenhower on the ballot.[74] The chairman and his allies on the committee took the president's proposal as an opportunity to advance their priority—90 percent of parity loan rates—and hitched it to the Soil Bank.[75] To make matters worse, the committee's design for the Soil Bank was too restrictive on corn acres. Midwestern Senators argued that the bill would leave Corn Belt farmers with little option but to opt out.[76] For the committee, this was a shortsighted strategy at best. Combined, 90 percent of parity and more acreage problems for corn would make for easy targets on the Senate floor. The narrowly passed bill provided corn interests the motivation to retaliate.[77] The farm coalition was achieving new levels of dysfunction.

Midwestern retaliation against the committee bill began on the Senate floor with a fight over loan rates and an amendment by former secretary of agriculture Senator Clinton Anderson (D-NM) to strike 90 percent of parity. Anderson argued that Congress should not be in "the position of reducing production with one hand, through the soil bank, and stimulating it with the other, through rigid high price supports" and that there was "no reason why we should throw money away on the soil bank" if Congress is going to restore that policy.[78] Some senators recognized that high loan rates were doing more harm than good, especially to cotton. They pointed out that cotton had been "deliberately . . . committing suicide" with price supports.[79] Senator Aiken added bluntly that the only thing that had ever helped prices and incomes was "war—cold, merciless, bloody war. . . . We paid pretty dearly for the high farm prices of those days."[80] Senator Aiken argued that the Senate was wrong to place so much emphasis and political capital on the high, fixed price support policy. Senators should, instead, focus on farm income and how Eisenhower's Soil Bank might be beneficial; not only had the 90 percent of parity policy failed to help farmers, returning to it invited a veto from President Eisenhower and jeopardized the bill.

Chairman Ellender and his allies latched on to the technological revolution as the cause of surplus rather than 90 percent of parity. Ellender explained that "for every major crop there has been a steady upward swing in yields per acre over the years" thanks to new "varieties, better cultivation methods, more efficient fertilizers, insect-and-weed-control formula, mechanical harvesters—these and other 20th century innovations have played their part in increasing the quantity of food and fiber that can now be produced from an acre of ground."[81] He also argued that flexible loan rates were to blame for the surplus and acreage competition because they led to lower prices. Farmers always sought to offset low prices with increased production. Thus, farmers were producing more crops on the same or fewer acres and not expanding acres because of price supports.

Southern Democrats were on the defensive, resorting to attacks on Secretary Benson and retreating to claims that they were merely seeking a temporary measure. Chairman Ellender, for example, claimed that Secretary Benson was "either woefully ignorant of what is going on today in American agriculture, or he is talking through his hat" and accused the secretary of failing to faithfully execute his duties.[82] Ellender and his allies bolstered their argument with threats of punishment at the voting booth, but the Senate again agreed to strike 90 percent of parity. Still, defeat did not put the issue completely to bed. Near the end of the farm bill debate, wheat-state senators attempted to revive 90 percent of parity for their farmers. Defeating the policy required Vice President Richard Nixon's tie-breaking vote and an agreement to provide USDA discretionary authority to implement the two-priced certificate program on a trial basis. Midwest senators were concerned that the certificate program was bad policy and a new threat to corn because it would dump cheap wheat into the domestic feed market or result in foreign retaliation, but they couldn't defeat it.[83]

Matters degenerated further for the coalition. Chairman Ellender complained that "the corn grower will get a bonanza . . . able to have its cake and eat it too" because of a lack of controls for the crop that had been the "fair-haired boy" and that had "always had a high standing in the price support program."[84] Senator Richard B. Russell (D-GA) joined him, complaining that "the corn farmers have occupied a privileged position ever since the inception of the program" and that he was "sick and tired" of it.[85] Midwestern senators Bourke Hickenlooper (R-IA) and Edward Thye (R-MN) instigated the fight when they sought to "make available to the corn farmer . . . more actual cash support value than he possibly could receive under the present law" or the committee bill.[86] Senator Young, the leader for wheat, made clear that crop's allegiance: "I admire the Senators from the cotton states because they will stand up here and help the wheat farmers get fair treatment."[87] At one point in the debate, Senator Hubert Humphrey (D-MN) accused senators of creating an "election-package . . . nothing more or less than an actual taking of money out of the Federal Treasury in an election year in an effort" to harvest votes with a "Trojan horse" meant to destroy the farm support programs completely.[88]

Federally diverted acres presented the major dilemma fueling these regional hostilities. Twenty million wheat acres had been cut out of the Great Plains, with some shifting to other states and countries. At the same time, cotton was losing market share to synthetic fibers and foreign production as it diverted seventeen million acres. Coupled with diverted wheat acres, many of the diverted cotton acres were going into direct competition with corn for the domestic feed market and pushed that crop into a surplus that otherwise would not have existed.[89] For example, USDA data indicate that during the years 1953 to 1957, cotton and wheat acres decreased by roughly twenty-nine million acres but acres in sorghum, barley, and oats (competing feed grains) increased by as much as twenty million acres. This made corn farmers "the direct victim of an unsound Government policy" because these diverted acres caused the feed grain surplus.[90] In response, corn farmers began abandoning the program in large numbers. Opponents argued that these acreage policies were failing across the board because they also caused some production to be "frozen in uneconomic producing areas simply because the producer [was] trying to protect his production allotment."[91]

The Soil Bank held potential as a work-around for these compounding complications. It could be a "place to put the diverted acres" in return for "some just compensation" so that the farmer could "reduce his plantings and at the same time not go bankrupt."[92] Senate supporters of the Soil Bank were asking for a lot out of the program, especially given the open warfare among the regions over acreage diversions. Midwestern attempts to fix corn's treatment in the program nearly opened a "Pandora's box of troubles" in the debate.[93] Senator Hickenlooper reminded his colleagues that both cotton and wheat had acreage reduction floors even though the commodities were in surplus. He pointed out that wheat had an acreage floor of 55 million acres but that the floor should have been set at 18.6 million acres; cotton should have been restricted to 6.8 million acres, but the floor had been set at 17.4 million acres. Pushed to the brink, senators managed to forge a compromise on corn acres that was more acceptable than what the committee had written.[94] The Senate also included a limit on acreage diversion known as cross compliance, viewed as some measure of protection against more acres going into competition with corn. Cross compliance was introduced in an amendment that the wheat-cotton alliance was not strong enough to defeat.[95]

The Senate managed the two most challenging matters for the 1956 farm bill debate, but it took nearly a month for senators to pass the bill; they ended up voting for it overwhelmingly.[96] The Senate floor debate had delivered strong signals to the coalition. First, partisanship was increasing the friction along the regional fault line. Second, high, fixed loan rates did not have a path through the Senate because wheat and cotton lacked the votes to overcome entrenched opposition. Third, restrictive acreage policy, especially as applied to corn, by now clearly held the most destructive potential for the coalition and the bill. Finally, while the cotton and wheat alliance held strength in the committee, it was far weaker on the floor, where its strength waned over contentious issues that had been the subject of previous battles.

Lurking underneath all this was a warning that support for helping farmers was weakening. This can be seen most clearly in the debate to include limitations on the total amount of federal assistance a farmer could receive.[97] Even when united in opposition, the farm coalition no longer had the vote strength to stop an amendment that had appeal outside its members. Thus, the Senate easily agreed to limit Acreage Reserve Program payments to $25,000 per person and reluctantly placed a limit on loans, but not before pushing the limit up to $100,000.[98]

Success on the Senate floor created complications for conference with the House bill from 1955. The House had once again continued 90 percent of parity, but President Eisenhower had threatened to veto it. Cotton interests held the upper hand thanks to their control of the House and Senate Ag Committees. They thought the Senate bill was too favorable to the Corn Belt. For example, Representative Poage best demonstrated this view when he complained that the administration was "obsessed with the idea that a certain group of farmers should receive a treatment that is not accorded to the general masses of farmers . . . whether they are Republican or Democratic farmers, whether they are farmers in Texas or farmers in Iowa, . . . we ought to treat them alike."[99]

Stacked in favor of 90 percent of parity, conferees produced an agreement to continue that policy in addition to Eisenhower's Soil Bank program.[100] According to Chairman Ellender, the "most difficult problems faced by the conferees concerned the provisions with respect to corn and feed grains," but the Midwest lost in conference; whatever gains they had made on the Senate floor were reversed.[101] Conference had reinstituted the compulsory requirements and acreage reductions, despised by midwesterners and guaranteed to provoke a fight with them and the president.[102] Conferees also removed the payment limitation provisions from the Senate-passed bill but kept the limits on Acreage Reserve Program payments, while adding the two-priced support program alternatives for wheat and rice. Congress passed the bill over Midwestern objections and sent it down Pennsylvania Avenue to be vetoed by President Eisenhower.[103]

The conference agreement darkened the partisan political cloud that had been hanging over the bill. It forced the president to veto a bill that included his proposed Soil Bank and take a difficult stand against farm interests during an election year.[104] In fact, midwestern Republicans accused southern Democrats of having designed the bill for that purpose.[105] It may have been that Ag Committee leaders were willing to gamble on the bill. President Eisenhower, however, called their bluff. His veto demonstrated that he was unconcerned about any political repercussions. He reportedly called the legislation a "private relief bill for politicians," and the House override vote was unsuccessful.[106] This outcome had been clear from the start. Ag Committee leaders either suffered from an unwillingness to acknowledge reality or thought they had little choice. It is possible they believed they needed to demonstrate to their farm constituencies that they had fought as hard as they could and thus that a return to 90 percent of parity was unattainable.

The veto was not the last word from President Eisenhower. Secretary Benson stepped in with a concession to corn farmers who had rejected existing programs. The Department of Agriculture permitted loans at a reduced rate without requiring compliance with acreage allotments. Roughly 41 percent of corn acres were in compliance with allotments in 1955, and this number fell to 24 percent in 1956 and then 14 percent in 1957; loans were made available at twenty-five cents below the loan rate of farmers who complied.[107] Southerners were livid. To them Secretary Benson had crossed a crucial, philosophical line on farm policy by disconnecting price supports and production controls.[108] Representative Poage, for example, blamed widespread misunderstanding about "the philosophy of our whole farm program" and attacked the Eisenhower administration for having "abandoned" this philosophy by giving "corn growers . . . support on every acre of corn they could grow, on every bushel they could produce, without regard to their acreage allotment."[109]

Although it further escalated the conflict, the two-part response by the Eisenhower administration had its intended effect. The Ag Committees quickly reported a new farm bill that removed the objectionable provisions in the vetoed bill, including 90 percent of parity.[110] Chairman Cooley grudgingly accepted that "the time has come when all of us must be realistic" about returning to 90 percent of parity.[111] Congress set aside the previous disputes to work quickly on a revised bill that the president would accept.[112] The power of the president's veto pen had defeated recalcitrant lawmakers, and the 1956 debate effectively ended the 90 percent of parity loan rate policy. The new bill provided flexible loan rates and included President Eisenhower's Soil Bank with specific funding amounts for each of the major crops.[113]

Results at the voting booth continued to be read into the farm bill debates. After the acrimonious 1956 debate, the issue moved with the members to the campaign trail. Democrats attacked Republicans and the president over the 1956 farm bill debate and veto.[114] Voters, however, responded in mixed fashion. President Eisenhower won reelection convincingly, but Democrats retained control of the House and Senate, gaining two seats in the House and one in the Senate.[115] Those wins were notable. J. Floyd Breeding won the seat vacated by thirty-year veteran and former House Ag chairman Clifford Hope, making Breeding the first Democrat to hold that seat since 1916; George McGovern won in South Dakota, making him the first Democrat from that state since 1936.[116] Subsequently, a special election in Wisconsin was held to replace Joseph McCarthy after his death, and voters flipped the Senate seat to Democrat William Proxmire.[117] Later, in February 1958, August H. Andresen (House Ag ranking member) died, and that special election was won by Democrat Albert Quie.[118] These results kept Republicans on the defensive in rural districts.

When he signed the farm bill, President Eisenhower expressed high hopes for the Soil Bank, calling it a "concept rich with promise for improving our agricultural situation."[119] Secretary Benson went to work on the Soil Bank immediately, and even with significant acreage committed by farmers, it failed to reduce the surplus.[120] Benson blamed the vast increase in technology on the farm for making "it virtually impossible to curtail agricultural output with the type of controls acceptable in our society."[121] The continued problems of surplus and acreage diversion were made worse by the failure of the new policy. The corn industry was reaching a breaking point. Despite nearly five million corn acres going into the Soil Bank, corn production increased and corn farmers rejected the 1956 acre program in a referendum.[122] The Soil Bank also failed to slow down diversion of wheat and cotton acres into feed grains. The acreage reductions that were required post-referendum were so severe that large numbers of corn farmers simply gave up on the program and took the lower loan rate offered by USDA.[123]

The 1956 farm bill was also not working for wheat or cotton, but parochial politics restricted Secretary Benson's options.[124] Technology was driving yield increases, and the acres Congress had locked in for wheat (fifty-five million acres) may have been four times the acres necessary to produce what the market demanded.[125] The Soil Bank was expensive, and some in Congress became concerned that it would be unable to cover the costs of all the cotton acres offered.[126] Irony piled on top of failure. While there was too much cotton, there was too little quality cotton for milling needs, and southern members of Congress sought to increase or freeze acreage.[127] They were worried this would create a severe shortage of quality cotton for the milling industry while continuing to produce a surplus of cotton overall.

With farm programs sinking deeper into a morass, the Ag Committees undertook yet another effort to fix them in an election year. The American Farm Bureau and corn interests called for new directions, including a payment-in-kind system to cut surplus and a market-based loan rate calculation.[128] Southerners on the Ag Committees—almost as if they couldn't help themselves—first tried to freeze parity levels for a year as a "stopgap measure" that they blamed on Secretary Benson's operation of the programs.[129] This added fuel to the regional and partisan fires, escalating the political rhetoric. For example, Representative Robert Michel (R-IL) attacked "the extraordinary benevolent attitude the Congress has taken with respect to such crops as cotton, tobacco, rice and peanuts . . . we are still being asked to pay reparations to the South for the Civil War."[130] President Eisenhower, of course, vetoed the bill.[131]

The second consecutive veto of farm legislation apparently drove the point home for southern senators, and they settled for compromises on corn and cotton. In summary, they ended corn allotments and shifted price supports to 90 percent of the three-year average price but with an absolute floor at $1.10 per bushel. They encouraged more voluntary cotton acreage reductions through higher loan rates but kept a minimum allotment of sixteen million acres, while putting in place a transition to 90 percent of the three-year average with an absolute floor price as well (thirty cents per pound floor).[132] The deal represented a tentative truce between corn and cotton, but potentially more important, it represented some movement away from the unraveling parity system. Wheat-state senators refused to go along, however, and some claimed the president sought to "divide and conquer" the commodities by purchasing cotton votes.[133] Some Democratic senators had concerns about both moving away from the parity system and setting floor prices in the legislative text because set prices could not move with inflation.

Senate success could not be easily replicated in the House. Chairman Cooley and his allies in the South wanted to preserve parity and help their farmers, but the politics on the House floor posed significant threats.[134] The committee had reported its bill unanimously, but Cooley requested that the House suspend the rules and agree to the committee-reported bill as a substitute to the bill that passed the Senate. The committee wanted to avoid amendments attacking the additional benefits they provided to southern farmers and to block efforts to add foreign food aid, food stamps, and changes to corn policy.[135] Food aid, food stamps, and the like were ominous signs for southern Democrats who reportedly "had increasingly hardened into sullen intransigence" largely in response to civil rights and other northern liberal legislative efforts.[136] Urban Democrats were looking for retaliation against southern opposition to food stamps. Midwesterners opposed the bill because it "takes care of the major commodities of the South . . . and does nothing worthwhile for the family-farms of the Midwest."[137] The combined opposition meant the committee was unable to muster the two-thirds majority necessary to suspend the rules and pass the bill. Unable to secure the procedural protections he needed, Chairman Cooley was forced to back down on cotton and permit corn farmers the option to vote themselves out of the parity system.[138] Notably, the bill did not address wheat, leaving the 1954 act policies in place.

More procedural gimmicks followed. House leadership refused to appoint conferees, thus blocking a conference negotiation on the two bills and leaving the Senate with only one option: pass the House bill. Chairman Cooley wanted to preserve parity for cotton, so he took no chances on conference and the Senate provisions. It was a gamble but one that paid off, at least in the short term. The Senate acquiesced and, in the words Senator Karl E. Mundt (R-SD), put "faith in our friends in the South, who, if we come to their rescue now, will come to our rescue next year, when we shall need" their votes; the senators' faith was backed with the threat of further dividing the farm coalition into "a series of warring camps representing various commodities."[139] It was a tentative truce. Corn farmers, given the option a year later to leave the parity system, voted overwhelming to do so. Subsequently, corn began reclaiming acres from the other feed grains.[140]

In his final year in office, President Eisenhower warned that the whole farm support system was at risk of a collapse; surpluses and farm income problems were intractable.[141] Eight years of trench warfare over farm programs had taken its toll, and the president had been unable to get any cooperation out of Congress.[142] The CCC was accumulating mountains of surplus crops while the farm interests engaged in self-destructive feuding. The surplus demonstrated that farm programs were failing at an astonishing rate, and the results of this failure were blatant and understandable to nonfarm voters and interests. There were far more nonfarm interests than farm interests, and the imbalance grew during this era.

Demographic changes in the nation were remaking Congress, as voters continued to move out of rural and into urban and suburban districts.[143] A large-scale shift from rural areas to urban and suburban centers had been going on for decades, but by the early 1960s, this shift would force redistricting and create a different vote-counting reality in the House. For example, population loss in rural America was reported to be 10. 6 million in the 1940s, 10.5 million in the 1950s, and 7.4 million in the 1960s. Farm districts went from 38 percent of House seats in 1950 down to 12 percent in 1960; nineteen rural House districts were redistricted into cities in that time.[144] Redistricting was subsequently pushed further by two Supreme Court cases that addressed proper apportionment in House districts, leading to the one-person-one-vote requirement.[145] All of this caused great consternation in the ranks of farm interest leadership but did little to reduce the coalitional infighting.

There was, of course, another significant element to the nation's demographic changes. The migration out of rural areas was especially pronounced in the South, where African Americans, including many former cotton sharecroppers, had left that segregated region for northern cities. Combined, these demographic upheavals represented tectonic-scale movements within the American political system. And they held real risk for farm policy. Southerners magnified these liabilities with vehement opposition to civil rights legislation and other northern social reform efforts. These metastasizing fights compounded already weakening political support for farm programs.

In a somewhat ironic twist, these very changes would provide a lifeline to the fractured farm coalition when losses brought an end to its internecine feuding in the sixties. During the New Deal efforts to combat the Great Depression, USDA had created a program that used surplus commodities to provide domestic food assistance.[146] It was known as food stamps. Throughout the 1950s urban liberal representatives in the House introduced various food stamp bills but faced substantial resistance in the House Ag Committee.[147] For example, when southerners sought support from urban members for the 1956 farm bill, Representative Leonor Sullivan (D-MO) "urged the conferees to bring some sense and some reason to the surplus food disposal program by adopting as part of the compromise bill a food stamp plan."[148] Food stamp interests may have been blocked in the committee, but they had increasing voting power on the House floor and were building strength and making inroads in Congress. By 1959 this coalition succeeded in attaching a pilot food stamp program to foreign aid reauthorization, but Eisenhower refused to implement it.

Escalating Conflicts Lead to Breakdown

The farm bills of the Eisenhower years featured a descent into seemingly irreconcilable positions that were further fortified by partisanship. President Eisenhower vetoed two farm bills written by southern Democrats. His efforts at reform also tilted the debate in favor of policy more aligned with corn; all efforts failed to make progress on the surplus. By 1961 the federal government owned over $4.25 billion worth of feed grains and was spending $1.175 billion to handle and store it.[149] In November 1960 Senator John F. Kennedy narrowly defeated Eisenhower's vice president, Richard M. Nixon, in the presidential election. The election elevated an urban, East Coast Democrat—with little farm policy experience or loyalty to any side in the conflict—to the White House. Kennedy did, however, owe some measure of his electoral success to southern Democrats and Lyndon B. Johnson.[150] President Kennedy made Orville Freeman, the former governor of Minnesota, his secretary of agriculture.[151] Midwestern corn interests, aligned with the Republican Party, were without a center of power in Congress or the administration for the first time in a decade.

The new administration had little time to get its footing on farm policy before southern Democrats acted. Representative Poage led a House effort considered to cause the "bitterest farm bill brawl in memory."[152] The premise was an emergency attempt to combat surplus. The method involved using payments in kind (PIK) to entice feed grain farmers to reduce acres. Payments in kind meant that bushels owned by the CCC because they had been forfeited under loan rates that were above market prices would be given back to the farmer to sell in place of crops grown. The farmer could sell the PIK grain, which had implications for an already oversupplied market.

Supporters argued that the policy would use the "surplus we now have to pay for reduced production" as a method for reducing that surplus.[153] Midwestern Republicans saw PIK as "a dangerous precedent for the Congress of the United States to peg and fix the support price on corn at a fixed figure."[154] The policy was also viewed as merely a different route to a similar southern-preferred end: fewer acres planted to corn in the Midwest with punishment for those that didn't go along.[155] Even House Ag Committee chairman Cooley thought the bill went too far and "would create more problems than it would solve."[156]

The House passed the bill in spite of the opposition. The Senate and conference toned down the more egregious and punitive provisions in the House bill. Midwestern Republicans continued to oppose it, but the revised bill garnered enough support to make it to President Kennedy for signature.[157] Rushed through to get ahead of planting, this first farm bill of the new administration was a noticeable step backward on feed grain policy.

The surplus was not easily abated, especially when farm programs were underwriting it. By 1962 the federal government had a $3 billion investment in feed grains and a half-year supply in storage, as well as $2.5 billion in 1.3 billion bushels of wheat and surpluses costing an estimated $2 billion per year.[158] Some from the Midwest tried to argue that the emergency programs were working to reduce the surplus, but they were met with concerns that the costs were too high to be sustainable.[159] Farmers continued to divert acres to feed grains, which by this point southerners such as Senate Ag chairman Ellender acknowledged, stating that there "is no doubt that the cotton farmers of the South have increased the production of crops which used to be produced as cash crops principally in the North."[160] Congress also had continued a policy that mandated too many wheat acres through minimum allotments, ignoring the fact that yields had doubled.[161]

Here were clear indicators acreage policies had failed. Acres diverted from cotton and wheat were adding surplus problems for feed grains. To cut acres when yields had doubled would have required acreage allotments too small to be tolerated by anyone elected from a cotton and wheat state or district. More pitfalls of federal acreage policies were mounting along with the piles of grain in federal storage. Allotments had become capitalized in the land, which discouraged farmers from adjusting crop mixes for fear of losing valuable allotments.[162] Of even greater political hazard, USDA had become embroiled in a scandal over fraud and abuse of cotton acreage allotments and storage contracts involving Texas businessman Billy Sol Estes.[163] The response from southerners in Congress, however, was to move from emergency measures to punitive. Thus, the 1962 effort marked the nadir for the farm coalition with two unprecedented defeats, one of which would not be repeated for fifty years.

Southern cotton, midwestern corn, and western wheat interests in Congress had been fighting over acreage policies as a response to uncontrollable surplus for a decade. By 1962 there were no minds changing on the issue.[164] Southerners viewed controls as a necessary return for receiving federal support, one they believed had worked well for their crops. Midwesterners viewed it as failed policy that was counterproductive and opposed by a large percentage of their farmers; "one man's medicine can be another's poison."[165] Southerners, however, controlled Congress and proceeded to force acreage reduction policy on farmers who opposed it.[166]

Inexplicably, they intentionally made matters worse because they were unwilling to apply the same medicine to their own farmers. The Senate Ag Committee included a special loophole to exempt southern farmers from the very acreage reductions required in the Midwest.[167] The result was a predictable fight on the Senate floor among members of the committee, backed by concerns that farmers would reject mandatory controls.[168] This opened minute and esoteric details of farm policy to various points of attack, risky in the larger political arena, where the farm coalition was politically weakest. At one point midwestern senators were trying to require simultaneous farmer referendums on all crops combined with cross compliance to force southern farmers to live with their referendum votes and limit the spread of surplus through diverted acres.

The experienced hand of Chairman Ellender, with backing from the administration, succeeded in forcing the policy through the Senate.[169] In doing so, Ellender escalated the quarter-century feud over farm policy to dangerous levels of dysfunction and animosity.

When the House Ag Committee tried to follow the lead of their southern Senate colleagues, they ignited political dynamite on the fault line.[170] Markup in the House Ag Committee had been extremely contentious; the process was bogged down because of the mandatory acreage reductions, and it took Chairman Cooley four tries to report a bill.[171] By the time the bill got to the House floor, Chairman Cooley could not hide his frustration. He addressed his "colleagues from the South," requesting that they "be willing and eager to give the farmers of the Great Midwest and elsewhere the same opportunity that we have with crops we produce, to determine in free and democratic referendums whether they want, as we have done, to place controls on surplus production and enjoy stable prices, or to go it on their own, with no check on production and no stability of price."[172] The chairman's comments were an inflammatory call to arms for southerners to forcefully apply their preferred policy options on farmers that had long opposed and resisted the policies, including to the point of abandoning the programs.

As had to be expected, midwestern opposition was intense. The midwesterners were upset that southerners were forcing detested policies on their farmer-constituents when diverted cotton acres contributed to (or caused) the surplus. Republicans from the Corn Belt argued, "Every time those farmers cut back their production of that commodity they did not cut total production, they put those acres into feed grains, so they were still raising crops from fence row to fence row."[173] They viewed it as a "harsh and vindictive measure" divorced from the realities of corn and wheat crops.[174] They considered the entire acreage allotment system bad policy that damaged farmers and the land; prone to abuse, it should not be restored. As one representative argued, "if you have mined your land from fence to fence, you get a decent acreage allotment," but if "you have farmed wisely, rotated your crops, tried to prevent surplus production, you are penalized."[175]

Southerners in the House seemed almost oblivious to the increasing instability around the fault line after more than a decade of fighting. They escalated the conflict further through what midwestern Republicans viewed as a "phony politically motivated committee amendment which would exempt feed grain farmers" in the South "from cutting down on their production."[176] Protecting southern farmers from the acreage reductions they required of midwestern farmers provoked "profound disgust at the actions of cynical advocates of high price supports and rigid crop controls, who blandly urge these principles for all farmers and then proceed to exempt from controls their particular friends and constituents by little understood amendments of the law."[177] Midwestern members saw the entire effort as an attempt to use the power of the federal programs to take acres from their farmers and hand them over to southern farmers.

Opposition and anger was not limited to the Corn Belt. Representative Bob Dole (r-ks) attacked the policy and its sponsors. He accused southerners of asking "the western wheat producer to cut his acreage and then give these acres to someone else," adding "it does not take any mental giant or an economist to figure who takes the brunt of the farm bill here."[178] Losing wheat support should have been sufficient warning, pitting the entire farm coalition against the South at real political peril. Even some southerners agreed that such a program "should not be shoved upon them."[179] These warnings went unheeded. Unable to remove the offensive provisions from the bill, midwestern Republicans counterattacked. In doing so, they resorted to the most explosive political issue for southerners and wielded it against the farm bill. This too had been foreseeable and could have been avoided; that it wasn't is puzzling.

President Kennedy and Secretary Freeman had initiated the 1962 debate with a comprehensive farm policy proposal that sought to better balance supply and demand through already controversial compulsory controls.[180] They were also interested in foreign food aid that would use the surplus to advance geopolitical goals for peace and freedom. But it was the conservation component of the president's proposal that provided the political dynamite.[181] President Kennedy had proposed turning unproductive farmland into recreational areas for urban residents, which included golf courses and swimming pools. Because the recreational facilities were linked to federal assistance, there were civil rights questions about whether these facilities would be integrated in the South.

The fuse was lit when northern Republicans in the Senate sought to guarantee that any facilities created in the South would be integrated and that "the use of Federal funds to develop racially segregated parks and similar recreational facilities" would be prevented.[182] These guarantees provoked strong objections from segregationist southerners. Senator Strom Thurmond (d-sc) attacked "this policy of forced integration . . . used in the past to deprive the South of their right of freedom of choice" by compelled "social intermingling in public recreational facilities, irrespective of local views, customs, and traditions."[183] Senator James O. Eastland (d-ms) added that "the Federal Government has not been able to induce the South to accept racial integration," and he was unhappy "that this provision is an attempt, by the use of Federal funds, to bring about integrated swimming pools . . . and other recreational facilities" in the farm bill.[184]

The president's proposal for recreational uses of farmland allowed the extremely volatile integration issue to be used in the long-running conflict over farm policy between the South and the Midwest.[185] Southerners tried to frame their concerns as a response to what they saw as broad authority to condemn private property, but they could not escape their segregationist baggage. Acreage reduction policy was itself burdened with its historical usage to remove poor black sharecroppers from cotton farms. Subsequently residing in northern cities, many former sharecroppers and their progeny were further influencing the politics of civil rights.[186] The path in 1962 could not have been more treacherous politically for southern farm interests, but they charged ahead. Shifting the focus of the debate toward the "grandiose scheme" helped Senate Democrats narrowly escape a further civil rights battle in that chamber.[187] The most explosive of issues had been ignited, however, and handed to the House.

The counterattack by midwestern Republicans in the House was effectively designed to split the urban and southern factions in the Democratic caucus. They used integration as a wedge against mandatory production controls on a bill that had been in a precarious position coming out of committee, before Republicans launched the counteroffensive.[188] Politically, 1962 was also a midterm election year, the first under President Kennedy. This typically made for tough partisanship and difficult defense for the party in power.

Republicans demanded party unity in opposition.[189] Northeastern Democrats attacked the wheat certificate provisions as a "bread tax," and scandals at USDA provided further complications.[190] In fact, Democrats postponed the debate because their initial whip counts indicated they did not have the votes to pass the bill.[191] Lobbying efforts, including extraordinary efforts by Secretary Freeman and the administration, turned up the already intense political heat.[192] But issues of race and integration proved too much for what remained of the farm coalition. These issues escalated partisan conflict and provoked angry objections from southern Democrats that it was a Republican sabotage effort.[193] Although the amendments requiring integration failed, the political damage was done, and the entire effort quickly unraveled. Led by Representative Paul Findley (R-IL) and midwestern Republicans, the House defeated the farm bill and sent it back to the Ag Committee.[194] It was an unprecedented loss for farm policy. The coalition lay in tatters on the House floor.[195]

President Kennedy and both chairmen were reluctant to proceed after the defeat, but Secretary Freeman pushed for a compromise that dropped the mandatory feed grain program; chastened southerners conceded.[196] Chairman Cooley acknowledged, "I realize that this Congress is in no humor to enact strict control legislation.... I have been in this body long enough to know that there comes a time when we must be realistic."[197] Conciliation in defeat allowed Congress to move the bill through the House and on to conference after the Senate was forced by House procedural tactics to pass the bill yet again.

During conference farmers fired another warning shot over acreage controls. In August they nearly defeated the wheat referendum, but conferees failed to heed this latest warning.[198] While they dropped controls on corn, they left in place another acreage control referendum for wheat in 1963. Conferees were concerned enough that farmers would vote down controls for wheat in a referendum they softened the impact on wheat loan rates that would result if the farmers did. Conferees also added once-controversial compensatory payments for corn and wheat farmers.[199] Still, the conference product was barely able to pass Congress. Corn Belt Republicans were unhappy with the conference compromises and nearly killed the bill, which passed the House only after votes were switched by leadership.[200] The bill limped to the desk of President Kennedy, who was "exasperated by the evident tendency of farm constituencies to insist on more than they were capable of obtaining" but still signed it into law.[201]

After experiencing an unprecedented defeat, the 1962 farm bill continued to present considerable risk to farm policy in terms of the 1963 wheat referendum. The combatants from the congressional arena responded by moving the fight to the farmers who would vote in it. Farm Bureau attacked under the banner of "Freedom to Farm," which resonated with farmers fed up with years of acreage controls and bureaucracy.[202] The administration campaigned for the referendum, and Congress tried to further entice farmers to vote for it, but the efforts were to no avail because farmers overwhelmingly rejected the wheat program.[203]

Rejection was not well received in Congress or by the administration. It left many urban Democrats feeling burned after they had been pressured by leadership to vote for a bill that farmers rejected. Chairman Ellender added his displeasure, "Democracy has spoken and wheat farmers have voted themselves out of a program. I wish them well."[204] These words rang hollow when wheat prices declined. As one historian wrote, "Congressmen aren't designed to withstand the pressure of $1.25 wheat."[205] In the fall of 1963, wheat district members returned to voluntary acreage reductions before history intervened again: first in Russia, then in the Senate, and finally, tragically, in Dallas.[206] Even in the wake of tragedy and turmoil, efforts to fix farm policy continued.[207]

Breakdown Leads to a New Coalition and Détente

By 1964 cotton and wheat were trapped by costs, prices, and damaged markets. Decades of failed policies, infighting, southern baggage on civil rights, and wheat baggage over the referendum had sapped political power.[208] For example, trying to address cotton's world market competitiveness issue, Congress in 1956 had passed a law to provide export subsidies, but these subsidies ended up harming the domestic industry because they made U.S. cotton exports cheaper for foreign textile industries, especially when coupled with aid programs to help build up foreign economies. By 1964 a frustrated Chairman Ellender lashed out at demands for more cotton assistance, stating, "I have heard that argument for the entire 27 years that I have been a member of the Committee on Agriculture and Forestry . . . that unless we do this or that, as they request, the cotton growers will be put out of business."[209]

Cotton was demanding that subsidies to the textile industry be added to its pile of existing supports while wheat recycled the two-price program, ignoring that policy's damages to cotton.[210] Politically, the two joined forces, threatened the Midwest with further diverted acres, and sought to buy votes in the Northeast with textile industry payments.[211] In the New Year, senators tacked on wheat to the cotton bill that had initially passed the House, but southern resistance to civil rights legislation nearly derailed the bill in the Senate.

The long-building breakdown of farm policy appeared complete, the path through Congress impassable. Back in the House, Republicans put up formidable opposition in the wake of the 1962–63 bill's failures and problems.[212] Desperation in this bleak moment led to opportunity. Democratic leadership in the House coordinated votes on the cotton-wheat bill with the Food Stamp Act of 1964 and forged a new coalition between urban and rural members.[213] Deal making between farm interests and nonfarm, urban and suburban interests proved to be a successful strategy for both. Despite continued vigorous Republican opposition, the House passed both food stamps and the cotton-wheat bill; farm policy crossed an important milestone.[214]

The Food Stamp Act had experienced its own arduous path during the 1950s owing to strident opposition in the House Ag Committee and lack of acceptance by the Eisenhower administration.[215] The policy had been pushed by urban liberals, led by Representative Leonor Sullivan, but House Ag Committee members resisted. They generally raised concerns that it would be charged against farm supports. House Ag held hearings in two successive Congresses but did not report a bill until 1958, when it was then bottled up in the Rules Committee. In 1959 House Ag reported out a pilot food stamp plan, and Sullivan successfully attached it to the foreign food aid extension, but the Eisenhower administration refused to implement the program. Southerners, followed by Republicans, proved to be the most resistant to food stamps. The two had joined together to defeat President Kennedy's request for legislation in 1963.[216]

The southern position on food stamps was compromised by farm supports, however, and urban members retaliated against tobacco legislation.[217] Retaliation against a southern crop delivered the appropriate message, and House Ag reported out the Food Stamp Act. It was pending on the House floor when the cotton-wheat bill needed help. The two bills were not formally connected, but President Johnson and House leaders made it clear that the deal was rural and southern votes for food stamps in exchange for urban votes on the cotton-wheat bill.[218]

The following year found yet another congressional effort to write farm legislation because "we are again at the crossroads on cotton."[219] Opponents pointed out that cotton policy was "a costly failure" that encouraged production "at a time when the surplus is near a record high."[220] By 1965, farm programs had "blossomed into absurdity" and were prohibitively expensive; one member humorously compared them to a farmer who got drunk while shearing sheep and accidentally grabbed a hog, "I got very little wool but a lot of squealing."[221]

Unlike its recent predecessors, however, the 1965 bill was omnibus legislation that brought the main commodities together in an attempt to repair the much-damaged coalition.[222] Representative Findley called it the "biggest legislative package in the history of the Congress dealing with agriculture" and one providing for a "longer than average period of time."[223] The farm interests would need all the help an omnibus bill could provide because they faced prominent opposition. Members were concerned about the cost of the farm programs and the wisdom of the policies the bill contained. For example, Representative Jamie Whitten (D-MS), the powerful chair of the Agriculture Appropriations Subcommittee, was a notable critic of the bill's programs. He bluntly blamed cotton policies for the commodity's problems.[224]

American cotton was drowning in a rising tide of foreign cotton production and a huge increase in synthetic fibers. The 1964 act proved to be another in a long line of failed policies; payments to the textile industry to offset export subsidies on domestic cotton helped less than they hurt. Cotton was also running low on allies and in desperate need of new policies.[225] Decades of diverting cotton acres into production that competed with other commodities, especially feed grains, had taken its toll politically.[226] Within cotton itself the allotment provisions created a dispute between the southeastern and southwestern (especially Texas and California) states.[227] Southern agricultural leaders had to work out a deal on cotton acreage allotments to avoid the situation deteriorating further.[228] The deal was meant to make the acreage cuts "a little less burdensome" but included reductions that were going to cause pain in hopes of bringing the cotton industry back to health.[229]

The wheat program had similar problems. It faced strong opposition in the House from the milling and baking industries, as well as consumers, because they saw it as a bread tax. For wheat, the two-priced program was the problem, and attempts to increase the domestic price by fifty cents were met with intense resistance. Democratic leadership scrambled to fix the problem by shifting the increased cost to the CCC.[230] By comparison, the Senate faced little opposition to the wheat program as Senator Young guided it through the process.[231]

The 1965 farm bill mashed together three different programs for corn, cotton, and wheat. It included a market-oriented approach for setting loan rates (especially for cotton) and moved further in the direction of compensatory payments.[232] It continued what four years ago had been the emergency feed grains program with land taken out of production devoted to conservation. The wheat certificate program provided an estimated 100 percent of parity on domestic production ($2.57 per bushel), but additional certificate costs were covered by the CCC and not the industry; the exported portion of production was supported at a lower level to remain competitive on the world market. The bill also ended many of the production control policies that had dominated the previous debates and split the coalition. Congress shifted more of the burden for cropland adjustment to conservation and the Soil Bank.[233] The unprecedented losses in 1962 and 1963 appear to have brought the warring commodity parties to some form of détente. In the words of one historian, there existed "a renewed spirit of cooperation across party lines" in 1965.[234]

The change was most evident in cotton when some southerners in the House and Senator Herman Talmadge (D-GA) in the Senate pushed market-oriented loan rates (90 percent of the average world market price), supplemented with compensatory payments; however, they continued to include acreage reductions and a sixteen-million-acre minimum.[235] The Senate Ag Committee's old guard had the toughest time accepting changes, but they could barely move a bill out of committee. Senator Talmadge successfully pushed the market-oriented changes on the floor over their objections, including those of Chairman Ellender. The chairman raised concerns about shifting acres out of the Southeast to the West and added, "The small rural communities in the South which are surrounded by small farms will become ghost towns."[236]

Payments to farmers drew the most concern and strongest opposition from the southerners because they didn't want farmers dependent on federal checks and appropriators. Mostly, they foresaw that payments would lead to more efforts to limit the benefits an individual farmer could receive. Limiting benefits was understood to impact southern farmers the most. For example, Representative Findley pointed out that cotton payments for large producers could exceed $950,000 in a single year.[237] Limiting benefits on price-supporting loans was also debated in the Senate, with an Arkansas company held up as receiving millions from the government, but southerners were able to beat the efforts with arguments that limits could not be applied to loans without destroying the program. This argument was not available to payments, however. Despite these challenges, the new direction for cotton policy won out—a victory thanks to decades of failures and the destructive intra-coalitional fight with corn that had broken the old system's grip. The last stand for the old cotton policy was in the Senate on September 10, 1965.[238]

After saving farm programs in 1964, food stamps would be tested in the closing years of the sixties, but the program continued to gain popularity with members of Congress. Because conservative southern Democrats and midwestern Republicans dominated the House Ag Committee's membership, most of the problems were found there. Resistance was generally led by Representative Poage, who had succeeded Representative Cooley as chairman and was less amenable to the program.[239]

William Robert (Bob) Poage was born in Waco, Texas, at the turn of the twentieth century and was raised on his family's cattle ranch.[240] He had earned his undergraduate and law degrees from Baylor and practiced law in Waco before he entered public life. He first served in the Texas legislature (1924 to 1930) and then was elected to the U.S. House of Representatives in 1936, along with Lyndon Johnson from a neighboring district.[241] The two worked together to steer New Deal funds to help their Texas districts. Unlike Johnson, however, Poage remained a staunch opponent of civil rights. He voted against the 1964 Civil Rights Act, a position that likely cost him his chance to be secretary of agriculture under President Johnson.[242]

Food stamps were up for reauthorization in 1967. The House and Senate took different approaches to reauthorizing food stamps with, respectively, a one-year and a three-year extension. When House conferees caused a deadlock in conference, food stamp supporters retaliated against peanut legislation that was pending in the House. They also relied on the Senate to help bypass the House Ag Committee and conference, flexing their strength on the House floor and negating the influence of southern Democrats.[243] Again in 1968 food stamp supporters in the House pushed for a four-year reauthorization and held extension of the 1965 farm bill hostage. Representative Sullivan made it clear to the House Ag Committee members in testimony that she wanted an open-ended extension and that she had the support of a majority of the House. Food stamp supporters had repeatedly requested an open-ended extension, but each time "you tried instead to kill the program, then reluctantly let it continue on a short string. . . . If we have to have another fight, let's have it. But let's make it clear now what the issue is going to be: [if you] won't let us use this method to assure adequate diets for all needy Americans, wherever they live, then many from urban areas are simply going to withhold our votes on farm legislation until we get another 'deal' as we had to do in 1964."[244] Although they were angered by her blunt threat, the opponents of food stamps reluctantly retreated. They apparently had learned their lesson after repeated tough tactics concentrated on their crops.

The program was growing in usage and popularity, which further strengthened its supporters in Congress.[245] The political benefit of aligning with food stamps was becoming clearer to House Ag members in an increasingly challenging congressional environment. Thus, conferees in 1968 agreed to extend food stamps through 1970 with additional funding. By 1969 even House Ag chairman Poage had been converted; he sought to combine food stamps and farm policy in a single bill. This time, however, the Senate and President Nixon were not interested.[246] This was a clear signal of how much the circumstances had changed in the House.

The tentative coalition between farm and food stamp interests created yet another fault line and one that developed sharp edges quickly. Partisan politics, of course, played a role. Examples included Senator Robert F. Kennedy's high-profile efforts to combat hunger in the late 1960s and the push by President Johnson to increase food stamp funding.[247] Food stamp participation expanded quickly while, at the same time, the number of farmers continued its decades-long decline. Food stamps served a very different constituency, which, among other things, highlighted the substantial disparity in assistance. Individual farmers received far more in federal payments than individual food stamp recipients, even as representatives in Congress fought against limits on farmer assistance. Difficult examples were made into political weapons. For example, reformers pointed out that Chairman Poage's district received over $5 million for 400 farmers but only $224,000 for around 140,000 food stamp beneficiaries; Senator Eastland, who was a cotton planter, had received over $200,000 in cotton payments but continued to fight against food stamps.[248]

The 1960s ended in a swirl of cultural and political storms, but it was relatively quiet for farm policy after 1965. On October 11, 1968, Congress extended the 1965 farm bill through December 31, 1970.[249] The rest of the national political scene was embroiled in unrest over civil rights and the Vietnam War. President Johnson declined to run for reelection in 1968. Martin Luther King Jr. and Senator Robert F. Kennedy were assassinated in 1968. President Johnson's vice president Hubert Humphrey lost the presidency to Richard Nixon in 1968. The 1965 farm bill and food stamps, however, continued through these storms into the Nixon administration and the seventies.

Concluding Perspectives on a Troubled Era

From the Agricultural Act of 1949 to Chairman Poage's conversion on food stamps in 1969, it was a consequential twenty years for farm policy. Of all that took place, the Food and Agriculture Act of 1962 was arguably the pivotal event for the farm bill—although it was obscured by more significant historical events, including the Cuban missile crisis and President Kennedy's assassination. The 1962 bill was the pivotal event because it was voted down on the House floor—an outcome that would not be repeated for fifty-one years—and wheat farmers rejected it in 1963. The farm coalition had been building toward these defeats for years with internecine feuding that worsened each time Congress worked on commodity legislation. The 1962 effort was pivotal because this feuding coincided with a major political transformation in Congress and the nation, bringing farm policy colliding with social reform efforts in Congress that included civil rights. The twin defeats contained in the 1962 effort required a different direction for the farm coalition and opened the way for the coalition with food stamps.

The fundamental question of the era was how to make the needed corrections and dismantle the parity system. The Agricultural Act of 1949 can be seen as the culmination of the New Deal efforts to construct a policy infrastructure out of high price floors and intrusive acreage controls. It was built for the emergencies of a different era (i.e., Depression and war) and proved counterproductive as agriculture's technological revolution rendered acreage controls ineffective. The massive surpluses in federal storage provided a clear verdict of failure. At the time a squeeze on farm incomes gave the impression of crisis in the booming postwar economy. It was more likely that technology adoption raised the farmer's costs at the same time surplus kept crop prices relatively low. So tangled were these matters that the course for the policy was not easily corrected.

Congress, pressured by President Eisenhower, tried foreign food aid and limited reform of loan rates in 1954. Using surplus for geopolitical purposes to combat famine, buttress world peace, and enhance export market development produced limited outcomes, and partisan politics complicated loan rate reforms. Congress tried conservation payments in 1956 with Eisenhower's Soil Bank, but the bank failed; surplus worsened and so did the regional conflict. Corn farmers voted themselves out of the system after 1958, but southerners, with President Kennedy's support, attempted to buy back feed grain acreage reductions in 1961.

All these policies were ineffective at reducing or controlling surplus because they were half measures at best. Each amounted to a different version of acreage-based control, and no matter the version, acreage-based control ignored (or avoided) the core realities of commodity production. Surplus resulted from legacy decisions combined with technological advancements. It was made worse by harming export competitiveness and by subsidizing cross competition within American agriculture. None of these issues were amenable to solution by any federal policy with so many internal contradictions and problems.

Experience with acreage controls taught that farmers diverted acres into competing production or removed their least productive acres. This was also true with conservation because the least productive acres were the cheapest to rent with limited federal funds and the most likely to benefit from retirement. Furthermore, federal assistance, whether loans or payments, was underwriting the cost of inputs, mechanization, and better seeds—investments to increase the intensity of production on acres not retired. Acreage reductions were further disruptive because they inflamed the regional fights in the farm coalition while setting up the collision between farm policy and civil rights.

Acreage reductions did serve a purpose: they forced cotton farmers to diversify their crops. Price supports, however, were incapable of returning a profitable price given the problems overtaking the industry. Combined, the policies enacted harmed U.S. cotton's competitiveness with foreign production or synthetic fibers.[250] This was a self-feeding cycle made worse by increasing cotton production costs. Increased costs and relatively depressed prices demanded greater efficiencies, cost-saving measures, and better economies of scale. The traditional cotton areas of the Southeast were affected the most, yet for at least three decades, southerners engaged in these efforts even as the problems compounded.

Politically, acreage reductions appeared to be attempts to benefit southern farmers at the expense of midwestern farmers; diverting acres to other crops was part of the formula regardless of whether it spread surplus and held down prices. To keep the efforts going, southerners needed to lower corn and wheat acres to make room for diverted cotton acres. Limiting corn production in the Midwest by blaming surplus added insult to injury. All of this inflamed the regional conflict because southerners were perceived as using federal policy to advantage their farmers in the marketplace with little regard to the effect on other members of the farm coalition.

Focusing on the damage to the farm coalition misses more profound changes underway and overlooks the conspicuous timing. The American electorate and Congress were undergoing a major political transformation that required a stronger, unified farm coalition, not one that was tearing itself apart. A large-scale population shift from rural areas to the urban and suburban centers had been going on for decades, but by the early 1960s, it had forced redistricting (including court-ordered) and created a different vote-counting reality in the House.[251]

Earlier efforts to move sharecroppers out of farming and consolidate cotton land into larger, more efficient operations were contributing to these changes. Thus, cotton acreage policy had not only created competition with corn, but it had emptied the rural South of many poor, mostly black sharecroppers, many of whom migrated to the North.[252] Sharecroppers pushed off the farm often ended up on "relief and welfare rolls," which increased the need for programs like food stamps. The migration out of the South also further transformed voting blocs and power dynamics in Congress.[253] It precipitated a refashioning of the policy focus for the Democratic Party, increasing the emphasis on social reforms and, most notably, civil rights. The rural South was losing political power within Congress but also within the Democratic caucuses. Civil rights and diminishing political leverage appears, however, to have only hardened southern opposition to change and resistance to helping those left behind by acreage reduction policy and the economy.[254]

All of which only adds further intrigue to the 1962 farm bill effort. It is striking that experienced policymakers proceeded with a bill that combined these volatile issues. Compulsory acreage controls split the farm coalition, and the president's proposal to put land into recreational uses provided the opportunity for midwestern Republicans to raise segregation concerns that split the Democratic caucus (northern urban versus southern rural). President Kennedy was already wrestling with civil rights. He had made strong remarks about civil rights in his State of the Union but delayed action on integrating federally supported housing because of concerns about southern Democrat votes in the election. By September 1962, however, pressure built to the point that the president used federal troops to integrate the University of Mississippi.[255] Southerners in Congress had been on the defensive over the issue for some time. Yet they all either missed or ignored the farm bill's implications.

Some of this may have been another example of liberal intentions being confounded by southerners.[256] Reportedly, Secretary

Freeman's desire for a "rural renewal" was "a major reason for including recreation and food provisions in the new farm bill." The secretary saw a "growing demand for recreation, wildlife, and simply open space in and around the cities of our increasingly urban Nation."[257] Maybe southerners gambled on the popular, urban president and his agenda.[258] Maybe they were too confident in a strong Democratic majority in the House.[259] They had made similar miscalculations before, such as allowing Republicans to craft a vote bloc with urban Democrats in 1954. Moreover, when southerners pushed for support for the 1956 farm bill, they were met with food stamp demands from Representative Sullivan.[260] Either way, they gambled on a long track record of relative legislative success but ended up with unprecedented losses.

Food stamps saved farm policy, or at least gave it new life.[261] The program added votes that might otherwise oppose farm support programs; in 1964 it did so at a particularly precarious time. An increasingly urbanized voting population and congressional makeup were looking more skeptically at farm programs in general. Pressure from food stamps had been building throughout this era. Food stamp supporters had the power to kill individual commodity policies in the House and demonstrated that they were willing to use that power in retaliation against southerners who blocked food stamp legislation. The altered balance of power meant that the farm coalition had to cut deals with urban, non-farm interests to survive.

The fractured farm coalition in the House and failed wheat referendum were farm policy at rock bottom. In the end the 1962 losses answered the era's fundamental question, forcing through discovery at rock bottom new directions for policy and an expanded coalition going into the 1970s. The 1962 bill achieved the end of the acreage fight, succeeding where flexible loan rates and the Soil Bank had failed. More important over the longer term, it also created the opening for food stamps in 1964. This consequential and contentious era ended in relative peace for the farm coalition as the policy approached its fourth decade and the start of an entirely new era.

5

The Commodity "Roller Coaster" and the Crash, 1970–1989

Introduction

From the beginning farm policy was largely a response to low crop prices, or at least prices that were low compared to costs that squeezed or damaged income. In that view, the difference between the previous era (1950 to 1969) and the one that is the topic of this chapter (1970 to 1989) could not be sharper (see appendix 1, graph 1). Average prices throughout the fifties and sixties were relatively flat, indicating the depressing impacts of the surplus and the extensive federal controls on farming. Comparatively, crop prices increased sharply beginning in 1971, and this increase was followed by significant volatility, particularly for wheat and cotton. The relatively flat price environment had produced damaging conflict within the farm coalition. The conflict contributed to the failed policies of the fifties and sixties. This chapter explores the interactions between farm policies and politics in a time of volatile crop prices.

The run-up of crop prices in the early seventies was, at the time, the only such price increase not tied to a major war event. Previously, prices had spiked with World War I and underwent a relatively gradual increase during World War II. The price increase in the seventies was sharper and more pronounced, and it resulted in significant policy revision. A second farm economic crisis followed in the eighties, and program costs increased in response. The demographic changes that had so influenced Congress, politics, and policies in the previous era were well established by the 1970s, but they continued to exert pressure for change. The political environment became more conservative in the 1980s, which added challenges for both farm and food assistance programs. Conservative ideology, as practiced in the Reagan years, collided with the farm economic crisis to define the developments during that stage of history.

The first farm bill of this era, the Agricultural Act of 1970, was transitional, but the coming changes to farm policy were evident. Congress had extended the 1965 farm bill to give the new Nixon administration time to formulate its policy positions. It may well have been that after two decades of internal warfare, the farm coalition wanted to continue the détente of 1965. Even more likely, the coalition lacked a clear path out of the tangle of loans, payments, and failed acreage policies. The bill itself quickly proved unpopular and has been credited with causing Secretary Clifford Hardin's resignation.[1] Earl Butz, dean of the College of Agriculture at Purdue University, replaced Hardin in 1971. Butz was destined to be a notable change because he was a critic of farm policy and was interested in getting the government out of farming and freeing the American farmer from the "yoke of bureaucratic control."[2] He also aggressively promoted American agricultural exports.

For the 1970 effort, food costs were increasing in an era of inflation, and consumers were increasing the pressure on farm programs, including for limits on payments. In response House Ag Committee chairman W. R. Poage (D-TX) claimed "that the great thrust of this bill is to help consumers."[3] This comment cannot be taken without a grain of salt because, to Chairman Poage, helping the consumer meant supporting the farmer. Further proving the point, food stamps were caught in a logjam in the House despite passing the Senate overwhelmingly. The other problem for the chairman was that to the extent that consumer interests understood farm policy, they knew that it paid farmers to reduce planting in order to increase prices, but that it had failed to control surplus for decades at great cost to the taxpayer. Farm interests, holding tenuously to eroded political power, were outnumbered. It was clear that farmers needed new policies, but Congress needed to avoid reopening the destructive regional feud.[4]

Despite the 1970 farm bill's transitional nature, Congress made notable changes to the price and acreage elements of farm programs that year. In part this was due to President Nixon's efforts to control federal spending.[5] President Nixon increased the influence of the management and budget functions in the executive branch, efforts that would lead to a reorganization and the creation of the Office of Management and Budget (OMB) a year later.[6] Reducing spending necessarily included farm programs, which complicated Secretary of Agriculture Hardin's role in farm bill development. One House member claimed that the 1970 bill was the first created by the Ag Committee rather than crafted from a proposal handed down by USDA.

The 1970 farm bill formally ended all production controls for the major crops, providing farmers with previously unheard of flexibility.[7] President Nixon advanced this change by proposing set-aside acres in place of allotments; acres set aside were to be put into conserving uses. Nixon's set-aside concept differed from past policy mostly in that it was a requirement for eligibility for any payments or price supports but the farmer would not be paid for the acres set aside. As part of his argument to consumer interests, Chairman Poage stated that supporters of traditional farm policy had "abandoned the old philosophy of paying for not planting."[8] Republican supporters called the bill "a step toward freedom in planting in American agriculture."[9]

In short, the policy provided more discretion to the secretary and more planting flexibility to the farmer, but it continued to work within the allotment system. Farmers were required to plant some of their allotment to the supported crop to maintain their acreage histories but with some flexibility to substitute crops.[10] The Ag Committees also pushed parity to the brink of extinction with direct compensatory payments triggered by fixed prices, although Chairman Poage noted that they "did not go as far toward tearing everything down" as maybe some had hoped.[11] The programs remained predominantly loans but supplemented with direct payments to reach price goals. Representative Thomas Abernethy (D-MS) proclaimed, "I am not so much interested in parity as I am in cold cash ... a specific dollars-and-cents return to my farmers."[12] Age-old regional conflicts flared up over concerns that cotton farmers were getting more benefits than wheat or corn farmers, but matters cooled in the Senate after the midterm election.[13] Senators were able to muster enough votes to send the bill to President Nixon.[14]

Finally, the 1970 farm bill contained payment limitation reforms that were far more difficult to resolve than minor program revisions.[15] Compensatory payments had long raised the specter of strict payment limits, and by 1970 those fears had been ushered into reality. Limits had been attached to farm programs before the 1970 farm bill debate, but each time they were amendments to appropriations bills. Because compensatory payments could be tracked to the person or entity that received the payment, opponents could latch onto large payments made to specific individuals and, in some cases, large companies such as Standard Oil of California.[16] The issue played on regional tensions because, as midwestern payment limit supporters acknowledged, "there is no question but what a payment limitation would hit cotton hardest and would hit feed grains, my type of area, the least."[17]

Both chairmen fought back against tighter payment limits arguing that going further would damage program operations, force larger farmers out of the programs, and result in overproduction. They also acknowledged that further limits were largely a cotton issue.[18] Both the House and Senate Ag Committees had included a $55,000 per crop limit on payments as a purely defensive move to protect against tougher reforms.[19] The tactic proved successful as both chairmen were able to protect their bills from tighter limits, much to the surprise and frustration of payment limit supporters.[20] This success demonstrated the strength of the compromise worked out in the committees and the difficult path for making controversial changes on the floor.

Policy on the Upswing: The Agriculture and Consumer Protection Act of 1973

President Nixon was easily reelected over Senator George McGovern (D-SD) in 1972, and he pushed for further reforms. His timing was good. World market challenges, especially for export commodities like wheat and cotton, were nothing new to American farmers, but the early seventies were different. Rising inflation had begun in the late 1960s and picked up speed into the early seventies, which increased farmer costs while an inflated dollar hurt export competitiveness.[21] In addition, the system of fixed currency exchange rates disintegrated. President Nixon devalued the U.S. dollar in 1971 and 1972. In 1973 he ended the ability to convert U.S. dollars to gold (i.e., closed the gold window) and moved the U.S. to a floating currency.[22]

These moves had profound impacts on American export competitiveness, including for agricultural commodities. The world market appeared to have become a much friendlier place for the major U.S. commodities, which enhanced optimistic views about export markets. This also meant that U.S. agriculture became "more closely integrated with the world economy," and prices were more dependent on it.[23] Exports fueled stronger prices and powered a sharp ascent up the commodity roller coaster, accelerating drastically after the summer of 1972.

That summer the Soviet Union suffered a massive crop failure. Instead of letting its citizens suffer as it had in the past, the USSR went out into the world market looking for grain. Soviet officials found Secretary Butz waiting. An aggressive promoter of American commodity exports, Butz orchestrated massive sales of U.S. surplus stocks of wheat to the Soviets.[24] The Soviet wheat deal cleared out CCC holdings, but it was a controversial move.[25] Topping it off, an El Niño event in the winter of 1972–73 decimated Peru's anchovy harvest, harming the supply of protein supplement for feed; a rush on soybeans ensued, and prices spiked.[26]

For the first time in a long while, food consumption and exports were increasing faster than production. Combined with the Soviet wheat deal, this trend was a large contributor to increasing commodity prices. Increased prices were, in turn, blamed for driving up the cost of food.[27] Consumer pushback built to the point that many were boycotting supermarkets and beef.[28] Responding to the rising pressures, President Nixon and Secretary Butz released tens of millions of acres back into production.[29] By 1973 the situation for farm policy was far removed from where it had been the previous two decades.

Inflation, spiking crop prices, consumer backlash, floating currencies, and export demand—this was undoubtedly a heady moment in farm policy circles. For those with a market-oriented conservative view, it provoked reaction that bordered on exuberance. For example, Secretary Butz proclaimed that agriculture had finally reached "the promised land" after forty years in the wilderness of low prices and failed, outdated policies.[30] The president announced that he wanted to free the farmer from government support with a three-year phaseout.[31] For those with a more traditionalist view, the situation provided great cause for concern. Among those traditionalists were Senate Ag Committee chairman Herman Talmadge (D-GA).

Herman Talmadge was born and raised on a farm in Georgia, attended the University of Georgia, graduated with a law degree, and served in the Pacific with the navy during World War II. He was governor of Georgia from 1948 to 1954, then was elected to the U.S. Senate in 1956, but served in the shadows of his father Eugene and Senator Richard B. Russell (D-GA). Talmadge's work on agricultural issues was also overshadowed by his opposition to civil rights legislation. He was part of a dynasty in Georgia. He had helped his father win a fourth term as governor, but his father had died before he could take office and anointed Herman as his successor. Talmadge served five terms in the U.S. Senate until he was shockingly defeated by Republican Mack Mattingly in 1980. He had been part of the effort to write Medicare and Medicaid legislation with President Johnson in the sixties before he chaired the Senate Ag Committee. He reportedly battled alcoholism and other personal scandals, such as divorce and a Senate investigation of his finances. But he was also known for his support of white supremacy and his opposition to desegregation in the South, although those views softened over time.[32]

Chairman Talmadge lamented that "drafting new farm legislation this year [1973] comes at what is perhaps the poorest possible psychological and political moment."[33] Reauthorization was in the hands of a politically altered Congress facing rapid changes in the economy. This dynamic was especially noticeable in the House, where southern power and influence had declined. During the sixties' battles over civil rights, the South drifted away from much of the Democratic Party and toward Republicans, which elevated business and industry interests in the stronghold of farm policy. In addition, demographic changes and redistricting had significantly dropped the number of rural districts and their influence in reaching a majority vote for legislation.[34] Such changes had major implications for cotton, although not cotton alone.

Change came quick and began in the Senate Ag Committee. Senators were "operating this year in a totally different arena than the one which existed prior to the adoption of the 1970 Act."[35] During markup of the Senate's farm bill, Senator Milton Young (R-ND) introduced the target price concept. He proposed that the government peg the wheat price at two dollars per bushel and make a "compensatory payment" if the market price didn't reach it. His fellow committee members liked the concept and adopted it for all supported crops.[36] President Nixon's proposal to phase out support was unacceptable, however, because the senators feared an inevitable collapse in commodity prices. Senator Young wanted a backstop in place for wheat farmers to help when prices fell.[37]

The policy was rooted in the long-maligned Brannan Plan, and Senator Young admitted as much, while acknowledging that he had opposed the plan in 1949. But farming had changed, he concluded, and the payment policy's "time had come" because reform was needed.[38] Young also reportedly had his eye on the increasing crop subsidies by European countries.[39] The target price policy had been a long time in coming, and its arrival required high prices caused by export demand and the Soviet wheat deal. More than anything, the policy was a recognition that the parity system had failed, and consumers were rising in opposition to it.

The target price concept was a different method for dealing with the old problem of low crop prices. Instead of a loan that acted as a floor price and risked forfeitures, the target price policy supplemented farmer income in times of low prices with payments. Senators considered it a market-oriented reform, but they also considered it a production incentive. It would help ease price inflation and encourage American farmers to capture growing export demand.[40]

The Senate Ag Committee report led off with a statement that the "purpose of this bill is to assure the production of adequate supplies at reasonable prices to consumers by insuring producers against losses if their expanded production results in prices below the target prices."[41] The committee also claimed that the new policy would benefit consumers and was "a radical departure" from the previous "four decades of Federal farm programs"; "farmers will no longer be paid for not planting crops, nor will they be guaranteed a Government subsidy regardless of what price they receive for their crop in the marketplace."[42] The committee said the bill was budget-conscious because if prices remained high enough, no payments would be made to farmers.[43]

The target price concept was built on the optimism of the moment, but the committee also understood that encouraging farmers to expand production carried great risk. Wary of the "promised land," senators were not willing to abandon farmers completely to "bear the entire risk of unforeseen developments that would be devastating to the agricultural community."[44] For that reason, the federal government was to share the price risk from expanded production. Senators referred to it repeatedly as a "share the benefit/ share the risk" policy.[45] Taxpayers would benefit if prices stayed high but would have to support farm incomes if prices fell too far. The committee members understood that a production incentive after decades of controls could have unintended and difficult-to-control consequences. They argued that the target price policy was the farmer's equivalent of the minimum wage concept familiar to urban and suburban interests, allies farm interests needed.[46]

Payment limits threatened to spoil the triumphal march into law for the new policy.[47] Senator Birch Bayh (D-IN) sought to lower the cap on payments from the committee-passed $55,000 to $20,000. He was seeking to target help to the smaller family farm rather than large corporate farm operations. Chairman Talmadge argued that the amendment "would gut the whole purpose of the bill" because "the thrust of the committee bill is to assure greater productivity of our Nation's farmers so that consumers will benefit from an abundance of food and fiber."[48] Southern Senators remained concerned that cotton would be hurt the most by lower limits, but they lacked the strength on the floor to defeat the Bayh amendment.

The switch to target prices was only part of the grand bargain to keep consumer interests allied with farm policy. The food stamps program was also up for reauthorization in 1973, and it had come under attack in the Nixon years. The Senate Ag Committee combined the two policies into a single piece of legislation, making formal the alliance begun in 1964.[49] On the floor senators increased spending for food stamps and eased enrollment with amendments that produced a little controversy and some resistance from Republicans and southerners.[50] Senators agreed to the farm bill by a large margin. The Senate advanced the most substantive changes to farm policy since the Agricultural Adjustment Act of 1933.[51]

After holding hearings on the target price concept, the House Ag Committee decided to go along with the "single most significant change" for farm policy.[52] The American Farm Bureau was concerned about European-inspired price-fixing, and USDA warned that target prices set too high would create "tremendous incentives" for expanded production that could "backfire."[53] Some Republicans argued that the target prices were established according to an anomalous price situation and opposed it because it was policy from the Brannan Plan rejected long ago.[54]

Chairman Poage and his committee advanced target prices to the House floor, arguing that the provision offered farmers more freedom, assured "an income which will keep many farmers on the farm," and provided "the incentives needed to encourage American farmers to increase their production to the level necessary to meet and sustain all of our domestic and export needs."[55] The House farm bill contained lower target prices and higher loan rates than did the Senate bill, as well as a higher payment limit ($37,500), but otherwise followed the Senate's lead.

The House Ag Committee had also included food stamps in the bill, but the new coalition could not provide the same level of protection on the House floor as it did in the Senate. Trouble began with amendments for tighter payment limits and attacks on cotton assistance.[56] Attacking cotton threatened to bring down the farm bill in a storm of combined southern and Republican opposition. Farm district allies led by Representative Bob Bergland (D-MN) pushed an amendment to strike the cotton provisions from the bill. If cotton was left out of the final bill, cotton policy would revert to the 1958 provisions. This was tactical; the goal was positioning for conference with the Senate on target prices and payment limits.[57]

While a clear indication of diminished southern political power on the House floor, the attacks may have also been short-sighted; they weakened the coalition with food stamp interests.[58] When the debate turned to food stamps, partisanship took over. Republicans fought to bar food stamps for striking workers.[59] They claimed the amendment was to address concerns from business about being able to settle disputes if strikers were able to receive food stamps. Democrats and the Nixon administration opposed the amendment, but the ban's supporters succeeded three different times on the floor, applying great stress to the coalition.[60] At one point Chairman Poage had to pull the bill and regroup. Republican efforts succeeded in large part because they received support from southern Democrats who were still smarting from losses on payment limits and cotton.[61] Dancing across both fault lines at once, the farm bill was in jeopardy. Democratic House leadership had to rescue it with a risky, complicated floor strategy.[62]

The conference committee was similarly unable to settle the fight over food stamps for striking workers.[63] Conferees had little problem settling on target prices and loan rates. To get them through, however, more procedural maneuvers were necessary. A clear, decisive vote by the Senate to reject the controversial provision was followed by parliamentary tactics in the House that blocked any attempts to reattach it.[64] In short, because the Senate had passed the farm bill as an amendment to the House amendment, Chairman Poage offered what is known as a message amendment that members would be reluctant to vote against. Under House parliamentary rules, no other amendments were in order until the pending amendment (i.e., Poage's message amendment) was voted down, but it passed and permitted passage of the bill, sending the Agriculture and Consumer Protection Act of 1973 to the president to become law.[65] The 1973 act formalized the partnership between farm policy and food stamps, but it had not been an easy journey.

The bill marked a clear turning point in farm bill history, launching the target price era combined with food stamps.

In retrospect, the proponents of target prices may have been at least partially ahead of themselves regarding the federal budget. In 1974, having fought with President Nixon over spending, Congress undertook efforts to reform itself in part by creating federal budget discipline.[66] Congress created standing Budget Committees, established the Congressional Budget Office (CBO), and formalized the budgeting process in order to provide it better control over fiscal matters.[67] A formalized budget process considered all federal spending together, estimated budget costs (known as scoring), and prioritized expenditures. This process necessarily pitted spending priorities against each other in a manner that would come to have a major impact on farm policy.

Prices for corn and wheat peaked in 1974 and began to decline shortly thereafter. Meanwhile, the Nixon administration unraveled in an even more dramatic fashion. On August 9, 1974, President Nixon resigned from office over the Watergate scandal.[68] His newly appointed vice president, former House Minority Leader Gerald R. Ford (R-MI), replaced him as president.[69] Nixon's resignation in August and Ford's pardon of him in September had a devastating impact on congressional Republicans in the 1974 midterm elections; Democrats gained forty-nine seats in the House and four in the Senate.[70] This was a different Democratic majority, however, because it was no longer the same southern stronghold that it had once been.

The 1974 elections upended the House Ag Committee. The new Democrats on the committee flexed their political muscles and voted to remove the gavel from Chairman Poage, replacing him as chair with Representative Tom Foley (D-WA).[71] Representative Foley, born in 1929 in Spokane, Washington, was elected to Congress in 1964 and served for thirty years. He had practiced law and been a prosecutor and assistant state's attorney general before he went into politics by working for Senator Henry M. Jackson (D-WA). Foley was known as a conciliatory figure in Congress and rose through leadership, becoming Speaker of the House in 1989.[72] The vote against Chairman Poage was largely due to his attitude toward consumers and food stamps, as well as his being from the South.[73] Poage's removal rather dramatically underscored the changed dynamics in the House.

Critics of the new policy argued that "setting target prices four years in advance in a highly uncertain world can only lead to trouble," and they proved correct.[74] Target price policy had been designed as a production incentive, and farmers responded to it. They planted fencerow to fencerow in 1974, raising record crops, but the promise of export markets did not hold up.[75] Geopolitics added to the problems because the United States imposed moratoriums on grain exports in 1974 and 1975.[76] Surplus reappeared, weighing down prices once again, but when Congress tried to increase target prices and loan rates, President Ford vetoed the bill.[77]

In 1976 President Ford was defeated by Jimmy Carter, the one-term governor of Georgia. Ford handed the new president a brewing farm crisis. An additional twenty million acres of U.S. wheat added to a worldwide oversupply of the crop, depressing prices. As usual, depressed prices harmed incomes, but in the late seventies, they were pushing many farmers to the edge of bankruptcy.[78] Candidate Carter had promised to improve farm prices so that they would cover the costs of production, but he also promised to balance the federal budget in his first term. On farm policy alone, he had a difficult task ahead of him.

Farm Policy on the Downswing: The Food and Agricultural Act of 1977

The first reauthorization of the combined farm and food stamp bill took place in 1977, as expanded production was coming home to roost. Another bumper wheat and corn crop was waiting to be harvested.[79] If farmers were "suffering the consequences of an unenlightened fencerow-to-fencerow farm policy," the answer in Congress was to increase target prices to "help get us off the roller-coaster of up and down prices."[80] In other words, the solution to the problem caused in part by target prices was to increase target prices. President Carter proposed to increase target prices, but not as much as Congress wanted. Food stamps gained participation, popularity, and political challenges; President Carter called on Congress to eliminate the purchase requirement.[81]

Despite new spending discipline, the 1977 farm bill was a grab for more, an effort by the coalition to consolidate its gains at a time of problems in the marketplace and with unemployment. Congress debated increasing target prices and reforming both food stamps and farm program payments. The final bill was a compromise product. Congress increased target prices but returned to set-aside authority and diversionary payments.[82] Congress also removed the purchase requirement on food stamps in exchange for a cap on total program costs and other reforms.[83] Senator Young called it "one of the best farm bills that has passed Congress during my service here," but House supporters were less enthusiastic, calling it "less than we need, more than we expected and about all we could get."[84]

The Senate was the more favorable turf for consolidating gains on both target prices and food stamps.[85] The Senate Ag Committee had decided that costs of production should determine how much to increase target price levels, but cost of production presented its own challenges, notably when it came to figuring out which costs would count.[86] To the senators on the Ag Committee, farmers had responded to demands for increased production, but the market had failed to hold up its end of the bargain, in part because embargoes had damaged exports.[87] The brief floor debate was a lukewarm standoff between farm state senators (mostly wheat) trying for higher target prices and those senators (mostly midwestern and eastern) wanting to lower them.[88]

The terrain in the House was less favorable. Members contended with budget discipline, parochial politics, increased target prices, and food stamp revisions. The disagreements began when the House Ag Committee engaged in a heated fight over raising target prices beyond the level President Carter had indicated he would accept. Chairman Foley was able to hold the president's line by a single vote. Chairman Foley acknowledged the "conflicting forces" of depressed prices and the budget deficit, explaining that the target prices in the bill were "an attempt to strike a deliberate balance."[89] Wheat district members, having lost at the committee level, pushed hard for higher target prices on the floor. They managed to secure enough votes to force the president to back down and the chairman to cut a compromise deal. Opponents cautioned that doing so could lead to another round of farm program failures.[90]

Farm program supporters had to again defend against additional farm program reforms or payment limits that went further than the committee had agreed. Added to their defenses were pleas on behalf of farmers sinking deeper in debt. The developing crisis weighed heavily on Congress.[91] Representative Paul Findley (R-IL) again led the effort to tighten payment limitations, but the committee compromise held on the floor. Other amendments seeking to reform farm program payments and operations reached similar fates, unable to move the committee compromise.

Once again southern Democrats and Republicans attacked food stamps but were mostly unsuccessful. They managed only to win an amendment that capped total spending on food stamps in any year.[92] Among the reforms to the food stamps program was elimination of the purchase requirement, which was opposed by Republicans and southern Democrats.[93] Republicans tried to restore the purchase requirement and picked up some southern Democratic support but were defeated overwhelmingly.[94] Concerns that the bill would make the farm problems worse and increase spending too much were unable to slow its path into law.[95]

The horizon darkened over Secretary Butz's "promised land" just four short years after he had declared it. Farmers were sliding back into painfully familiar problem territory—oversupplied and incomes squeezed between prices and costs. Many had gone deep in debt to expand production to meet the push from Secretary Butz and the 1973 farm bill. They were vulnerable to inflation, interest rates, monetary policy, embargoes, low prices, and surplus. In the late seventies, they were hit by a perfect storm. Target prices were contingent supports that provided no assistance until prices fell below the target price. Overextended and vulnerable farmers had to absorb the initial losses when prices fell.

Whatever the expectations for increased target prices and the 1977 farm bill, they were buried by events. The most visible included farmers driving tractors on the National Mall in protest and demanding more help from Congress.[96] In 1979 the Federal Reserve raised interest rates to help "curb spending, restrict credit, and halt inflation," but this maneuver crushed farmers carrying heavy debt loads.[97] Farming fencerow to fencerow also had entirely predictable consequences for natural resources, such as soil erosion and water quality degradation. This time farmers came face-to-face with an environmental movement that had grown in political power. Farmers would have to answer for the "promised land" not only in the markets and with their bankers, but also in the political arena. When the commodities roller coaster went off the rails, it wrecked the farm economy.

Of Budgets and Boll Weevils: The Agricultural and Food Act of 1981

The wreckage was already fairly severe by January 4, 1980, when President Carter made it worse. Under political pressure in an election year, the president imposed a grain embargo on exports to the Soviet Union in response to its invasion of Afghanistan in December 1979.[98] The embargo provided an odd counterpoint to the 1972 Soviet wheat deal. In the course of a single decade, much had changed for farmers. The cost of farming, especially energy and interest rates, had increased; export markets had brought increasing volatility to prices compounded by unexpected geopolitical problems, such as the Soviet grain embargo.[99] The situation developing in agriculture would provide political cover against additional drastic redirections in farm policy. It also formed the defining characteristic of eighties farm bills. Mired in crisis, the bills managed only incremental change.

In the 1980 elections, Ronald Reagan, the conservative governor of California, defeated President Carter in a landslide.[100] Reagan's coattails also brought the first Republican Senate majority in twenty-nine years and elevated Senator Jesse Helms (R-NC) to chair of the Senate Ag Committee.[101] Senator Helms was a controversial figure and polarizing politician who served a total of thirty years in the U.S. Senate. A former journalist and radio commentator, he switched parties from Democrat to Republican in order to ride President Nixon's coattails to an upset victory in 1972. He built a controversial record owing to his views on race, civil rights, conservatism, and religion.[102]

Change in control of the Senate was expected to bring real consequences to the 1981 farm bill effort.[103] Democrats retained their majority in the House, and Representative Foley moved to the House majority whip. He was succeeded as House Ag chair by Kika de la Garza (D-TX).[104] The 1981 effort would prove to be a tough initiation for the new chairman. Representative de la Garza was a Korean War veteran and the first Mexican American to represent his Texas district in Congress. He had dropped out of high school to join the navy in 1945 and served in the army during the Korean War. He went on to earn a law degree and was elected to the Texas House of Representatives in 1953 and to the U.S. House in 1964. He would hold the Ag Committee gavel until 1995 and help write four farm bills.[105]

In 1981 the full weight of federal budget discipline first came to bear on the farm bill. It added a significant new pressure on the traditional regional fault line and on the recently created fault line with food stamps. President Reagan and his allies in the Senate prioritized reductions in federal spending, programs, and taxes coupled with increased defense spending and a balanced federal budget.[106] Reagan's were tough priorities to reconcile under budget discipline and parochial congressional interests. Cutting taxes and increasing defense spending would need to be offset in other programs or priorities if the budget was to be balanced.

President Reagan's budget in March and farm bill proposal in April called for ending the federal role in farming, eliminating deficiency payments, and splitting food stamps from the bill.[107] According to the head of President Reagan's OMB, David Stockman, the strategy was "to come in with a farm bill that's unacceptable to the farm guys so that the whole thing begins to splinter."[108] They almost succeeded. They were tripped up by their own deal making and the need for votes to push their priorities. The declining situation in the farm economy also played a significant role.

Likely unforeseen when Congress created it in the seventies, the new budget reconciliation process provided President Reagan and his allies a powerful weapon.[109] Congress had created budget reconciliation to help enforce spending discipline throughout the appropriations process. The Reagan administration and Senate Republicans decided to use it as "a dramatic and revolutionary maneuver" to push through the economic, tax, and spending agenda with reduced opportunities for Democratic opposition.[110] The effort was so drastic and "extraordinary" that Senate Majority Leader Howard Baker (R-TN) attempted to negate its precedential value for future Congresses.[111]

Reagan and reconciliation created an unfavorable environment in Congress. It did not, however, force any major changes to farm policy. In fact, farmers came away from the process with target prices on an increasing schedule that would likely trigger more payments.[112] This juxtaposition of goals with outcomes can be explained by the rough realities that face ambitious proposals as they travel the legislative process. The realities are more difficult when the ambitions involve taking away benefits from powerful constituencies. Republicans' novel use of budget reconciliation was limited in its farm bill outcomes by political circumstances and the deteriorating farm economy.

The president enjoyed a majority in the Senate, but he was facing a Democratic House, albeit one with numbers reduced in the 1980 election.[113] This would prove critical because the path from calling for an end to farm programs to signing a bill into law that achieved it required three stages of legislation. First, Congress had to agree to a budget that included reconciliation instructions to cut spending. Second, Congress had to pass legislation that enacted those spending reductions and that would be signed by the president. Third, Congress had to write and pass a farm bill that reauthorized the programs but remained within budget limitations. Each one of these stages provided separate but interconnected opportunities for negotiations replete with regional, partisan, and parochial pressures and interests.

The Democratic majority in the House had been weakened in the 1980 election, but more important, it contained as many as forty southern representatives who were conservative and more likely to side with President Reagan than Speaker Tip O'Neill.[114] They were concerned with their own political survival in districts that voted heavily for Reagan.[115] These southern Democrats were critical to the president's victories in the first two stages of the process. They helped deliver a surprising victory on the budget resolution, rejecting the one crafted by the House Budget Committee in favor of the president and the promise that he'd consider granting "relief on other regionally important programs" in the reconciliation bill.[116] They were more critical to passing the reconciliation legislation that made the actual spending and program reductions.

On the reconciliation package, a total of twenty-nine Democrats, mostly from the South, voted with the president. This time the vote included rejecting the caps on food stamps that had been agreed to by the House Ag Committee in favor of specific cuts favored by the president and Republicans.[117] The trade was more specific. Southerners voted against Democratic leadership in return for President Reagan's promise to protect peanut, sugar, and dairy programs.[118] For example, Representative John Breaux (D-LA) said that his vote could "be rented" for budget and tax cuts.[119] Presumably lost in the deal making was President Reagan's goal for farm program reform, at least for southern commodities. Branded "Boll Weevils," these southern Democrats risked a backlash from their colleagues.[120]

The main victim in the Boll Weevil deal to save farm programs turned out to be food stamps.[121] The focus on food stamps was among President Reagan's goals for reforming welfare programs by tightening eligibility requirements and reducing benefit levels.[122] His strongest support was in the Republican-controlled Senate, where, for example, Senate Ag chairman Helms wanted to go even further than the president. For reconciliation the Senate Ag Committee made substantial reductions to food stamps, but Chairman Helms failed to get a reinstatement of the purchase requirement that had been eliminated in 1977. The chairman bragged that the "bulk of the reductions come in the two largest areas of the U.S. Department of Agriculture's budget—food stamps and child nutrition programs." He added, "I personally believe that additional reductions could be made in the food stamp area."[123]

In the spring both Ag Committees ignored the president's proposals for farm policy, but the bills were forced to wait. This was part of the administration's strategy: holding up work on the farm bill to get the tax cut bill completed. Reagan needed southern Democrats who did not want his farm program reforms. This intersection of interests created the potential for deal making on votes for the president's priorities.[124]

After the bruising fights over the budget and reconciliation bill, Congress still had to complete work on the farm bill. Reconciliation's wake created impediments, especially in the House, where many urban Democrats were angry with the Boll Weevils. Though diminished, the southerners remained a formidable bloc; seventy-two districts had more than 60 percent of their residents living in areas that could be classified as rural.[125] Adding to their anger with the Boll Weevils, urban Democrats had specific farm bill concerns about the impact of sugar, dairy, and peanut programs on food costs. These were the programs that the Boll Weevils were protecting in the deal with the president.[126] The Boll Weevil deal making therefore cut food stamps and child nutrition programs but helped southern commodities.

Additionally, farm state members and senators were frustrated that Congress was unable to do more to help struggling farmers.[127] Making matters worse, both the House and Senate versions of the farm bill exceeded agreed-on budget limits, and both would have to be revised to cut costs; what they had been able to put together wasn't enough for farmers but was too much for the budget.[128] Democrats in the House, especially, voiced their frustrations with the Reagan administration's influence on the farm bill and the way it used the budget to dictate farm policy. They could do little about it and were not alone. Some Republicans also saw the effort as unfair, especially given recent embargoes; many senators thought the budget constraints tied their hands at a time when farmers needed help.[129]

From the budget to the Boll Weevils, revising the farm bill on the Senate floor presented significant risk. Substantial pressure came from some farm state senators who wanted to keep or even increase target prices. Democrats blamed President Reagan and the Republican budget, but they couldn't win the votes.[130]

The Senate Ag Committee retained "the basic framework of

recent farm legislation" but under Republican leadership made "further advancements toward a market-oriented agricultural policy."[131] The higher target prices—concessions to commodity interest groups in committee—were estimated to increase spending because of lower price projections.[132] Because Congress fixed target prices in advance against unpredictable commodities markets, the program had the potential to trigger large payments. The committee's target prices had been written in the spring. By the time the bill was on the Senate floor, however, prices had fallen further, and the target prices turned out to be too high. Senator Richard Lugar (R-IN) argued that target prices were "as uncontrollable as anything in the Federal budget . . . ticking time bombs in the fiscal dilemma of the total budget of the U.S."[133] Chairman Helms was proving that point by trying to lower the targets because they broke budget limits.[134]

Target prices created more than budget problems. As Senate farm bill supporters tried to fix the budget problems, they had to contend with questions about whether some commodities were benefitting more than others—questions that played into blowback against the Boll Weevils. Senator Lugar informed his colleagues that many farmers "resent[ed] attempts by special interests in American agriculture to carve out special support of agricultural production which makes inefficient use of valuable energy and water supplies."[135] Target prices were not equitable across all commodities and setting them provided ample opportunities to play favorites.[136]

Senator Lugar attacked the peanut program first, arguing it permitted peanut farmers to lock in profits and shielded them from market signals, which, in turn, caused them to produce surplus crops and depress prices, triggering bigger payments.[137] He was unable to take down the peanut program against his own leadership and chairman. Senator Lugar did manage to draw attention to the Boll Weevil deals, and he created headaches for leadership's efforts to reduce target prices against strong Democratic opposition.[138] His efforts also caused a brief return of the regional dispute. Midwestern Senators expressed little interest in protecting peanuts, and southern Senators, in turn, threatened to bring down the entire farm bill.[139]

At this point in the *Congressional Record*, the institutional inclinations of the Senate appear to have taken over. Because a minority of senators can derail legislation, the Senate tends to favor negotiating bills to completion rather than having them collapse on the floor. This inclination was likely triggered by the deteriorating situation. Senators forged a delicate, complicated compromise and deployed procedural maneuvers to save the peanut program.[140] Saving the peanut program then carried over to help sugar and tobacco defend against similar attacks; although the tobacco program in particular raised uncomfortable comparisons with the treatment of programs like food stamps. Senators sidestepped the traps laid by Stockman's strategy—made more dangerous by the Boll Weevil deals—and found a way to pass the bill.[141]

In the Democratic House, the Reagan administration became more involved in the negotiations, at one point working with liberal Democrats to decrease farm program spending.[142] This appears to have complicated earlier deals with the Boll Weevils, possibly even breaking them outright. The efforts prompted Representative Billie Lee Evans (D-GA) to warn, "Beware of a Republican bearing a commitment."[143] The political waters were clearly more tumultuous. Farm district members were able to push through increases to the wheat and feed grain loan rates for those that reduced acres in times of oversupply but failed to increase cotton's loan rate, and the House voted to eliminate both the peanut and sugar programs.[144] Notably, representatives spared tobacco but only because Democratic leaders were concerned that they would lose seats in North Carolina—strong evidence that retaliation focused on the Boll Weevils.[145] By comparison, the amendment to reduce dairy and sugar supports to reduce the bill's projected expenditures passed overwhelmingly and withstood a number of subsequent challenges.[146] The House succeeded in passing the embattled farm bill in part because the bill helped food stamps. This held some urban members in line and produced a compromise on acceptable reforms to the embattled program.[147]

The debate also featured a return to prominence of conservation issues because expansion had revived old concerns about soil erosion and farming on marginal ground. Policymakers worried that America was exporting its precious soil resources.[148] The return of concerns about erosion coincided with a run of success by the environmental community in the seventies to combat large-scale environmental problems. Conservation received bipartisan support and generated little controversy.

Secretary of Agriculture John Block acknowledged the problem and led efforts to address it.[149] The Senate Ag Committee expanded conservation programs to include assistance to protect farmlands from development pressures, and conservation amendments to the bill were simply accepted by the Senate without recorded votes.[150] The House farm bill conservation title included assistance that was targeted to areas with serious erosion problems and focused on cost sharing, technical assistance, and long-term contracts. It also provided for a conservation loan program to landowners and farmers.[151] Tackling soil erosion dovetailed with issues common to the farm economic struggle, but the 1981 provisions were only the beginning of the response.

Conference began with the House and Senate entrenched in opposing camps, and the Reagan administration backing Senate Republicans. The sides were separated by $6 billion and deadlocked for six weeks until House Democrats backed down.[152] Secretary Block led the Reagan administration's opposition to the House bill, and the administration applied substantial pressure on conferees to lower target prices and reduce budget exposure. Conferees and the administration finally reached a deal in early December to set target prices for the major commodities that increased each year but did not leave the increases to an open-ended cost-of-production calculation. The deal included a one-year extension of food stamps.

The administration's demands put the bill at significant risk of being rejected by the Democratic House. Democrats in the House felt that the budget had sealed the farmer's fate as victims of President Reagan's misplaced spending priorities.[153] Chairman de la Garza called it "perhaps one of the most difficult moments I have had in my legislative career, which is some 28 years." The result, he added, was "not an easy bill to support" because it "falls far short of what I personally believe would be fair and reasonable."[154] Ranking Member William Wampler (R-VA) added, "Perhaps those who are claiming this will be the last 4 year farm bill that Congress will enact are right."[155]

Republican senators generally accepted the bill, with Chairman Helms proclaiming it "both sound farm policy and sound budget policy" that "makes further advances in the direction of a market-oriented agricultural policy."[156] It was "the first omnibus farm bill that has been drafted under tight, self-imposed budget constraints," said Senator Bob Dole (R-KS), unlike previous bills that were "developed using the 'mutual admiration society' approach to legislation—each commodity or special interest got pretty much what it wanted by supporting similar benefits across the board."[157] Republican senators expected that its improved market orientation would help correct the excesses and mistakes of the seventies.

Opposition came mostly from farm state Democrats who demanded more assistance for struggling farmers. Regional tensions remained over the special treatment provided southern farmers as a result of the deals with the administration. The deals were viewed as having helped one group of farmers but as having limited assistance to others.[158] Democrats in both chambers also concentrated on the fact that food stamps were reauthorized for only a single year. This contrasted with four years for the rest of the bill and the windfall for peanuts. This further highlighted deals between the Boll Weevils and the Reagan administration.[159]

The budget, Boll Weevils, and food stamps created a bipartisan coalition of dissenters that tried to defeat the bill. Democratic opposition was not enough to slow the bill, let alone defeat it, and the Senate passed it easily.[160] In the House the opposition came within two votes of defeating the bill.[161] The tough process appears to have convinced some food stamps supporters to be pragmatic. House Majority Whip Foley noted that he was voting for the bill because "it is the best bill that we can produce for now."[162]

Crisis and Conservation Produce the Food Security Act of 1985

If Congress expected the 1981 farm bill to reduce spending—and it is hard to believe the legislators did given that they were increasing target prices—those expectations were quickly dashed.[163] The only argument that could square increasing target prices with spending restraint would have been that they increased at a lesser rate than inflation. Farmers were hit from all sides: Europe subsidized its production and exports, increasing competition with U.S. farmers; domestically, interest rates and tight monetary policy to combat inflation reduced export competitiveness while damaging indebted farmers and land values. The farm economy slid deeper into recession, rivaling the 1930s. In response, the countercyclical farm programs produced a wave of spending as lower prices triggered larger payments under increasing target prices.[164] When prices fell below the loan rates, the program again caused forfeitures. Government-held stocks and associated outlays increased. By 1982 the farm economy was in deep trouble, and Reagan's USDA used set-aside acres to reduce production.[165]

Like control efforts before it, Reagan's move failed. Farmers "increased their per-acre yields and harvested even larger crops of wheat and corn in 1982," causing end of crop year CCC inventories to grow.[166] Congress tried to mandate further acreage reductions, but USDA estimated that 1983 expenditures would approach $19 billion.[167] In response, USDA dug further into the past and announced a PIK program in January 1983. Regan's PIK program was similar to the sixties version but applied to wheat, corn, sorghum, cotton, and rice. It was an additional measure—farmers could still receive deficiency payments and loans—that provided farmers with certificates for CCC-owned grain in return for not planting the crop. Commodities were to be transported to the farms or nearby storage facilities at government expense, and farmers could bid to remove their entire base acreage from production with payments set at 80 percent of normal yields (except wheat at 95 percent). It was an incredibly generous program, a sign that the Reagan administration was desperate to stop the deteriorating situation and concerned that set-aside policy was not effective.[168]

The PIK program succeeded in one respect: it diverted 77.9 million acres from production in 1983, idling "more crop land in the United States than all of Western Europe planted that year."[169] Especially for corn, 1983's drought reduced production further, but the total impact harmed the agribusiness industry (e.g., input and equipment) and gave away export markets to expanding foreign competition.[170] The PIK program's generosity to farmers was expensive, with costs reaching $10 billion for the 1983 efforts alone.[171] The otherwise budget-conscious Reagan administration treated PIK certificates as off-budget for government spending purposes, but real problems came about when PIK certificates took on a life and value of their own. Speculators got involved, and some farmers figured out how to profit from PIK.[172] The program was also counterproductive. It injected surplus commodities back into the market, further depressing prices. Lower prices required additional action by USDA and Congress, particularly for cotton.[173]

By 1985 a decade of export potential had been lost to a strong dollar, foreign subsidies, and U.S. farm policy that inflated commodity prices but left American crops uncompetitive worldwide.[174] Farmers were in a "depression such as it has not been seen since the 1930's . . . a flood of grain weighing heavily on our markets," Senator Jim Exon (D-NE) explained, because exports had "dried to a trickle" and incomes were being consumed by "excessive interest rates."[175] Farmers, especially in the Midwest, had suffered through three bad weather years and were underwater on loans from the excesses of the seventies. The eighties witnessed a debt-fueled crisis that infected the agricultural lending sector and nearly collapsed the Farm Credit System.[176] The crisis hurt not only farmers but the rural communities that depended on them; depicted in movies, it became a cause for Farm Aid benefit concerts.[177]

In the depths of crisis, farm program expenditures jumped sixfold between 1981 and 1986, reportedly making the programs the "fastest-growing item in the federal budget."[178] For the 1984 campaign, the Reagan administration temporarily paused efforts to cut farm spending even as it continued to seek cuts in other areas of the federal budget.[179] Reforming the programs in the midst of a crisis was politically problematic, so the president waited until after he won reelection in a landslide to propose drastic cuts and radical revisions.[180]

Looking to the 1985 reauthorization, Secretary Block and President Reagan wanted to cure the farm crisis through the market—lower prices and less assistance offset through greater exports.[181] This familiar ideological perspective was a tough sell after export market optimism in the seventies had left farmers in such a pre-dicament. Caught in a tangle of extremes, Congress had limited options. More money for farmers might not solve the problem, but many proposals did not seem adequate for the crisis at hand.[182] The president's proposal was on one end. On the opposing end was the demand for a return to failed policies from the past. And still, the relentless drive to cut federal spending clashed with burgeoning outlays.[183] The president pushed further on budget reductions, and Congress passed another anti-deficit law that required any new spending to be offset by revenue increases or cuts to spending else-where in the federal budget. The president was back in deal-making mode with key Democrats just as the Senate began its work on the farm bill, re-creating potential political traps. Consumer and urban interests especially took note of the disparate treatment for farm supports compared to other domestic assistance programs.[184]

The president's proposal also faltered on midterm election pol-itics and partisanship in his second and final term. The farm crisis put the fear in Senate Republicans who were defending twenty-two seats, many in farm states. President Reagan's farm bill pro-posal ended up dead on arrival in Congress. Its demise spawned more than 140 different bills and a fierce, difficult debate.[185]

At center stage for the 1985 reauthorization effort was the farm crisis.[186] Adding to the drama, the governor of Iowa declared a state of economic emergency that coincided with the opening rounds of the debate—a move that built more tension for the farm bill. Just over fifty years after the Great Depression, lawmakers rummaged through the detritus of history for old familiar policy tools, dis-regarding the failures they had long produced. Others were more concerned that farm policy was damaging export markets; prom-inent critics and academics from within the agricultural commu-nity concluded that farm policy was out of date and out of touch. Former secretary of agriculture Bob Bergland told the House Ag Committee, "We debate target price levels, while the realities in agriculture scream out for an entirely new approach."[187]

Expectations for substantial changes were high, which put farm interests in a real bind. They were caught between the demands from their farmer-constituents and the budget constraints for which many had voted.[188] The bill they produced was a strange hybrid containing steps forward and backward. It mixed market orientation with acreage-based production controls. It featured new efforts to improve American competitiveness in export markets while protecting farm income with payments. It was a complex support system with no shortage of potentially contradictory components. The anchor of the entire system remained deficiency payments issued directly to farmers through target prices, but Congress was, in some sense, throwing every policy option they could find at the problem.[189]

In the House the farm coalition was strongest when united in defense. It successfully defended against attacks on sugar, peanuts, and dairy that repeated the 1981 Stockman strategy to divide and conquer the commodities.[190] The more difficult challenges came from issues that actually divided the coalition. The most notable of these included a new marketing loan concept and demands for a return to supply controls.[191] These competing concepts encapsulated the forward- and backward-looking views in a Congress trying to respond to crisis.

Representatives Arlan Strangeland (D-MN), Pat Roberts (R-KS), and Dan Glickman (D-KS) pushed for the marketing loan concept. Representative Roberts argued that the president's proposals "sent the farmer out to do battle with one hand tied behind him" whereas the marketing loan would improve export competitiveness "without marching an entire generation of farmers into bankruptcy."[192] The concept sought to resolve problems that were as old as the loan program itself. It was designed to address free rider problems and permit prices to fall to competitive world levels while keeping the CCC out of the grain purchasing, storing, and handling business. It revised the price-support policy to avoid having the CCC act as a buyer of last resort when prices were depressed. If market prices were below the loan rate, it permitted farmers to repay the loan at the lower rate and keep the difference instead of forfeiting the loan commodities. With this change, the old price-support loan would operate more like target prices and deficiency payments. It was lost to a divide in the farm coalition over whether programs should benefit smaller operations more than larger operations, a concept at odds with southern views.[193]

By comparison, farm interests successfully sidestepped a destructive fight over acreage controls that could have reignited the regional conflagration of the fifties and sixties. The committee had narrowly agreed to provide farmers with a referendum on an optional program to reduce acres.[194] On the floor the referendum opened a tug-of-war between opposing views. One side was trying to pull policy back toward mandatory reductions and controls. On the other side were midwesterners and some Democrats who opposed both the committee's voluntary provision and the mandatory amendment. Midwesterners had the votes on their side; they defeated the mandatory amendment and struck the committee's voluntary provision.[195]

Compared to the House, the Senate farm bill faced far more adversity, in large part because of strong ideological differences over target prices.[196] Problems began in the committee, which deadlocked for months. The problems carried through to the floor in a grudge match over raising, lowering, or freezing target prices. Majority Leader Dole, who was also a member of the committee, emerged as the key player in the effort. His deal making rescued the bill both in the committee and on the Senate floor.

In committee Senator Dole's deal gave Democrats enough votes to defeat Chairman Helms on target prices, but in return Dole provided protections for tobacco to buy peace with the chairman.[197] The "somewhat complex panoply" of farm programs in the reported bill set up a tough floor debate.[198] The committee had also rejected President Reagan's reforms for target prices, which were backed by the chairman, but it had included the marketing loan concept and continued acreage set-aside authority.[199] By the time it got to the floor, the bill was estimated to break its budget limits by $9 billion.[200] For opponents of the bill on both sides, this fact was a handy indicator that the policies were failing.

The budget problems fueled intense opposition to target prices led again by Senator Lugar, who proposed to freeze target prices for a year and then reduce them by 5 percent each year over the life of the farm bill. He was supported by Chairman Helms.[201] Senator Lugar argued that target prices were a product of the seventies boom and created an incentive to overproduce because they provided "guaranteed minimum prices . . . [that] trigger Federal outlays that are as uncontrollable and unpredictable as U.S. agriculture itself."[202] The Senate bill contained "target price levels" that Chairman Helms pointed out were "some 50 percent higher than actual market prices."[203]

Opposing Lugar were Democratic arguments that under Republicans and President Reagan "the U.S. Government is pulling the plug on those farmers who are today remaining."[204] Farm state Democrats argued that the bill did not do enough because Republicans were too focused on its costs. Farm state Republicans walked a fine and difficult line, balancing between the challenges presented by the market and economic situation and the impact of decisions like embargoes and monetary policy. Market-based arguments generally appealed to Republicans, but farmers were struggling because of obvious market failures.

Drama and deal making ensued. First, Senator Lugar's amendment lost by only three votes and sent shock waves through farm state senators and leadership.[205] Senator Dole undertook extraordinary measures to save the bill. He deployed procedural tactics to force negotiations for proposals that he admitted "may be a little smoke and mirrors."[206] Negotiations were protracted and difficult. Senator Dole was seeking an outcome that senators could vote for but that would also leave them negotiating leverage in conference and with the president.[207] Out of necessity, Majority Leader Dole crafted a confusing, contradictory compromise that managed to both freeze and reduce target prices and that included additional benefits negotiated for wheat, rice, sugar, and soybeans.[208] The resulting legislation was ugly—and valuable only because it permitted the Senate to complete the bill. The compromise worked.

Although the floor debates had been tough and President Reagan came in trying to enforce spending restrictions, conference turned out to be relatively smooth.[209] The calendar likely provided an assist as the holidays loomed. Additionally, the Reagan administration had little leverage in the negotiations. For example, southerners managed to secure more favorable treatment for cotton and rice—enough for Chairman Helms to call the bill "the great train robbery."[210] More important, the necessity of food stamps had been made clear to Republican senators and the administration; no farm bill would pass the House without the program.[211] The final version continued and expanded food stamps, avoided controversial changes, and preserved the coalition with urban votes.[212]

At the end of the effort, hard-fought compromises produced "a hold-the-line measure that will help buy time for agriculture to work its way out of its current crisis."[213] Some Republicans were concerned the bill did not reform farm policy enough, with farmers planting "for Government programs rather than for the market," but others recognized it was the best they would get and took comfort in the progress toward competitiveness and market orientation; almost all supported it.[214] It was simply the best that any one side could get—a balance "between the economic problems on American farms and the budget deficit problems of our Government."[215] The bill prioritized improving American competitiveness in export markets, carrying a distinctive market-oriented, Republican imprint.

The most vocal opposition came from some farm state Democrats frustrated that "the long shadow of the President's veto pen" shaped the outcome.[216] Their opposition covered significant ground. Some saw it as the wrong policy solution for the crisis. It would neither increase commodity prices nor decrease the surplus and was a bad deal for farmers because it would reduce prices in a time of crisis.[217] Others were frustrated that it would allow the largest farmers to take home the most in payments, while others wanted a more forceful return to mandatory controls. Happy or not, many Democrats voted for it because "legislation is the art of the possible," and they recognized it was the best deal they were likely to get under the circumstances.[218] The Food Security Act of 1985 passed easily in the House but by a narrower margin in the Senate. It cleared Congress on the same day and just in time for the holidays.

A focus on the farm coalition fights over target prices and acreage controls sells the 1985 farm bill short. It remains a landmark bill in the annals of farm policy because it marks the beginning of modern conservation policy. Congress created conservation programs designed predominantly to address conservation challenges rather than farm support and acreage reductions.[219] Expanded production in the seventies contributed to the farm economy's collapse but also recycled soil erosion and other environmental concerns about commercial farm production practices. By 1985, however, the environmental coalition was empowered by public support, as well as major legislative and political victories.[220] The conservation title was generally noncontroversial and nonpartisan. It garnered effusive praise from its authors as "historic reforms in conservation policy."[221] Congress took major actions to advance conservation of soil and other natural resources. The first dovetailed with the farm crisis and a need for reduced acres, the second was a deal with the environmental community that was even more consequential.

Congress re-created the Conservative Reserve Program's (CRP) long-term land retirement contracts but with a heavy emphasis on removing environmentally sensitive land from production. The policy was a direct link to the 1956 Soil Bank that could retire up to forty-five million highly erodible acres into conserving uses.[222] The more jarring policies were the compliance provisions known as sodbuster and swampbuster. In the depths of the farm crisis, as members were fighting to provide more generous payments to farmers, they also acceded to demands that receipt of those payments should be conditioned on conservation measures.[223] They agreed that "no Federal subsidies will be paid to farmers who employ unsound conservation practices generally, and specifically includes sodbuster, swampbuster, and cross-compliance provisions."[224] Compliance was a significant turning point in farm policy, fifty years after the 1935 act first enshrined farm conservation policy.

Such were the problems for the nation's soil and waterways that a farmer could be kicked out of the farm support programs for breaking out new ground or draining wetlands to farm them. And more was required: those farming on highly erodible land had to have approved conservation practices to remain eligible for payments. Conservation programs in the past had been primarily designed to reduce planted acres by helping defray the cost to the farmer. By comparison, Representative Glickman called the 1985 provisions "a very dedicated and considered effort to try to reduce the amount of highly erodible land in this country."[225] He worked to strengthen them further and succeeded.

In 1985 conservation policy crossed an important threshold. It was no longer cover for reducing acres or a desperation move to park acres that would otherwise be diverted to competing commodities. But it also had not completely escaped its past either. The CRP provided a convenient partner in that it removed up to forty-five million acres from farming at a time of low prices and severe depression. But Congress also included tough medicine meant to help heal the excesses of the seventies. These outcomes were likely made possible by the farm bill's struggles in the Reagan era, struggles that had necessitated building a coalition with environmental interests.[226] Forming an alliance with environmental interests completed a broad-based farm bill coalition that spanned interested communities from farm commodities to the environment and low-income food assistance.

The 1985 farm bill was an expensive rescue mission for struggling farmers. The Department of Agriculture reported expenditures on the major row crops above $15 billion each of the first two years and over $50 billion for the 1986 to 1990 period.[227] The bill generally froze target prices, but it also lowered loan rates, increasing the spread between the two price triggers.[228] The larger spread produced larger deficiency payments when prices fell; larger payments multiplied over millions of farms meant larger federal budgetary exposure.[229] From set-aside acreage authority to target prices, PIK, loans, and the new CRP, the 1985 farm bill also increased the government's role in farming.[230] One measure of that role can be found in the fact that program payments represented nearly 60 percent of the value of the corn and wheat crops—payments exceeded the value of the entire rice crop.[231]

Republicans lost control of the Senate, and Democrats took a 55 to 45 majority in the 1986 midterm elections.[232] In some states the continuing farm crisis was given credit for helping Democratic candidates overcome the campaign efforts of a still-popular president.[233] The partisan and difficult nature of the 1985 debate provided Democrats with plenty to attack Republicans, and they used the issues to their advantage.[234] The farm crisis had not yet run its course, however, and the new Democratic majority inherited the near-collapse of the entire Farm Credit System, which required a $4 billion bailout.[235]

These problems were another symptom of the seventies' excesses. The system had lent heavily to farmers during the expansion only to be caught when inflation and interest rates hammered its borrower-farmers. System banks were left with bad debt that chased out good borrowers. A collapse threatened to send shock waves through the economy.[236] The Farm Credit System bailout, strangely enough, also provided a marker to the end of the second great farm crisis.

President Reagan did not give up the push for farm policy reforms, however, and in July 1987 he opened a new front in his attack. He turned to international trade negotiations known as the General Agreement on Tariffs and Trade (GATT).[237] He proposed complete elimination of all farm subsidies by the year 2000. World-wide experiences in the eighties included surpluses and subsidized fights for export markets. Nations had also provided extremely expensive assistance to farmers. These failures were seen as pro-viding potentially receptive minds for reform.[238] But the president's proposal faced strong resistance. He pushed aggressively for farm support reforms when the Uruguay Round of negotiations began in 1988. Although it would take many years to bear fruit, Reagan's effort would eventually come to make an impact on the develop-ment of American farm policy.

Having cast their lot with export markets, American farmers benefitted when exports improved in the late 1980s, but they were also helped by drought and acreage set-asides; supplies fell back in line with demand. Farm exports picked back up from 1986 to 1990, helping improve prices, but the target price program remained expensive, if declining. Payments totaled $26 billion in 1986 and $22.4 billion in 1987 and then dropped to $10 billion in 1989 and between $7 and $8 billion in 1990.[239] The 1985 farm bill also spent significant funds to improve and expand export markets around the world. As the recovery strengthened prices and improved farm incomes, it helped improve land values. The eighties, which had provided such pain and grief for farmers and for policymakers ended on an increasingly improving farm economy. The 1985 farm bill would expire in the new decade, and the stage was set for the introduction of the modern era of farm policy.

Concluding Thoughts on an Era of Crisis and Change

Critical building blocks for modern farm policy and politics were put in place during the years spanning 1970 to 1989. First, the commodity support system switched from predominantly loans to predominantly payments with loans as a secondary backstop. Second, modern conservation policy took hold in the landmark provisions of 1985. Third, the complicated partnership between food stamps and farm programs was formalized and tested. And finally, congressional budget disciplines were enacted and then strengthened, setting in place the single biggest element in program development and political difficulties going forward. The ground under farm bills shifted, encompassing concerns about the consumer's pocketbook and then those of the taxpayer's. The 1973 farm bill was written during a commodities price spike and an era of food cost inflation; the 1985 farm bill written in the depths of crisis and intense budget-based demands for reform from Reagan Republicans.

If, as discussed earlier, the 1949 act was the culmination of the New Deal parity system, the Food Security Act of 1985 represented another peak for policy, this time for the two-part support system of payments and loans. The 1985 system resulted in billions of dollars in federal outlays to rescue farmers, lenders, and natural resources from the crisis that unfolded from 1973.[240] The cycle of expansion and collapse that defined the seventies and eighties was certainly not unique. It compares with the World War I era from which farm policy originated. The 1973 farm bill unleashed farmers who had labored under production controls since World War II. High crop prices and export enthusiasm proved to be powerful incentives to expand production, but so were federal policies from target prices to the suspension of acreage controls. The consequences were severe and expensive.

Senator Milton Young, given credit for creating target price policy, claimed that it saved farm programs. It was policy innovation at a critical juncture with strong prices and consumers in open opposition.[241] It was nearly billed as a panacea. The production incentives were a share-the-risk federal backstop that was market oriented. They were to finally free farmers to reap great gains in the "promised land" and reduce federal expenditures. But panaceas do not exist in policy. This is especially true when the policy itself helps fuel problems. When the bottom fell out, target prices were the primary pipeline for funneling billions to struggling farmers.

While target prices continued supporting a select group of commodity farmers, they *were* a different policy mechanism. They provided support on a contingent basis, meaning the farmer had to suffer some estimated price-based loss before he or she received assistance. To a certain degree, target prices also removed the federal government from the grain handling and storage business because it issued payments directly instead of taking on forfeited crops. Farmers were thus free to sell the crop at the best price they could get and take the payment as supplement. Arguably, grain companies benefitted even more. Having consolidated and grown more politically powerful, they opposed the production controls and struggled with CCC lending and forfeiture. By the early seventies, these larger, more powerful companies had become more involved in the policy process.[242]

Direct payments to farmers came with significant liabilities, however. Payments made to individuals and entities left a trail that was easy to understand, compare, and criticize; payment limitation reforms were obvious and necessary follow-ups. As Senator Lugar repeatedly argued, when prices fell, they exposed the federal treasury to immense outlays. The irony was that this policy change coincided with new efforts to control federal spending. Tied to production, direct payments also provided incentives to farmers to plant for the payment even when market prices were too low.

As a policy innovation, target prices were very much dependent on the rosy price and market predictions of the moment. They did, however, allow farm interests to reset the debate. A reset was needed after years of failed policy that had become impossible to defend. The timing was especially important because agriculture's influence in general had waned.

More consequential was formalization of the partnership with food stamps, consumer interests, and the urban vote bloc in the

House. It exemplified the benefits, challenges, and perils associated with building a large, diverse policy coalition to pass legislation through Congress. Success required a deal between cotton and food stamps that could hold together enough Democrats while adding farm district Republicans who needed to support a bill. This was no small challenge. Urban and suburban Democrats were more aligned with non-southern Republicans against cotton on tough issues like payment limits. Republicans and southern Democrats were aligned when it came to controversial food stamp issues like striking workers. Attacking benefits for striking workers was politically potent, driving a large wedge into the center of the Democratic caucus between the South and labor and touching on core issues such as poverty and race. Moreover, Republicans who would support a farm bill against their leadership were not in the same position on food stamp votes. Some saw an attempt to rebuild the Roosevelt coalition between southern farm interests and northern labor in a Democratic Party fractured regionally over civil rights and social policies.[243]

The coalition between farm interests and food stamp supporters was tenuous from the start and burdened with the troubled legacy of removing African-American sharecroppers from cotton farms. By 1974 the number of black-operated farms in the South had fallen 95 percent compared to 1920.[244] The misleadingly benign "surplus labor" and resulting "efficient reshaping of farm resources" in the South had pronounced and profound human consequences, compounded by racism, segregation, and the fights over civil rights.[245] All of this was inextricably embedded in the food stamps program. It became further complicated by budget disciplines and the partisan drift of the South to the Republican Party. This partisan realignment in the South permitted overtly race-based antagonism over food stamps to become more covert, cloaked in concerns about federal spending.

The election of Ronald Reagan and a Republican Senate in 1980 amplified budget disciplines, which had been enacted by Democrats in response to perceived abuses by President Nixon. Farm support and food stamps had grown considerably, and Reagan's budget-based strategy laid bare just how precarious the entire coalition was.[246] In the end, budget disciplines proved powerful in shaping a farm bill but not insurmountable.[247] Deal making in 1981 helped protect farm interests, especially southern, but harmed food stamps. It also ended up harming many of the Boll Weevils in the longer run and failed to reverse partisan trends in the South.[248] By 1985 the farm crisis was severe enough that it proved too big a political obstacle for more drastic reforms.

The impact of conservation and the alliance with the environmental community was also vitally important. It added yet another nonfarm coalition partner at a time when Reagan and the Republicans were on the attack against both food stamps and farm programs. The deal appears to have formed around the farm sector's need to remove acres from production and the environmental community's concerns about erosion. Farmers received CRP and protection on target prices in return for environmentalists' demand for conservation compliance. The Conservation Reserve Program and compliance constitute the most significant reforms of the two farm bills in the Reagan era, which is ironic given CRP was a new expenditure and compliance was a new form of government intrusion.

Of the two, compliance was the more striking because it had the potential of ending payments to some farmers. It was also quasi-regulatory, providing some federal oversight of farming practices. Compliance would have been generally controversial but presumably more so in a time of crisis. It also ran contrary to the views espoused by the president and many of his partisan allies. It was part of a deal, however, that provided new allies in the fights with the president and the budget. In that way it both made policy and political sense and provides important lessons.

The substantial costs of the Reagan farm bills call into question the president's success in achieving his goal for reforming farm policy.[249] Whereas the switch to target prices in 1973 constituted a major change in direction, the farm crisis limited the eighties bills to a form of incremental change or slow evolution.[250] The actual achievement may lie more in what was avoided than what changed; high target prices transferred large sums to the farm sector, but they were not Depression-era price supports and production controls. Moreover, Congress continued to push the loan provisions in a market-oriented direction, lowering loan rates and using moving averages to calculate them. And Congress introduced the marketing loan concept to permit repayment instead of forfeiture. The PIK programs and set-aside acres were the only significantly dissonant policy features; PIK, at least, was temporary.

It is likely that farm policy avoided a more substantial reversion to New Deal–era policies in the eighties crisis because neither farmers nor consumers nor voters had an interest in a return. An urbanized nation and Congress played a role, pushing policy evolution and reforms in a time of crisis. Evolution was also driven by a better understanding of the linkages between the interests of the farmer and consumer—and society in general—combined with continued commercialization in the ag sector.[251] Critics were growing louder, and demands for reform became stronger and more bipartisan. Larger, more sophisticated farmers had less need for, or interest in, the intrusive policies of the past. Writing farm bills was no longer the sole province of farm interests, and whether admitting it or not, farm interests were more reliant on nonfarm alliances in a growing coalition.

In the history of American farm bills, target prices and food stamps in the Agriculture and Consumer Protection Act of 1973 mark one of the most distinct turning points in farm policy development. The 1973 bill offered a potential way out of the self-destructive feuds over parity policy that had consumed the previous two decades. But the changes in farm policy and perspective set up many for substantial economic pain, driving more farmers out of farming and nearly collapsing the Farm Credit System. The Food Security Act of 1985 likely helped with the farm crisis and benefitted many struggling farmers. It also charted a new course for conservation and the coalition. The Reagan farm bills were squeezed between two eras of drastic change in the seventies and nineties. The two decades distilled to target prices and food stamps; these two policies dominated farm bill debates and politics for at least forty years, including the effort to write what became the Agricultural Act of 2014.

6

Revolution and Reform Launch the Modern Era, 1990–1999

Introduction

The eighties farm crisis was receding in history's rearview mirror as the new decade got underway. Some Republican senators made certain to credit the 1985 farm bill for having saved farmers, but they also had to admit that exports drove record net farm income in 1989.[1] Democrats, particularly those hailing from wheat country, were concerned that export enthusiasm and President Reagan's push to eliminate farm supports in the GATT negotiations would cause the U.S. to "unilaterally disarm" and leave farmers at the mercy of foreign governments.[2] Overall, export markets increased demand for American commodities; farm policy in the nineties promised to be influenced by trade and exports. The decade experienced strong growth in Asian market demand, especially from China, as well as major free trade agreements, such as the North American Free Trade Agreement (NAFTA), that helped open Mexican markets for American feed grains.[3] The impact of this had yet to be felt when the 1990 farm bill was written.[4]

The nineties opened with political status quo. Ronald Reagan's vice president, George H. W. Bush, was elected president in 1988. He won handily, but Democrats gained seats in Congress and remained in control of the House and Senate.[5] President Reagan had used the budget to reinvigorate the Republican perspective on farm policy, setting up conflict between conservative ideology and the demands from farmer-constituent interests. He had pushed reform and a more limited federal role in farming, consistent with his Republican predecessors Eisenhower and Nixon. Following in this tradition, President George H. W. Bush pushed for more flexibility in making planting decisions and better responsiveness to market signals.[6] He also sought assistance to restore wetlands because the issue had gained prominence with voters and environmentalists.

In Congress Senator Patrick Leahy (D-VT) assumed the chair of the Ag Committee in 1990. Senator Leahy was elected to the Senate in the post-Watergate 1974 election after he had served eight years as a state's attorney.[7] He replaced George Aiken, who had been in the Senate for thirty-four years and who, as chairman of the Ag Committee, had taken leading roles in numerous farm bills. Senator Leahy has served as chair of the Agriculture and Judiciary Committees and held lead roles on the Appropriations and Intelligence Committees. In farm bill debates, he has been a powerful voice on dairy, conservation, and food assistance issues.

His ranking member in 1990 was Senator Richard Lugar (R-IN), who had fought against target prices throughout the eighties. Senator Lugar, an Eagle Scout and Rhodes Scholar, had served in the U.S. Navy before he returned to Indianapolis to join the family food product machinery business. He was elected as mayor of Indianapolis in 1967 at the age of thirty-five and as a U.S. senator in 1977. He served in the Senate for thirty-six years.[8] He grew up in Indianapolis but also lived part-time on a farm his parents purchased the year before he was born. Senator Lugar continued to own and work on the farm during his years in the Senate. He has said that his "soul is on that farm."[9] Widely respected, known as a true statesman and expert on foreign policy, Lugar left the Senate in 2013 with a legacy that includes the Nunn-Lugar Cooperative Threat Reduction Act, passed in 1992 to safeguard nuclear weapons.[10]

In the House, Chairman Kika de la Garza (D-TX) remained at the helm of the Ag Committee with two farm bills under his belt. He was joined by Representative Ed Madigan (R-IL) as ranking member. Representative Madigan was a ten-term congressman from Lincoln, Illinois, who went on to serve as George H. W. Bush's secretary of agriculture in 1991.[11] He was a member of the Illinois House of Representatives from 1967 to 1972 and elected to the U.S. House of Representatives in 1972.[12] In a sign of things to come, Madigan lost a close and difficult race against Newt Gingrich (R-GA) for the House whip position in 1989 in what has been explained as a clash of generational and style differences within the Republican Party. Madigan was more of an establishment Republican who sought consensus and deal making with Democrats, whereas Gingrich wanted confrontation in service of building a Republican House majority.[13]

Congress wrote two farm bills in the nineties. The debates followed more than a decade of cross pressures from crisis, budget battles, environmental concerns, and a popular president's ideological demands for reform. While the 1985 farm bill had helped with environmental concerns, the crisis had largely cancelled out Reagan's reform agenda. Target prices were decreasing in 1988, 1989, and 1990 as prices increased, reducing the baseline for the 1990 farm bill but not criticism or budget pressure.[14] Another price upswing and relief from crisis decreased pressure to deliver assistance on par with that of the eighties. The budget pressures remained. Budget reconciliation, which had been introduced to the farm bill process in 1981, would feature prominently in the decade's two bills.

High crop prices had delivered major revisions in 1973 but proved expensive when long pent-up pressure, released on the upswing, went too far and led to crisis. Assistance was coupled to production, providing incentives for farmers to make planting decisions according to expected payments. Congress used CRP and set-aside authority to attempt a counterbalance to this incentive, but problems remained. When the policy debate shifted from crisis to reform, it was logically defined in terms of spending through target prices and acres; the focus was on production incentives and cost exposures from coupled farm payments.

Reform also involved controversial questions about who received payments and how much they received. For example, as part of the Omnibus Budget Reconciliation Act of 1987, the House Agriculture Committee prevented payments to persons who were not actively engaged in farming.[15] This reform was meant to address concerns that certain farm entities were legally circumventing payment limitations, a problem that had been exposed by the Government Accountability Office (GAO).[16] Reformers attacked what they saw as absurd abuses of the farm programs through "schemes [that] have been developed that allow passive investors to qualify for benefits intended for legitimate farming operations."[17] Changes in the last decade of the twentieth century would make bright the spotlight on these matters.

Another export-fueled climb in prices was a key ingredient for the substantial revisions produced by Congress in the two farm bills of the nineties. Better prices combined with budget pressures, partisan upheaval, and demands for reform to create a powerful political mixture. The decade launched the modern era of farm policy. Congress moved in the direction of market-oriented programs with unprecedented flexibility for the farmer.

Prelude to Revolution: The Food, Agriculture, Conservation, and Trade Act of 1990

The 1990 farm bill was legislated in two parts and marked a third consecutive effort consumed by budget discipline. Congress had to pass both the farm bill and a budget reconciliation bill that included a $13 billion reduction to that farm bill. That reduction was achieved by creating a policy called "triple base," crafted out of marathon negotiations over target prices, loan rates, and payments.[18] The policy involved "real pain and real hurt and real cuts" that left some Democrats opposed to the bill, placing blame on the Bush administration for the final outcome.[19]

Budget pressures had repeatedly limited congressional efforts to increase target prices while Republicans pushed the emphasis on exports and better market orientation. The 1990 effort went further, effectively removing 15 percent of acres from the programs. It also contained modest revisions to farm program payment limits and eligibility requirements, as well as revisions to the crop insurance program. Congress continued conservation compliance authorities with modest revisions and increased CRP.[20] Finally, the bill continued food stamps with little change, but the budget negotiations had forced the House to abandon its efforts to increase benefits.

Most of the action took place in the Senate, where the Ag Committee markup had been difficult. On the Senate floor, Chairman

Leahy cautioned senators that the committee had "wrestled with at least three cardinal issues in the bill: The loan rate structure, the target price structure, and the set-aside acreage proposition."[21] He repeatedly argued against amendments that threatened to unravel the delicate compromise struck by the committee, claiming that upsetting the balance would bring down the entire bill. The committee fought over authority for the secretary to lower loan rates, with Republicans and enough southern Democrats holding off plains Democrats. The coalition between southern Democrats and Republicans ran into trouble over target prices and whether farmers who set aside additional acres should get a higher target price. The committee had also struggled with conservation compliance, but deal making between Chairman Leahy and Ranking Member Lugar brought the sides together to report the bill on a strong bipartisan basis.[22] The chairman was committed to the committee's work to protect "the water we drink and the wetlands our wildlife depends on" and to stop "the decline in target prices and loan rates."[23]

The most significant challenges to the committee compromises would come from farm state senators, however. Their inability to get the changes they sought demonstrated the power of the deal struck in committee. Farm state Republicans targeted the set-aside authorities because they wanted more flexibility for farmers to make planting decisions and improved market orientation.[24] Their view also aligned with that of the grain industry, which played an increasingly larger role in farm policy debates.[25] Set-aside authority was core to the compromise because each side of the deal held strong views and even minor changes would have major political consequences. The committee compromise had to hold in total or fall apart. Such stakes reinforced the Senate's institutional tendencies and helped the chairman protect the deal.[26]

Similarly, plains state senators (primarily Democrats) demanded higher target prices for small farmers.[27] Senator Tom Daschle (D-SD) wielded the populist perspective of the plains on behalf of smaller family farmers who "cannot exist on economic dialogue and trade theories. . . . We simply cannot sacrifice family farmers on an altar of imagined free trade."[28] His was an attack on the cardinal issue of target prices, and he wanted to increase them for small farmers at lower levels of production. Target prices were supposed to help farmers compete with subsidized foreign production, especially European, but farm state Republicans accused their colleagues of trying to "give farmers more than farmers apparently want" and maybe more than they deserved.[29] The focus on helping small farmers caused political problems with southerners that Republicans exploited to help defeat the challenge.[30]

Price-supporting loan rates were the third of Chairman Leahy's cardinal issues and marked an unwelcome return of this long-controversial component of farm policy. Loan rates had taken a backseat to target prices beginning in 1973. The 1985 farm bill went further in the direction of market orientation, in part by calculating loan rates using market average prices. The Senate Ag Committee had reinstated a fixed price floor under the market-oriented calculation.[31] Farm state Republicans wanted to prevent a return to loan rates that made U.S. commodities uncompetitive with world market prices. Senator Rudy Boschwitz (R-MN) represented the view that this was one of the most important features from 1985 because 30 percent of corn, 50 percent of soybeans, and 64 percent of wheat were exported.[32] He and like-minded Republicans were not comfortable with having a floor back in the program.[33] Here again, the committee agreement held mostly because those who personally preferred more competitive loan rates, like Senator Lugar, understood the need to avoid changing one of the cardinal issues.[34]

Issues on the periphery of the main committee compromise were open to being amended on the floor. For example, crop insurance was debated and amended after the Bush administration had proposed eliminating it but the Ag Committee disagreed.[35] Congress was generally frustrated with crop insurance by 1990 because participation remained too low and the program was not actuarially sound.[36] Farm state senators such as Senator Daschle and Senator Kent Conrad (D-ND) were concerned that the effort to force reforms would backfire, arguing that crop insurance was "the best method for farmers and the Federal Government to manage the risks of natural disasters" because disaster assistance payments were costly and inequitable.[37] Crop insurance was not, however, a cardinal issue of the committee compromise, and senators were able to work out reforms on the floor.

Chairman Leahy effectively wielded the committee compromise to fend off challenges on the Senate floor in a demonstration of the power and effectiveness of hard-fought deals at the committee level. The bill had been attacked from the market-oriented perspective over acreage set-aside authorities and the loan rate floor. It had also been attacked from the more populist perspective on target prices. These key components were balanced against each other to provide maximum political effectiveness. A senator who wanted to vote for higher target prices (more funds to their farmers) knew that the political price would have been more backsliding into acreage reductions and fixed loan rates that hurt international competitiveness and the grain trade. A senator who might otherwise oppose the set-aside and loan floor provisions understood that they were the price to be paid to keep target prices in check. This was the source of the deal's strength, but the bill also carried great risk because a loss on any one piece would unravel the entire deal with more revisions, challenges, and possible defeat. The uncertain budget situation also helped protect the cardinal issues of the compromise.[38] The Senate compromise was tentative because the farm bill had to clear the House and remained at the mercy of a yet-to-be-completed universal budget deal.[39]

In the House the Ag Committee struggled with the same disagreements. Democrats wanted to increase loan rates, but Republicans were opposed, concerned about the damage higher loan rates would do to U.S. competitiveness.[40] The Bush administration's demand for spending discipline limited efforts to provide more generous assistance, much to the frustration of many farm-district members, including some Republicans.[41] Frustration carried over to the floor, where Republicans won minor revisions to loan rates and target prices to reduce the costs of the bill but little else. Democrats were unable to increase assistance levels.[42] The debate was muted by the budget shadow hanging over the entire effort. Everyone understood that a pending budget summit would be where the real cuts were made.[43]

The most contentious moments in both the House and Senate involved reforming farm program payments.[44] Notably, the Senate farm bill initially did not contain a nutrition title (and food stamps) because it had been caught up in ongoing budget negotiations. This denied farm policy the protective benefits of that coalition. Debating a farm bill without food stamps coincided with continued budget challenges for the bill that required the committee managers to fix it on the floor.[45] Opening up the bill presented opportunities for reformers that risked the deal. The farm coalition would have to defend payments on its own. Senator Harry Reid (D-NV) argued that payments "lavished on wealthy farmers must first be taken from other citizens," and he fought unsuccessfully for means-testing farm payments, blocking them from farmers with high incomes.[46] Although he was defeated, Senator Reid had put farm program supporters further on the defensive.

In the House a coalition of urban Democrats and conservative Republicans led by Representatives Chuck Schumer (D-NY) and Dick Armey (R-TX) pushed a similar amendment that exposed a deep vein of frustration with farm programs from nonfarm district members. They considered it unfair to pay farmers who earned more than average workers.[47] Some were particularly frustrated with past abuses that had left the impression that "the swindlers and the parasites" were taking advantage of the programs while others suffered.[48] Reformers clashed with farm district members, especially southerners, who wanted to preserve the status quo and fought against the perception that the payments were welfare for farmers.[49] They thought this misrepresented what they considered assistance that helped ensure an adequate food supply for consumers and the nation.[50]

In the House the bigger concern for farm policy supporters was that the amendment was an ideological attempt to destroy farm programs rather than reform them. Representative Armey's involvement confirmed this concern for many. As a result, farm district members who supported the need for reforms were convinced that the amendments would do more political damage than good.[51] Reform efforts had made a noteworthy push but were defeated on both sides of the Hill.

After having achieved success in 1985, environmental concerns shifted their focus to wetlands protection and restoration. In the Senate the policy remained popular and bipartisan, with minor amendments generating little controversy.[52] In the House the addition of environmental interests in 1985 continued to strengthen the coalition. Environmental interests wanted to preserve the conservation compliance provisions and improve efforts to restore wetlands.[53] Preservation of their top priorities proved enough for them to join farm interests and help to defeat reformers on the floor.[54] Similarly, amendments to cut food stamps were also no match for the voting bloc, and the farm bill passed easily.[55]

As expected, budget negotiations upended the effort in conference. Conferees were pushed to cut $13 billion from the bill. There was little appetite for cutting farm payments so close to midterm elections and on the heels of the eighties farm crisis.[56] Additionally, the Bush administration demanded increased flexibility and spending reductions. These forces led to a relatively creative outcome.

Instead of lowering target prices to achieve savings, policymakers simply reduced the acres that were eligible to receive payments. The result was triple base; 15 percent of a farm's base acres were ineligible for payments.[57] The farmer received almost complete flexibility to plant on those acres, and the change moved farm programs a step further into territory that left traditionalists uncomfortable.[58] This created three categories of program acres: (1) acres that received program payments; (2) acres that were required to be set-aside from production in order to receive payments; and (3) the 15 percent of acres ineligible for program payments of any kind that the farmer had freedom to plant to almost any crop. Supporters of market orientation viewed the new policy as progress. A more drastic revision to farm policy was on the horizon, but it would require political upheaval that was still a few years off.[59]

Republican Revolution and Freedom to Farm, 1995–96

Four seismic political shifts took place before Congress next returned to the farm bill. First, Bill Clinton, the Democratic governor of Arkansas, defeated President George H. W. Bush in his 1992 bid for reelection in a three-way race largely defined by a struggling economy.[60] Second, President Clinton inherited the Uruguay Round of trade negotiations, which included the far-reaching Agreement on Agriculture.[61] Third, President Clinton negotiated and Congress approved NAFTA in 1994.[62] Fourth, Republicans took control of Congress in the 1994 midterms for the first time in forty years, carrying promises of conservative reform to the federal government.[63] For the farm bill, all four shifts carried substantial implications. They elevated budget and reform priorities with a market-oriented, conservative ideological perspective and an emphasis on exports. The changes in world trade did not fall equally on all commodities; export dynamics specifically separated corn and wheat.[64]

The Republican victory in 1994 also flipped the traditional reform approach. Instead of a Republican president pushing reforms against a (largely southern) Democratic Congress, it was now a reform-minded Republican Congress working reforms against a reluctant Democratic (southern) president. Arguably more important, when the House flipped to Republicans, it was on a wave of votes that defeated many southern stalwarts of traditional farm policy. The result was that both southern commodity interests and conservative reformers were on the same side of the partisan line.

Longtime champion of farm program reform, Senator Lugar took over as chair of the Senate Ag Committee. In the House, Representative Pat Roberts (R-KS) became chair of the House Ag Committee. It was a challenging time to assume the chair, but Representative Roberts took the gavel with a significant amount of farm bill experience. Born in Topeka, Kansas, and a graduate of Kansas State University with a degree in journalism, Roberts served four years in the U.S. Marine Corps and worked as a reporter in Arizona.[65] In 1967 he joined the staff of U.S. Senator Frank Carlson (R-KS) and moved to the staff of Representative Keith Sebelius (R-KS) two years later. In 1980 he was elected to the House of Representatives, where he led efforts to develop the marketing assistance loan program in 1985. He went on to be elected to the U.S. Senate in 1996, and in 2015 he became the only member of Congress to chair both the House and Senate Ag Committees.

The farm bill reauthorization would face the familiar sting of budget discipline, but from a very different perspective. Republicans had taken over the House under the Contract with America, a pledge to cut spending, balance the budget, and enact reforms to get the federal government out of people's lives and business.[66] Congress—particularly the House—was wielding the budget axe, and the $56.6 billion farm support system would not escape unscathed.[67]

Congress crafted and passed a budget in June 1995 that included reconciliation instructions to achieve $894 billion in federal spending reductions over seven years.[68] This included a difficult $13 billion reduction in farm program spending. The House and Senate Ag Committees were then required to produce a bill that achieved those reductions. The budget reconciliation process, put to unprecedented use in 1981 with the intention it be limited to those extraordinary circumstances, came into play once again.

The two Ag Committee chairmen took decidedly different routes. Senate chairman Lugar was at his core a reformer, and he went all in. He informed farm groups that American farming had changed but the policy had not kept pace. He made clear his view that "every agricultural program will be called on to justify its existence and continuation."[69] He wanted to phase out deficiency payments by lowering target prices year over year. This would make the target prices less likely to trigger payments and more likely to avoid repeating previous failures to control expenditures.[70]

His proposal, however, was burdened by two heavy political liabilities. First, it would result in disparate outcomes for commodities depending on the spread between market prices and target prices. The larger the spread the longer it would take the target price reduction to affect payments; a smaller spread for corn would end payments in four years whereas the larger spread for rice would take nine years.[71] Even more consequentially, the proposal led directly to the elimination of farm program payments. The proposal therefore put Chairman Lugar in a direct collision with southern Republicans, led by Senator Thad Cochran (R-MS), and all committee Democrats.[72] This combined opposition was too much, and Chairman Lugar had little choice but to abandon his proposal or face certain defeat in his committee.

The direction taken by Chairman Roberts was difficult in its own right and required considerable negotiating and legislating skills.[73] He negotiated his way through budget challenges with House leadership and navigated a turf war with the Appropriations Committee.[74] He also faced challenges from urban Democrats. During this difficult period, Chairman Roberts appears to have strategically held back on making a specific farm program proposal. He waited out the budget and appropriations processes before he introduced a farm bill proposal; he dropped "Freedom to Farm" just as Congress broke for its long August recess.[75] His Freedom to Farm proposal sought to completely eliminate all remnants of acreage reduction policies and replace price-based assistance (target prices) with fixed, annual payments on a declining basis that were completely decoupled from prices and production.[76]

Chairman Roberts's Freedom to Farm proposal constituted an unprecedented move for reform by a House Ag chair and required a rather delicate political balancing act. The appearance of drastic reforms to decades-old programs rooted in the New Deal held important appeal for House leadership but also provoked strong concerns from farm program traditionalists. The 1994 Republican victories in the midterm had shifted a large number of southern seats into the House Republican caucus, making it home to both sides of this divide. Southern farm program stalwarts were in direct conflict with their aggressive leadership over this small but symbolic slice of federal spending. Chairman Roberts offered something for both sides but only if they could see their way to it. Moreover, he could not count on any help from committee Democrats opposed to drastic measures by Republicans.[77]

Crop prices were on the rise, and Roberts was proposing to discard any potential exposure to large expenditures from price-based assistance. Freedom to Farm was also legislative sleight of hand in the budget discipline era. High crop prices reduced expected expenditures from deficiency payments because market average prices were expected to be above target prices.[78] By comparison, the fixed payment structure would capture farm program baseline in an increasing market.[79] Understanding the interaction between CBO scoring and the way it treated fixed payments compared to deficiency payments contingent on market prices was vital to the chairman's plan. He had to get his committee to go along with him, however. The fixed annual and declining payments jettisoned almost the entire farm program infrastructure they had long known and defended. The plan faced substantial political hurdles, starting with southerners who united in opposition to a concept that was simply anathema to them.[80] Southern senators had already defeated Chairman Lugar's reform proposal, and their House counterparts took up Senator Cochran's proposal rather than acceding to major changes.[81]

After the August recess, the Roberts Freedom to Farm proposal, a Democratic alternative, and the southern proposal offered by Representatives Bill Emerson (R-MO) and Larry Combest (R-TX) all collided in an intense markup.[82] While Democrats lacked the votes, farm interests were confident that the southerners could defeat the chairman. The chairman and his allies, including Representative John Boehner (R-OH), predicted that Freedom to Farm would prevail because failure would put the farm bill's future in jeopardy.[83] Both were wrong. Committee Republicans turned against each other in a long, difficult debate that neither faction could win.[84] The split resulted in a narrow defeat for the southern alternative followed by a "dramatic roll vote" that rejected the chairman's proposal because four Republicans voted against it.[85]

The defeat was reported as having stunned everyone at the markup. A Democrat on the committee remarked, "No chairman has ever lost a bill like this of this magnitude."[86] The loss threw the farm bill in jeopardy and created political challenges for Chairman Roberts, who adjourned the markup. Hard feelings and real fear about farm programs' future had to have hung heavy over the farm interest community and the House Ag Committee. But Chairman Roberts would not be easily defeated, and he had a backup plan with House leadership. He had secured an agreement from leadership that they would include his Freedom to Farm proposal in budget reconciliation if the committee could not complete its work.[87] This deal produced no small amount of political tension within the Republican caucus.

House Republicans moved forward with budget reconciliation, including Freedom to Farm. Chairman Roberts argued that it was the proper policy for the changed farm economy. Existing programs were rooted in the New Deal, he said; they "no longer achieve their original goals and have collapsed as an effective way to deliver assistance to producers." He added, "Government outlays under current programs are the highest when prices are lowest . . . [which] has had the effect of encouraging production based on potential government benefits, not on market prices."[88] He wasn't just cutting spending; he was attacking the very foundation undergirding farm policy.

Logically, Democrats attacked Freedom to Farm as the destruction of farm policy, jammed into budget reconciliation without hearings and after the bill had been rejected by the Ag Committee. The proposal would "unilaterally disarm farmers in the international market place" and was a recipe for farm failures, not freedom.[89] Democrats were also angry with the heavy-handed tactics Speaker Gingrich and his leadership team used to bring reluctant southerners along to pass the bill in late October.[90] Democrats were pushing for less drastic changes and cuts to farm spending, but their remarks carry no small amount of desperation in the face of not only the drastic revisions to farm policy but also their powerlessness against the budget reconciliation process. The debate displayed the markings of the devastating loss for Democrats after forty years of House domination.

Chairman Lugar, having pulled his proposal, sided with Senator Cochran's bill to reduce spending by expanding triple base acres, but the bill remained a heavy lift against Democratic opposition. A final compromise barely received committee approval at the reconciliation deadline.[91] Democrats were powerless to stop reconciliation in the Senate, and the bill moved to conference with the House and Freedom to Farm.[92] Considering that it was trying to balance the federal budget in seven years with cuts across numerous areas and committee jurisdictions, the entire reconciliation bill was of immense size, scope, and complexity. Ag Committee conferees would be tasked with working out the differences in their two bills, the Senate/southern decrease in payment acres versus the House/Roberts Freedom to Farm. This would have been challenging in a normal conference situation, but the farm program debate was just one among many challenges in a massive effort.

Conferencing the farm program provisions within the larger budget reconciliation conference had to have altered the negotiating dynamic significantly. It appears to have ceded significant negotiating responsibility (and power) to House and Senate Republican leadership, which ultimately favored Freedom to Farm and the elimination of target prices.[93] Under pressure Chairman Roberts cut a deal with southerners that helped cotton and rice, and then Senator Lugar switched his support to the Freedom to Farm proposal. It was included in the final conference agreement.[94] Leadership and budget reconciliation forces helped bring about the final agreement, but commodity prices also continued to improve. The higher prices, and expectations for continued strong prices, were also instrumental in getting otherwise opposing sides in Congress to agree to the reform program.[95] In short, the idea of locking in the CBO baseline grew more and more attractive. Deals struck, Congress passed budget reconciliation in late November.[96]

Budget reconciliation created multiple political problems for the farm bill beyond farm program payments. House Republican leaders were intent on revising social safety net programs across the board. They sought to block grant funding and reduce overall funding for food assistance, such as food stamps and the school lunch programs.[97] Disparate treatment for farmers compared to poor children or families was a politically charged move in a highly partisan atmosphere. Senator Leahy, ranking member on the Senate Ag Committee, made clear the dangers when he informed the Senate, "Because some farm groups have proposed taking food from the needy to subsidize wealthy farmers . . . trying to repeal a decade of legislation that has brought harmony between agricultural and environmental policies. Let me make my position clear—very clear. If farm programs become the enemy of the hungry and the environment, I will not support them. Indeed, I will join those on the floor who want to dismantle them."[98] He had allies in the Senate from the likes of Chairman Lugar and some southern Senate Republicans who resisted drastic cuts to food stamps and school lunch assistance.[99]

For President Clinton, the budget reconciliation effort was a political gift. He opposed Republican spending reductions and pushed back against their proposals, forcing Republicans to stand alone on budget cuts and fight among themselves over how to achieve them. House Republicans, by seeking to cut food stamps, block grant funding to the states, and reduce all assistance to the poor, especially schoolchildren, made the president's strategy simple and politically effective.[100] House Republican political vulnerabilities were largely self-inflicted. President Clinton, Secretary of Ag Dan Glickman, and congressional Democrats wasted no time attacking them. They took advantage of the poor optics for Republicans and reminded farm interests that food stamps had political value to them as well.[101]

The political trap was set when reconciliation passed. President Clinton vetoed it after Thanksgiving, which forced a partial shutdown of the federal government just before the holidays and dragged Republicans into a budget summit that went nowhere.[102] The entire fiasco was to the political and public detriment of Republicans, particularly House leadership.[103] The 1990 authorities had technically lapsed on September 30, 1995, and farm policy risked reversion to the 1949 act permanent law parity provisions. Budget negotiations left everything in limbo, as political repercussions resulted in failure for the Republican leadership.[104]

High crop prices continued to change the political calculations for Roberts's proposal in the New Year. Both he and Chairman Lugar introduced Freedom to Farm as stand-alone legislation in January 1996.[105] It would have to survive a difficult regular-order process in a presidential election year without the protections from reconciliation, but momentum was in its favor. Farm groups and committee members calculated that Roberts's proposal would pay more than existing price-based polices. The lure of known benefits delivered to farmers from fixed payments was working. Most of the committee Democrats, however, remained opposed to the bill.

Chairman Roberts pushed ahead as spiking commodity prices put wind to his back. The second time around in committee was better for him as he fended off Democratic amendments and reported the bill with three Democratic votes.[106] Freedom to Farm cleared the hurdle that had tripped it up the year before and managed to pick up a measure of bipartisanship. These were remarkable achievements for the controversial, drastic changes to long-established policy.[107] The achievement was even more notable because Chairman Roberts's strategy included a commodities-only farm bill and he maneuvered to keep out nutrition and conservation. He also eliminated the 1949 act as permanent law to preserve his negotiating leverage in conference.

Chairman Lugar moved next. He bypassed his committee and took the bill straight to the Senate floor.[108] The Senate farm bill, similar to Freedom to Farm, was centered on the fixed annual payment program. Reflecting process realities in the Senate, however, Lugar included conservation, nutrition, and the other farm bill titles. Chairman Lugar needed sixty votes to break a filibuster and pass the legislation, as opposed to a simple majority for budget reconciliation. This also strengthened Minority Leader Daschle's hand for his alternative proposal benefitting smaller farmers. But Chairman Lugar had a few options to get through the Senate. He could deal with Senator Daschle and midwestern Corn Belt senators or with Ranking Member Leahy. Reportedly, Lugar was close to finalizing a deal with Daschle before leadership pulled him back.[109]

Chairman Lugar argued that the bill was a simple, straightforward reform of farm policy that would provide farmers with the certainty and flexibility (freedom) they needed to survive in the world economy.[110] He pushed for this new direction in farm policy from the farmer's perspective: the need for market-oriented policies that allowed a farmer to plant for market demands—especially those from China and Asia—rather than farm programs.[111] The bill had gained support from across the farm coalition and the Senate Republican caucus, but it was lacking support from key Democrats.[112] Farm state Democrats, especially from the South and the northern Great Plains, appeared inalterably opposed.[113]

Senate Minority Leader Tom Daschle attacked Republicans for slashing numerous assistance programs but giving "farmers huge payments whether they do anything to farm or not."[114] He called it the "most radical farm bill in 60 years."[115] Senator David Pryor

(D-AR) likened the bill to "going on a cheap drunk" under high crop prices.[116] It was "the death knell for farm programs," Senator Kent Conrad proclaimed, because making payments in times of high prices and without regard to whether the person farmed was a serious mistake and "a scandal waiting to happen."[117]

Chairman Lugar acknowledged the partisanship that had enveloped the bill, but he had real leverage. Republicans had consolidated support behind it, including from southern Republicans who were now in favor of the guaranteed payments. The bill locked in baseline, which would provide for future farm bills under budget rules. This was an important feature given that the bill was a transition to an unknown future for farm policy.

Democratic opposition was enough to initially block the bill's progress but not enough to derail it completely.[118] Moving forward required additional Democratic votes through a deal with either Ranking Member Leahy or Minority Leader Daschle.[119] Negotiating with Senator Daschle proved more difficult because of his strongly held philosophical opposition to Freedom to Farm policy. But it also appeared to have a lot to do with the level of partisan politics. Senator Leahy was in a different negotiation position. He was willing to accept Freedom to Farm in exchange for dairy, food stamps, and conservation provisions.

Senator Leahy's willingness to deal on the Republican's priority won out, and when Senator Daschle tried to push his compromise proposal as an amendment, he was soundly defeated. The tense negotiations produced a deal between Senators Leahy and Lugar that preserved both sides' top priorities: Freedom to Farm along with conservation and food stamps.[120] Senator Daschle's Democratic faction had lost the ideological fight, but some accepted a final concession as sufficient to allow the Senate to pass the bill.[121] Specifically, Senator Daschle was able to rescue the 1949 act's position as permanent law. In the last-minute deal making, Louisiana's senators secured additional benefits for rice farmers.

Here was a clear result of elections. Upheaval at the ballot box in the 1994 midterm elections stood on the precipice of delivering substantial changes to farm policy. Freedom to Farm had cleared the U.S. Senate and the House Ag Committee. It was the most far-reaching change in the policy, surpassing target prices 1973. Its fate rested on the House floor.[122]

The only thing that seemed capable of slowing the House farm bill at that time was, ironically, the weather—a February snowstorm paralyzed DC.[123] When the farm bill came before the House, its reforms were protected by a rule that limited the amendments and the amount of time for debate.[124] The closed rule served the farm bill well, protecting it from traditionally difficult amendments on cotton, dairy, peanuts, and sugar. The House easily defeated an amendment to return to (and then phase out) price supports, as well as one to eliminate marketing assistance loans for cotton. The peanut and sugar programs were also subject to attacks, but they appeared perfunctory in spite of the close votes to protect them; only dairy was moderately revised.

Chairman Roberts opened the debate stating, "Farmers and ranchers know, boy do they know, the current farm program is outdated and in need of reform."[125] The Agricultural Market Transition Act would allow farmers to compete for export markets instead of watching foreign producers "steal our market share." The old farm policy system had "collapsed as an effective way to deliver assistance to farmers." Freedom to farm would "get the Government out of farmers' fields" so that the farmer could follow the market with more environmentally friendly crop rotations.[126]

The House Ag Committee's report provided the Republican blueprint and philosophy for reforming farm policy. Reform was needed because of long-running budget pressures that had rendered the old policies ineffective.[127] It promoted a transition to market orientation and away from failed New Deal–style policies that were ill-suited for the modern farm economy.[128] While the report stopped short of proclaiming arrival in the "promised land," it contained echoes of 1973. At its core were optimistic forecasts for export demand that farmers needed to be able to capture free of the yoke of federal controls. It was a new version of farm policy built on old arguments and another big bet on the markets. This time fixed payments would smooth the path and capture baseline.

In Chairman Roberts's words, the fixed payments "must be looked at from a new perspective . . . a transition to full farmer responsibility for his economic life, a risk management account."[129] Roberts and his allies can be seen walking a fine line on the issue of transitioning farm policy, however. Principled claims that the bill was buying the federal government out of farming were balanced against taking credit for capturing the CBO baseline for future farm bills.

High crop prices had bought peace among Republicans on the House floor in 1996, but not among Democrats. Farm district Republicans were united behind the new policy direction; seven years of fixed payments presented an attractive option.[130] To Democrats, a flawed process had produced bad policy. House leadership was again forcing this policy on America's farmers in spite of its having been rejected by the Ag Committee.

Ranking Member Kika de la Garza countered that the policy represented "a sudden and dramatic abandonment by the Government of its role in sharing the farmer's risk. . . . The Federal Government will withdraw completely from its partnership with the producer in providing the food security of our Nation."[131] The Democrats were skeptical of the optimism about exports and cautioned that the bill was tantamount to disarming American farmers in an unpredictable international marketplace against heavily subsidized foreign competitors.[132] And they derided a bill that provided a "$35 billion . . . entitlement to people, whose requirement will simply be that they have been the owners of the land at a certain period and in the program" but not a penny for food stamps.[133]

Chairman Roberts added conservation provisions on the floor. Specifically, he added the Environmental Quality Incentives Program (EQIP) to the growing suite of conservation programs in the farm bill, along with CRP and the Livestock Environmental Assistance Program (LEAP). These additions, however, were not enough to appease Democrats. House Republicans passed the bill without them.[134] Chairman Roberts had successfully steered through the House the most substantive revisions of farm policy to date, against long odds and substantial resistance. It was no small accomplishment.

A brief conference sealed the deal. The final negotiations accepted the reforms in Chairman Roberts's Freedom to Farm, with payments that were decoupled from planting decisions, production, and market prices. That conferees also provided significant victories for Democrats was evidence of Senator Leahy's leverage and skill in the negotiations.[135] With Chairman Lugar's help, Senator Leahy was able to deliver on conservation policy and, at least partially, on food stamps.[136] For conservation, the farm bill continued the CRP created in 1985 but lowered the acreage cap to 36.4 million acres. The bill also included EQIP funded at the Senate's higher billion level. Finally, it kept the conservation compliance provisions from 1985 but removed the linkage to purchasing crop insurance. As for food stamps, the bill provided a two-year reauthorization of the program as compared to seven years for farm and conservation programs. This kicked any further fighting over the program to the comprehensive welfare reform efforts underway in Congress.

Senate Democrats, led by Minority Leader Daschle, represented the last pocket of resistance. To them, the bill represented a repeat of past mistakes over market reliance, especially exports, and they warned of a repeat of the seventies. Much of their opposition was based on making payments irrespective of prices. The bill would "phase out the partnership the government has had with agriculture," trading a farm safety net for welfare payments.[137] Democrats also continued to level accusations of hypocrisy when comparing fixed payments for farmers with the treatment of food stamps.[138] But their ideological opposition lacked sufficient power in the new Congress.

The final bill eliminated acreage set-asides for the major commodities in return for seven years of fixed payments. To receive the payments, farmers were required to keep the land in agricultural uses and in compliance with conservation requirements. They were paid using the base acre system that remains in use today, a system that provides farmers with maximum flexibility. Avoiding a complete break from the farm policies of the past, the bill retained the nonrecourse marketing assistance loans and continued the peanut and sugar programs, dairy export incentives, and Senator Leahy's Northeast Dairy Compact.

Chairman Lugar credited House chairman Roberts for "tenacity

[that] led to reforms that a short time ago were clearly unthinkable . . . a revolution of consequence, perhaps the greatest in 60 years . . . because we are now in a situation in which the market-distorting target price system is replaced."[139] Chairman Roberts celebrated the "major departure from the past and a bold plan in regard to the future" with passage of the Federal Agriculture Improvement and Reform (FAIR) Act of 1996.[140] The new Republican Congress had passed, and a Democratic president had signed into law, the most substantial, sweeping changes to farm policy in its sixty-year existence.[141]

With a few exceptions, farm policy had traveled from the parity system of price-supporting loans and acreage controls. It had left behind the target price system of income supports with set-aside acres. After two contentious, partisan years, it landed on a fully decoupled system of fixed payments. Under the new bill, farmers simply received a yearly check from the federal government for having historic acreage under contract with USDA. If prices fell, the payments did not change, and the only backstop became the marketing loan program (which included loan deficiency payments [LDPs]). When he signed the 1996 farm bill into law, President Clinton mentioned reservations about the impact of the new bill if prices declined; by fall 1996 market prices were doing just that after peaking in May and July.[142]

The 1996 Reforms Crash in Asia

History delivered a tough verdict on the 1996 farm bill, casting doubt on its revisions.[143] The full impact of the 1996 bill remains unknown because prices collapsed soon after it became law. Although agriculture avoided a post-1973 expansion, it sat on a foreign market bubble.[144] Republicans in Congress had placed a big bet on the volatile world market, especially in Asia. The markets once again did not cooperate.

The Thai financial crisis in 1997 set off a series of events that quickly spread throughout Asia and harmed exports, dragging down American commodity prices.[145] Prices fell more drastically than they had after 1973, especially for cotton, and continued to fall through 1997, hitting bottom in 1998.[146] By 1998 the situation had deteriorated to the point that President Clinton vetoed an agricultural appropriations bill specifically because it did not provide enough emergency financial assistance to farmers. Congressional Republicans responded by increasing disaster assistance and rolling this aid into an omnibus appropriations bill that was too difficult to vote against and veto.[147] By the late 1990s, the 1996 farm bill "and the export-led growth strategy upon which it was based" was being written off as a "massive failure."[148]

History has demonstrated that Congress does not weather low crop prices well, and the late nineties proved no exception. The 1996 farm bill had been a difficult, highly partisan fight, and lower commodity prices added significant fuel to the partisan flames. When prices fell, Democratic opponents launched an all-out attack and demanded that the bill be reopened to provide price-based assistance to farmers.[149] They used the price collapse to re-litigate the 1996 act, largely made up of Republican policies. The weakening farm economy plus natural disasters added pressure, but Congress had few options available.[150] Democrats elected to go after an increase in the loan rates. They also complained that crop insurance was ineffective; concerns were also rising in the South. Without acreage control (or set-aside) authority, making price-based payments could encourage production into depressed markets. Re-creating such assistance would also require opening the seven-year farm bill just a few years into it.

The debates over how to respond to the depressed prices were very partisan, highlighting the different perspectives between Republicans and Democrats. Democrats and the president were quick to lay blame at the feet of Freedom to Farm and the Republicans in Congress.[151] Republicans, however, blamed the Asian financial and economic crisis for a loss of export markets and the resulting damage to commodities prices.[152] Senator Pat Roberts (R-KS), the former House Ag chairman who had led the effort on the bill, defended his far-reaching revision of farm policy. He argued against the view "that the wheels have fallen off and sent agriculture policy crashing into a wall. . . . They are regional problems . . . caused by weather and crop disease and the 'Asian economic flu' . . . but not the 1996 farm bill."[153] Democrats savagely attacked the 1996 farm bill while pushing for disaster assistance. A reelected President Clinton backed them in a showdown with congressional Republicans intent on preserving the reforms they had muscled through.

Democrats could not roll back the 1996 provisions, however, nor could they return to price-based assistance. The only option that emerged was to make additional, or supplemental, contract payments. Congress appropriated billions in additional payments— known as market loss assistance payments—in 1998, 1999, and 2000.[154] Low prices had threatened to drown out the 1996 reforms but in the end managed only to indirectly reestablish the connection between payments and market prices.[155] Importantly, congressional Republicans were able to hold the line on decoupling payments from production. Democrats temporarily settled the policy dispute by throwing additional money at farmers, doubling the contract payments, and building a political issue for campaigns.[156] The immediate threat to the 1996 farm bill had subsided, but questions about its long-term viability remained as farm policy crossed the threshold into the new century.

Concluding Thoughts on the Nineties Reforms

Congressional efforts in the 1990s produced the most jarring and significant revisions of farm policy to date by decoupling farm income support policy from planting decisions and market prices. Triple base in the 1990 farm bill was a mere prelude to changes wrought by the 1996 bill, which included fixed contract payments known as production flexibility contract (PFC) or agricultural market transition assistance (AMTA) payments. These payments were the first, and so far only, commodity assistance program that Congress designed without some form of price-based mechanism. That Congress drastically revised farm policy in 1996 is beyond doubt. Whether these revisions constituted reform is more debatable, as are questions about how substantial were any reforms.

Drawing a continuum for reform would place status quo on one end and something radical or unorthodox on the other end. The most obvious example of the latter being complete, immediate elimination of all programs and payments to farmers. Somewhere in between would fall more incremental or transitional revisions such as a buyout or phaseout of the support system.[157] An element of time may also be important to consider. For example, some of the once-radical farm policy concepts incorporated in the 1933 AAA had been debated throughout the twelve years leading up to it. Secretary Brannan's radical concept for issuing direct payments to farmers was soundly rejected in 1949 but reappeared in full twenty-four years later. The PFC concept first appeared in the 1985 debate in the Senate Ag Committee. It was quickly rejected but became law eleven years later and with only one law (1990) in between.[158] This constitutes fairly rapid adoption of a new policy.

The 1996 farm bill did not eliminate assistance to farmers; it did not eliminate price-based programs nor was it a buyout. Subsequent history made this all the more obvious, from the market loss assistance payments of the late nineties to the farm bills that have followed thus far. This conclusion was also apparent at the time. The bill continued the marketing assistance loan program, including LDPs, as the catastrophic price floor for commodities. This was largely a negotiating concession to southerners, and it also helped decrease Democratic opposition.

The 1996 bill also continued to include the permanent legislation provisions that ostensibly meant programs would revert to the 1949 act policy if Congress failed to act upon expiration in 2002. These provisions are designed to force Congress to write a new farm bill; inclusion was a negotiating concession to Senator Daschle and his Democratic allies to get the bill through the Senate. Most important, the 1996 contract payments were *not* intended to be a buyout. They were a clever method to capture CBO baseline and preserve estimated funding. Capturing baseline meant that the Ag Committees had funding available for writing the next farm bill, and the permanent law provisions would force them to do so. Falling short of elimination or buyout, however, should not be the full measure of reform.

Analyzing the 1996 revisions in terms of reform begins and ends with the two versions for decoupling farm support. The PFC payments decoupled income support payments from planting decisions and production through the use of contract acres. Specifically, payments were made on these contract acres, calculated using historic planting records for the farm. The payments were also decoupled from market prices by making a fixed payment amount to contract holders each year, albeit on a declining basis.

Separating federal support from the farmer's planting and production decisions was the more significant revision because it helped address a fundamental conundrum for farm policy. Supporting commodity farmers had always carried the potential that the assistance would encourage farmers to plant for the payments or loans. Federal assistance can easily create a cycle that is self-defeating when payments (or loans) in response to low market prices encourage farmers to plant more of the crop. This directly contradicts the market signals that less of the crop is needed. Planting for the government assistance in contradiction to the market's signal further depresses prices but increases payments, along with government outlays and political problems. History and experience had clearly taught that no program or policy could effectively control commodity supplies, especially if the controls were applied to the acres planted. Weather and farming practices were the more influential factors, and lawmakers were reluctant to make deep cuts in acres planted in their states or districts.

After the 1996 farm bill, farmers could plant any of the program crops on their farms without affecting their federal assistance. Payments would be determined solely by the crop for which they had contract acres. This was the point of decoupling, hence the phrase "freedom to farm." Going forward, market signals, not federal assistance, were to be the predominant factor in a farmer's planting decision. Farmers responded mostly by expanding soybean acres and largely at the expense of wheat.[159]

Decoupling from production represented a leap in an otherwise long, evolutionary process. It was rooted in the effort to remove the government from farm decisions, which had been the long-standing policy position of Republicans in Congress, especially those from the Corn Belt. In fact, the phrase "freedom to farm" dates to the American Farm Bureau's efforts in the heyday of midwestern power in the fifties and sixties. It was most notably deployed in the campaign to defeat the 1963 wheat referendum. Decoupling thus evolved out of the fights over acreage allotments and diverted acres through set-asides, CRP, and triple base. In 1996 the acres in the program were called contract acres, effectively the base acres from the 1990 farm bill; that bill calculated base using the average planted to the crop in the five years prior (1986–90). Thus, decoupled acres were a historical record of what had been planted on the farm, averaged over multiple years and at a partial remove from actual planting history. Most important, they were designed to avoid influencing planting decisions in the future.

Decoupling was also part of the market-oriented shifts that were most notable in 1973 and 1985. Target price policy was a necessary step because decoupling required an assistance program that was not a price-supporting loan. Loans are made on the actual commodity harvested. The harvested crop serves as collateral for the loan, and thus, the assistance is fully coupled to production. By comparison, target prices issued payments and could be separated from the actual crop produced or acres planted. Congress did not take that step (decoupling payments from planting) for twenty-three years.

Target prices connected to planting decisions also provided a production incentive, which had been one of the goals of the 1973 program and the Nixon administration. Production incentives tend to work far better than production controls—problems exposed by the eighties farm crisis and the huge federal payments made to famers. Experience with target prices demonstrated the impact a coupled price floor can have on farm production, market prices, and federal outlays. Separating the federal support from the farmer's planting decision was an important policy response to an age-old challenge, albeit within the chain of policy evolution up to that time. Acreage decoupling has been the more enduring revision and continues to be used in farm program payments. Contract acres will be redefined as base acres and come to constitute the whole of decoupling.

By comparison, decoupling payments from price-based mechanisms was the most radical revision in the 1996 farm bill because it was a complete break from all policy designs before and since. Fixed payments were a method of budget control, which is not necessarily the same as reducing expenditures. In fact, they were designed to capture CBO baseline. They were also a method for buying political support for budget discipline and decoupled acres, part of a package deal. Deficiency payments were triggered depending on market prices, incorporating market volatility and risk, which rendered the payments inherently unpredictable for budget purposes. High prices might result in no payments, but low prices would produce massive outlays as demonstrated by the 1981 and 1985 farm bills. Fixed payments removed this uncertainty because there was no potential for increased spending if prices fell, but it also meant farmers received payments even when prices were strong. The latter weighs against considering these payments as reform.

In addition, supporters of the bill were clear that the fixed payments were intended to capture the CBO baseline.[160] Export optimism fueled forecasts for continued high prices, which, if continued, meant that CBO would estimate little to no outlays when it came time to reauthorize the farm programs in 2002. Under CBO scoring rules, the authorities in the final year of the bill are assumed to continue through the next budget window, and expenditures are estimated based on that. Fixed payments, by comparison, provided over $40 billion in baseline for 2002.

Capturing baseline was more than simply using the CBO rules to the Ag Committee's advantage at a time of high prices. It was also political protection for farm spending in a Republican Congress determined to cut federal spending. The fixed payments were part of a seven-year contract with farmers and were therefore guaranteed. Future Congresses could not cut the expenditures without being legally liable for breaching the contracts.[161] In that way the fixed payments in the 1996 farm bill were designed to save farm policy from the combined effects of Republican efforts to cut spending and continued high crop prices, placing them well short of reform.

Finally, fixed contract payments were also a form of political buy off for acreage decoupling.[162] Southerners and Democrats opposed decoupling, but southerners had acute problems with both base acres and fixed payments.[163] Decoupling carried risks for southern crops if market forces pushed farmers to plant less cotton, rice, and peanuts and to replace them with corn, soybeans, or wheat. This would potentially decrease political support for cotton, rice, and peanuts but also create localized problems for gins, rice mills, and peanut shellers who would lose production for their operations. Strengthening crop prices and political pressure to cut spending helped make the payments more attractive and thus an effective method of buying support from southern farm program stalwarts.[164]

The guaranteed payments distributed federal farm spending in a very visible method. Each commodity interest received a certain share that could not change even if planted acres shifted. The commodity-by-commodity allocation was something members could take credit for back home and farm interests could use to demonstrate the success of their efforts to influence Congress. But this also meant that the bill locked in place payment acres with the potential for allocations determined more by political influence than farm realities. Dividing up expenditures in this way shifted farm coalition competition, away from acres and toward capturing a share of the CBO baseline. Future changes that altered the 1996 division of expenditures resulted in winners and losers—a result clear to everyone. The 1996 bill also altered the politics outside the farm coalition, which would struggle to defend making payments when prices were high and incomes strong.

Comparing the two versions of decoupling clarifies the reform question. Decoupling payments from production represents the most important revision. Its continued use in farm program payments adds to its status as substantive reform. By comparison, decoupling from prices was the most radical or unorthodox revision, but it is difficult to consider reform. Payments to buy political support and capture baseline provided little in the way of policy reform and the subsequent turn to additional assistance when prices fell bolsters that conclusion. Those payments, however, may have been necessary to achieve acreage decoupling reform.

Political consequences and ideological changes delivered the revisions to farm policy in 1996, but none of it would have been possible without the increase in crop prices. It may be trite to say that elections have consequences, but that doesn't discount its accuracy; the 1994 midterms provided a stark example. They were considered a partisan revolution because Republicans took over both the House and Senate after forty years in the House minority. The election also empowered conservative ideological goals for reducing the federal footprint and spending, with farm policy an ideal candidate.

The 1996 farm bill may well be representative of the larger ideological and philosophical battle between Democrats and Republicans. Farm supports were rooted in the New Deal, created by Democrats and long defended by them. Without the 1994 Republican victory, it is difficult to conclude that decoupling would have happened in 1996. A Democratic Congress would not likely have moved as much, if at all, in that direction even under political and budgetary pressures.[165] Democrats would have been less likely to make major changes with improving prices reducing farm program spending.

But the political question runs deeper. Historically, reforming or changing farm policy had been a battle within the farm coalition, especially between midwestern Republicans (corn) and southern Democrats (cotton). This is where the 1994 elections may have had the biggest impact. Republican victories meant that the South was no longer a Democratic stronghold.[166] Southern Democrats were no longer in charge of the Ag Committees. They lost seniority and the outsized power that a network of senior positions throughout Congress had long provided. The partisan change in Congress altered the regional power dynamics, which cleared the way for Republicans, led by the Midwest, to decouple farm supports.

Democrats opposed and attacked the bad optics of paying farmers when prices were high and regardless of what they planted. For one, old policy habits die hard, and Democrats had long fought for price supports and production controls. It is, however, impossible to disentangle Democratic opposition to the 1996 farm bill from partisan realities. After forty years in charge in the House, Democrats were suddenly in the minority, and total opposition was likely the most convenient—if not the only—mode of operation in response. But Democrats lacked the power to defeat the Republican policy revisions, especially in the House. In the same vein, traditional farm policy had lost its southern Democratic firewall and was more vulnerable to changes and reform. The South was now home to a party that possessed an ideological perspective that appeared at odds with federal farm support. While the traditional fault lines remained, they had undergone significant changes.

The new Republican Congress delivered decoupling, but it took high prices to give them the room to make that move. This was the second time that a run-up in crop prices resulted in substantive revisions of farm policy. Until prices spiked in the early seventies, the parity system went through minor but difficult changes, mostly around the edges. Midwestern Republicans succeeded mostly in helping corn avoid the policy's more draconian aspects and eventually fight its way out of the system. High prices pushed Ag Committee members to change policy and save farm supports from consumer blowback in 1973. Similarly, in 1995–96, escalating crop prices made fixed payments acceptable, especially to the South, which permitted acreage decoupling. It was a clever legislative move to save farm support from Republican efforts to cut federal spending, bridging budget and parochial farm program priorities.

In hindsight the 1996 farm bill is a bit of a paradox. It dampened the production incentive problems of federal support to farmers but set farm policy adrift from actual farming and risks. The bill appears to have gone too far by decoupling support from low prices. High prices have never lasted long and are consistently followed by relatively low prices. It was a big bet that a fixed, declining payment would suffice when the historically inevitable happened. This, too, was familiar from 1973, when prices fell soon after the farm bill passed but had to fall below the target prices to trigger payments. After 1996, prices would have to collapse below the loan rates to trigger payments beyond the fixed PFC payments. Fixed payments would have to suffice to compensate for low prices, or Congress would have to be willing to step in and help farmers more after paying them when times were good. This was clearly the argument made by Democrats who opposed Freedom to Farm, which added political risks to the risks of historically volatile commodity markets. As to be expected, Democrats sought political advantage when prices did collapse.

Both the 1973 and 1996 farm bills provide similar lessons. High prices are a necessary component for big policy changes. They also serve as cautionary tales about the substantial risk inherent in export markets and in basing policy changes on high prices, especially if optimism about foreign market demand is involved. The late seventies and the late nineties were very different times, however. When Congress started pouring more money into the farm sector in the late nineties because of low prices, it did so in a world that had moved under the purview of the World Trade Organization (WTO).

Spending more money on farmers was more problematic in this new environment, and these problems were magnified by cotton interests that also managed to re-create a special export subsidy for their crop. The rules had changed, and there would be consequences. Known as the cotton Step Two program, it had been eliminated in 1997 but was reinstated by Senator Cochran in the 1999 fight for disaster assistance. Coupled with the money allotted to subsidize cotton production, it created an easy target for WTO rules. Along the primordial fault line, victories for midwestern corn in 1996 presented cotton with an opportunity to rebound when prices collapsed. If the 1996 farm bill went too far by decoupling support from price risk, the congressional response risked going too far trying to correct it.

In multiple ways, some of them inadvertent, the 1996 farm bill set in motion events that would present new challenges for the farm coalition. Those challenges would lead to further changes in farm policy. The 1996 farm bill launched the modern, decoupled era, but it nearly ran aground on the same hazards that originated farm policy decades previously. A big question lingered in the aftermath of the nineties about whether 1996 signaled the beginning of the end for farm programs. The answer to that question continues to be sorted out, and subsequent farm bill debates will provide important clues. In the end the nineties added proof that policymaking is probably better suited to evolution than revolution and that, if nothing else, it is a long game.

7

Cotton, Ethanol, and Risk Management
Form the Modern Era, 2000–2010

Introduction

Farm policy's development in the new millennium begins with the rise of risk management through the federal crop insurance program. The disaster assistance pipeline Congress opened in 1998 sent more than $30 billion in additional funds to farmers by the time the 2002 farm bill effort began.[1] Crop prices remained depressed in the early years of the new century. Lower prices added pressure to the opening debates for the modern, decoupled era. Coming so closely on the heels of 1996's major revisions, the return of low prices compounded doubts about the changes and fueled opposition to them. The disaster assistance payments had effectively reestablished the link between farm support and market prices, but the payments remained decoupled from planting decisions. The $30 billion in extra spending cast harsh light on the 1996 revisions and raised big questions for the next farm bill. As discussed in the previous chapter, the AMTA contract payments were not intended as a buyout of farm programs. The disaster payments further drove home that point. The 2002 farm bill eliminated any doubt.

Before Congress turned its attention to rewriting the farm bill, it first worked to improve the risk management policies of the federal crop insurance program. The authors of the 1996 farm bill considered crop insurance to be a key component of a reformed farm support system; Congress had not followed through.[2] Congressional efforts on crop insurance were an important clue about the upcoming direction for farm policy.

Crop insurance is a different policy mechanism. The farmer pays some portion of the cost of the policy and receives an indemnity only when there is an actual loss on the farm. Crop insurance is directly linked to production as farmers insure only the crop they intend to produce and much depends on their actual production history. It is likely not a coincidence that the importance of insurance grew in the wake of decoupling farm payments from production. The crop insurance program also may have benefited from a growing sophistication in farming, one that included technological advances on and off the farm.

Crop insurance had been created in the Agricultural Adjustment Act of 1938 as an experimental program for wheat. It had long languished as an inconsequential element of the farm support system and of farm policy debates. Congress and USDA had repeatedly worked to increase program participation, but it continued to fall far short of expectations. The Federal Crop Insurance Improvement Act of 1980 was designed to place insurance as the primary form of disaster assistance and protection—a requirement to get disaster assistance. The act also provided assistance to the farmer to offset the cost of the premium (known as premium subsidy or discount) and transferred delivery of crop insurance to private industry.

Crop insurance failed to replace disaster assistance, as the late nineties demonstrated. In 1990 the Bush administration proposed eliminating it in favor of disaster assistance. In 1994 Congress revised crop insurance again to boost subsidies for farmers purchasing higher levels of insurance.[3] Congress was generally frustrated that participation remained stubbornly low, that the program was not actuarially sound, and that it was ineffective for the cost. It continued the program, however, and continued attempts to improve it.

The final years of the nineties produced a drastically improved federal fiscal situation. In fiscal year 1998, the CBO reported a budget surplus that it estimated would increase through 2000.[4] One result of this surplus was that it allowed Congress to provide additional funding to make crop insurance more affordable for farmers.[5] The main method for doing so was to increase the amount of a farmer's premium cost that the federal government would cover.[6]

Congress also sought to bring revenue-based insurance coverage onto equal footing with yield-based policies. Revenue-based policies insured a dollar-per-acre value for a crop that was calculated using a yield history and the actual market prices before planting. Many farmers could purchase coverage up to 85 percent of a per acre value. Revenue insurance therefore covered the two main risks for farmers within any one crop year, production risk (yield) and market risk (price). In that way it was considered a more robust insurance policy for the farmer and would prove to be very popular.

The Budget Committees had provided additional funding; the Ag Committees had to craft legislation spending it.[7] Recent years of low commodity prices added urgency, but the Senate Ag Committee faced some regional fighting that included continued attacks from Democrats on the 1996 farm bill.[8] The Senate Ag Committee effort also featured competing philosophies for crop insurance between Chairman Lugar and others on the committee led by Senators Pat Roberts (R-KS) and Bob Kerrey (D-NE). In short, Chairman Lugar wanted to pay farmers directly to choose among a variety of risk management strategies. The committee preferred the Roberts-Kerrey version that increased premium subsidy and codified revenue policies.

On the Senate floor, two regional perspectives squared off against the bill. Senators from the northeastern states were concerned that the bill did not do enough to help their farmers manage risks and disasters on fruit and vegetable crops. The bill also faced opposition from southerners who thought that crop insurance had "a built-in bias against Southern agriculture" and didn't work for their "capital-intensive crops."[9] Southerners lacked enthusiasm for crop insurance because farmers had to pay for policies and the premiums were rated based on the estimated risk of losses. In their eyes this made coverage too expensive for southern crops, especially as compared to midwestern crops.

Northeastern concerns were addressed in a deal to provide millions for specialty crops.[10] There was little senators could do to address southern issues with crop insurance, and ultimately, their opposition was incapable of stopping the bill.[11] It also helped that Congress added over $7 billion for disaster assistance in conference, quelling any further opposition.[12] The final law followed the Senate's blueprint, increasing premium subsidies to make crop insurance cheaper for farmers.[13] It also permitted premium assistance on revenue-based insurance policies.

These two changes proved to be the solution to crop insurance that Congress had long sought. Legislators reduced the out-of-pocket cost of crop insurance to the farmer and provided coverage for both price and yield risk. As a direct result, crop insurance participation took off drastically, and the program grew. It also began operating on an actuarially sound basis, which had been required by the bill.[14] The bill improved the disaster assistance program used by farmers of fruit and vegetable crops that are not covered by crop insurance policies. Risk management via crop insurance was the opening act of the modern era of farm policy; it became a defining characteristic of modern farm policy.

Big Budget Break and the Return of Target Prices in 2002

The 1996 farm bill was scheduled to expire in 2002, but the effort to replace it began early. Discussions for a new bill operated in the uncharted territory of a projected federal budget surplus. Out of that surplus, farm interests secured additional funding: first for crop insurance and then $73.5 billion (over ten years) above the CBO baseline for the farm bill.[15] Once again, the Ag Committees would need to move quickly to capture those additional funds. It was an unprecedented situation in the period since budget disciplines had been introduced in 1974. The Ag Committees had the green light to spend more, but it was fleeting. Failure to secure the funds in a new farm bill would send members back home to explain to their constituents why the money was lost.

The legislators faced the traditional regional differences, partisan concerns about spending the additional funds, demands for reforms, and tragically, the terrorist attacks of September 11, 2001. New spending added its own pressures on the farm coalition over how to divide the budgetary spoils. The most fundamental policy differences centered on price-based assistance. It was eliminated in 1996 but had been a significant focus of subsequent attacks. Some of this debate was based in reality; when prices fell, farmers did not see any assistance directly tied to the falling prices. But it was also partisan, with Democrats attacking Republican policy choices, and regionally centered in the South and the Great Plains. In 2001 and 2002, the House Ag Committee was led by Chairman Larry Combest (R-TX) and Ranking Member Charlie Stenholm (D-TX), two members from Texas with deep connections to the cotton industry.[16]

Chairman Combest was born and raised in West Texas. He was first elected to the U.S. House in 1984 and served until his resignation in 2003.[17] He had previously served as director of the Agriculture Stabilization and Conservation Service at USDA and was a legislative assistant to Senator John Tower (R-TX) for much of the seventies. Notably, Representative Combest had been one of the leading opponents of Freedom to Farm when it was proposed and pushed by then-chairman Pat Roberts (R-KS).

Representative Stenholm was born and raised on his family's cotton farm north of Abilene, Texas. He taught agriculture at a vocational high school while farming.[18] In 1965 he became the lobbyist for the Rolling Plains Cotton Growers Association and was elected to the U.S. House in 1978. A conservative Democrat, he had been a key member of the Boll Weevils, the group that cut deals with President Reagan on tax and budget cuts in the early eighties.

In the controversial 2000 presidential election, a decision by the Supreme Court paved the way for Texas governor George W. Bush to narrowly defeat Clinton's vice president Al Gore.[19] The election also produced an equally divided Senate with fifty Republicans and fifty Democrats, but in May 2001 Senator James Jeffords (R-VT) left the Republican Party to become an independent caucusing with the Democrats, shifting majority control to them.[20] With the change in majority, Senator Tom Harkin (D-IA) took over as chair of the Senate Ag Committee.

Senator Harkin was born and raised in a small town in Iowa and served in the Navy after graduating from Iowa State in 1962.[21] His career in Washington began in 1969, when he served on the staff of Representative Neal Smith (D-IA). Harkin graduated from law school in 1972 and was elected to the U.S. House in the Watergate election of 1974. He was elected to the U.S. Senate in 1984 and retired in 2014. His ranking member was Senator Richard Lugar (R-IN).

With more than $70 billion to spend, Chairman Combest set out to reverse the 1996 farm bill. In 2001 he faced a reluctant president, who questioned the chairman's policy demands, as well as the additional spending.[22] President Bush, however, needed Chairman Combest's support for trade legislation; he traded his opposition to the farm bill to get it. Chairman Combest traded his vote on trade promotion authority for the administration's agreement to drop opposition to his farm bill.[23] To further protect his commodity program priorities, the chairman and his allies also threatened food stamp interests.[24] The House Ag Committee reported its farm bill in late July, but not before Chairman Combest and Ranking Member Stenholm were forced to shift roughly $3 billion out of farm programs and into conservation and food stamps to get Democratic support.[25] Progress was briefly stalled by the August recess and then the tragic events of September 11.

The House took up the farm bill in early October. Chairman Combest claimed that a crisis was facing American agriculture and that the committee had produced a "balanced" bill to address it.[26] The claim that it was a "balanced" bill was the key to the chairman's argument and strategy. He used it to defend against those members who sought to change provisions or the distribution of the bill's additional funds. He and his allies argued that any changes would upset this crucial balance and jeopardize the entire bill.[27] It was a delicate balance given that the federal fiscal situation had changed significantly in the wake of a technology bubble crash and the September 11 attacks.[28] Protecting the bill's balance also militated against the threat that other member's priorities might be charged against farm payments. Combest used the argument to defeat amendments that sought to cut payments to fund renewable fuels provisions that would've primarily benefitted the Midwest.[29] It was most notably effective in defeating amendments that sought to reform farm payments, especially where changes would fall hardest on southern farmers. The effort led by Representative Ron Kind (D-WI) was the highest profile example.[30]

Representative Kind wanted to redirect some of the additional spending provided for commodity programs to conservation programs by tightening payment limits and "fundamentally reform[ing] agricultural policy."[31] Chairman Combest claimed that the amendment was "extremely inequitable" and accused its supporters of sowing regional discontent—picking a fight to shift money from farm districts (especially those in the South) to the less agricultural districts in the Northeast and Mid-Atlantic.[32] Labeling it a threat that federal funds would be redistributed away from farmers rallied the chairman's bipartisan farm district allies against an otherwise popular amendment.[33] The farm district position was strengthened by the fact that the committee had provided additional funding for conservation, allowing the chairman and his allies to argue the amendment went too far in cutting farm programs to pay for more.[34] More important, however, the amendment faced strong opposition from urban members worried about retaliation against food stamps. The threats had worked, and the House narrowly defeated the Kind reform amendment behind the farm–food stamps coalition.[35]

The defeat may have been a particularly painful lesson in coalitional politics because the chairman's perception of what was balanced and equitable about the bill can be seen as highly questionable. Not only was the additional spending tilted heavily in favor of farm programs, but within the farm coalition, cotton appeared to benefit the most. Of the additional baseline funds for the farm bill, the House farm bill spent nearly 70 percent on commodity programs; the bulk of those funds (74 percent) were spent on target price policy in the new countercyclical payments program.[36]

Because payments depended on the spread between market average prices and the target price, the new version of this program favored cotton. Cotton prices were the furthest below target price levels, which would trigger the largest payments under the program. The bill also continued the fixed, decoupled payments created by the 1996 farm bill and the marketing assistance loan program and LDPs.[37] The clear result of the additional baseline was to allow the committee to balance subsidies on top of subsidies. Direct payments every year, countercyclical payments if prices were below the target price, and marketing loan gains or LDPs if prices fell further.

Among the three major commodities, only cotton was in the

money for all three programs at the time the bill was written. This windfall did not go unnoticed. For example, Representative Sherwood Boehlert (R-NY) argued that the bill was tilted heavily in favor of cotton interests, which he likened to a banana republic that divided up the spoils of power.[38] The bill's balance also raised concerns about the impact on the federal budget and the risks of violating WTO commitments.[39] But Chairman Combest and Ranking Member Stenholm dismissed these concerns and moved ahead.[40]

The return to price-based payments was in part a reaction to the 1996 farm bill and the run of low prices preceding the farm bill debate. The committee managed, however, to preserve the decoupled nature of federal assistance that was the key reform from 1996. All payments would be made on base acres, including the countercyclical payments. The House avoided reattaching payments to planted acres. This preserved planting flexibility for the farmer, which had become popular in operation. Decoupling's survival was important given the return to price-based assistance and the additional spending. It lessened the potential that farmers would plant for the payments and helped avoid repeating some of the mistakes from the seventies and eighties.[41]

The bill did not spend all additional funding on farm programs. Conservation programs also received increased spending. The Environmental Quality Incentives Program, created in 1996, was continued, and its funding increased to $1.285 billion per year.[42] This program provided conservation assistance on working lands, as compared to a retirement program such as CRP. The Conservation Reserve Program was also continued and received an acreage increase to 39.2 million acres. In total, CBO estimated that conservation spending would increase by $12.5 billion, along with an additional $3.6 billion for food stamps. Combined, the conservation and food stamp increases bought peace and protection for the additional commodity support and allowed the House to easily pass the farm bill.[43]

The farm bill's toughest tests would be in the Senate. Key Senate Republicans opposed it and complicated Chairman Harkin's efforts. In fact, Chairman Harkin faced opposition from both ends of the farm coalition. Midwesterners and some Republicans opposed

additional spending and the return of target prices. Southerners opposed his goal to restructure farm policy with reforms and
an emphasis on conservation. Political realities forced Harkin to
back down on his more ambitious goals for farm policy reform.
He did manage a deal in committee that secured more than $20
billion for conservation.[44]

Chairman Harkin put much of the funding into creating the
Conservation Security Program (CSP).[45] This program was designed
as a "green payments" program that supplemented farm income
and encouraged natural resource conservation and stewardship.
Farmers who qualified for a contract would receive annual payments for implementing conservation practices on their farms.
This balanced commodity interests with the chairman's interests,
but the bill struggled on the floor.[46] Harkin was also proud that the
committee had added an energy title for the first time to a farm
bill. The title contained policies to encourage energy markets for
agricultural products such as biofuels and for farmers seeking to
produce wind, solar, or biomass energy.

The Senate farm bill mirrored the House's three-tiered support structure (direct payments, countercyclical payments, and
the marketing assistance loan / LDP), but the Senate had added
more to conservation and food stamps.[47] Senator Lugar, the ranking Republican on the Senate Ag Committee, immediately went
on the attack with an amendment that would "create a more
effective, market-oriented and broad-based safety net program
for U.S. farmers and ranchers" based on crop insurance, savings accounts, and limiting payments to farmers.[48] Senator Lugar
opposed "this strange preoccupation with the Agriculture bill"
in light of the terrorist attacks and a national recession.[49] The
debate, he noted, "has proceeded almost as if we were in a different world from the one in which there is war, recession, and
deficits. . . . Senators with a straight face have said . . . that there
was $73.5 billion above the current baseline . . . we are going to
claim it." The bill, he argued, "really has no particular philosophy" but consisted of "one subsidy piled on top of another."[50] He
blamed Majority Leader Tom Daschle (D-SD) and Democratic
senators facing reelection in farm states for the bill's bad poli-

cies and large payments to wealthy farmers when farm incomes were actually higher than in 1996.

When Lugar proposed cutting farm payments to replace funding previously cut from food stamp benefits, he opened a far more controversial line of attack that carried substantial political risk for farm bill supporters. The blueprint for defending the committee bill was also the claim that it represented a balance, a compromise among competing interests. Senator Lugar attempted to pit the two major interests against each other in a fight to lay claim to the additional spending in the bill.[51]

His amendment was an all-or-nothing situation in which taking from farm programs would lead to retaliation and likely more amendments aimed at farm supports. Senators argued that taking money from farm programs to pay for more food stamp spending would upset the bill and destroy both programs. Senator Lugar complicated matters for Democrats, however, given their strong support for food stamps. For Chairman Harkin, the tough reality was that appeasing farm interests left him vulnerable in his own caucus over food stamp and conservation spending. He argued that losing on farm programs would risk losing the smaller gains on food stamps and conservation that he was able to secure in the deal.

Senator Lugar argued bluntly that "the only well-thought-out aspect of the bill before us are thoughts as to how a Senator might be enticed by more money for particular crops for his or her State."[52] He focused his opposition on programs that "give incentives, strong incentives, to plant and produce every year . . . [and] drive the market price down further."[53] Returning to price-based assistance and target prices was seen as a step backward to policies that shielded farmers from market signals and ran up large outlays.[54] Lugar was imploring senators not to abandon the reforms and progress Congress had made on farm policy, warning that he would seek major reforms and hoped "it will be an educational experience Senators will enjoy."[55] Up against billions in extra funds and animosity for the 1996 farm bill, however, Senator Lugar's threats and his attempt to slow down the bill appeared to be make little difference.[56]

The tallest hurdle for the farm bill in the Senate was not extra spending, or its distribution, but rather partisan procedural fights over amendments that produced deadlock and blocked progress on the bill.[57] Not only did the procedural fights block progress on the bill, they allowed more time for opposition to the bill to build. The fights also created more opportunities for senators to craft alternatives that could have upset the committee's balance. Time did not appear on the bill's side, especially if a new year complicated the availability of the additional funding in the bill.

As the calendar approached the end of the year and the holidays, Senator Pat Roberts teamed up with Senator Thad Cochran (R-MS) on an amendment to reform the farm support system. The Roberts-Cochran proposal was relatively novel, proposing a "guaranteed direct payment to producers when they suffer a crop loss" through the "creation of a farm savings account, set up by the producer" and matched by the CCC.[58] The two were unlikely partners given their opposing views during the 1996 debate, but the alliance was likely formed by the voting realities in a Democratic Senate.

Senator Roberts opposed the committee bill "because I think it will be counterproductive, because I do not think it is going to work, that it will take us back to policy that does not fit today."[59] He thought the bill was a misguided attempt to lock in additional spending and buy political favors with commodity groups. He opposed reducing crop insurance, returning to target prices, and increasing loan rates. He also targeted Democratic senators with likely qualms about the bill. For example, he argued that the three-tiered support structure provided "a greater incentive to farm fragile land and use excessive chemicals and pesticides to improve yields."[60]

Democratic leadership remained intent on wiping out the 1996 farm bill and was not interested in new policy ideas. Moreover, Senator Roberts was burdened by the baggage of having created Freedom to Farm. In the end his proposal was unable to break the deadlock, and the level of partisanship increased. As the situation worsened, southerners attempted to push the House farm bill in what was seen as a political stunt designed by House Ag chairman Combest to cause problems and force Democrats to vote against a farm bill.[61] As it lingered, the farm bill was also subject to other policy fights that were notable but probably contributed only a little to the bill's struggles. Unable to move forward, the farm bill was pulled from the floor, leaving it in a state of limbo.[62] The Senate did not return to the farm bill until February 2002, but the new year had not altered the politics on the Senate floor.[63] Instead, a new threat emerged.

Senators Byron Dorgan (D-ND) and Chuck Grassley (R-IA) led a bipartisan effort to lower payment limits and apply them to the marketing loan provisions.[64] The amendment split the coalition along its regional fault line. It was largely the work of senators from the Midwest and the northern Great Plains; opposition was almost exclusively from the South. Underlying this fight was the long-held view that southern farmers benefited more from the farm programs and had long led the way in getting around any limits Congress put in place.[65] This view was magnified by a sense that the 2002 farm bill provided more favorable support to southern crops than midwestern crops. Many in the Midwest and northern plains raised concerns about the harm large payments to large farmers had on small, family-sized farmers and to farm policy generally.[66] They were also concerned that the concentration of payments to large, well-off farmers would cause a public backlash against all of farm policy. This issue had been elevated by the work of the Environmental Working Group (EWG), an environmental interest group that had published a comprehensive listing of farm subsidy recipients. Senators noted a "flood of newspaper articles" with harsh criticism of farm payments.[67] They were confronted with the bad optics of large amounts of federal dollars next to the names of individual farmers, and it had a powerful political effect on the debate.[68]

Southern senators viewed payment reform as a direct attack on them and their farmers, but the issue was quickly getting away from them. From the southern perspective, the amendment was unfair to southern farmers because they grew more "capital-intensive" crops.[69] Southern farmers, Senator Blanche Lincoln (D-AR) argued, had been forced to expand their farms to survive, and tighter limits would hit them first and hardest. Justifying more generous payments based on expensive crops and larger farms did not stack up well politically. This was especially true when the opposing view was that limits were needed to protect smaller farmers from being pushed out of farming by subsidized large farms. Southern senators fought hard but were defending the indefensible, and the Senate overwhelmingly rejected their attempt to kill the amendment.[70] They were reportedly stunned by this result but shouldn't have been.[71] The EWG database was politically devastating. From Billie Sol Estes to the "Mississippi Christmas Tree," a long history of scandal tied to program payments was rapidly catching up to them in the digital age.[72]

Tightening payment limits and adding more funding for food stamps was enough to help the farm bill finally clear procedural hurdles and pass the Senate.[73] The bill went limping into conference with the House, but the Senate position suffered a near-fatal wound when CBO concluded that the bill would cost billions more than previously expected.[74] As a result, House Ag chairman Combest held the upper hand, and he used it. Senate conferees were forced to cut spending on nutrition and conservation priorities, as well as reduce their higher loan rates. House priorities won out, and the final bill hued closely to the House-passed farm bill. Chairman Combest applauded "a strengthened safety net for farmers" that provided tiered assistance from direct payments, countercyclical payments (target prices), and marketing assistance loans (including LDPs).[75] Conferees did, however, stay the course on the decoupled base acre system reform that had been put in place by the 1996 farm bill.

The main story line of the 2002 farm bill was both the additional spending and the return of target price policy. The regional imbalance did not slow the farm bill once it came out of conference.[76] The final distribution was striking. The Farm Security and Rural Investment Act of 2002 was estimated to spend $73.5 billion over what CBO estimated that the 1996 farm bill would have spent had it been extended.[77] More than 70 percent of the additional spending went to the commodities programs, and more than 60 percent of that went to the countercyclical program alone; conservation received less than 20 percent and food stamps less than 10 percent.

The story within those numbers was cotton's haul. The debate had provided clear recognition that senators and representatives outside the South concluded cotton farmers were getting a better deal than the rest of the farmers. The numbers bear this out. Each of the major crops received a target price similar to what was provided in 1990. Also, all target prices were above recent market year average prices, but the cotton target price was far above market prices. As a result, payments per base acre quadrupled for the cotton base.[78] Cotton benefitted more from the return of target prices because it had lost more in the market. For example, average cotton prices in 2001 were just 42 percent of what they had been in 1995. By comparison, average prices for corn in 2001 were 61 percent of the 1995 prices. This spread between market prices and congressionally fixed target prices, however, had also previously been a large part of the argument against the policy.

The political story line of the 2002 farm bill was that the traditional regional farm program dispute was contained almost completely among Republicans in the post-1996 Congress. Midwest Republicans like Representative John Boehner (R-OH) concluded that a "closely divided Congress fighting for control in the House and the Senate, got into a bidding war as to who could be the biggest friend of agriculture."[79] Senator Roberts was unhappy with a final product that would "take four checks from the Government for producers to receive what they could have received from one check under a supplemental. . . . This is not market driven; it is mailbox driven."[80] Senator Lugar was concerned about the substantial cost estimates for the bill, especially the tens of billions for commodity programs. Senator Grassley added that "unless you are a big meatpacker or a cotton and/or rice producer in the South, the new farm bill is not what it is cracked up to be."[81]

Southern Republicans were abetted by Senate Democrats such as Chairman Harkin and Majority Leader Daschle. The distribution of additional baseline strained the Democrats over farm payments, food stamps, and conservation assistance. That strain can be seen in the final Senate vote.[82] It was not substantial enough to block passage or alter the outcome. The southern Republican–farm state Democrat coalition had enough political muscle to push the bill through Congress. Their efforts were also aided by just enough additional funding for conservation and food stamps to keep the peace.

In actuality, there was little balance to the bill, tilted as it was toward farm payments and cotton. Political liabilities reside within such imbalances, and the 2002 farm bill was vulnerable to attack for its spending and multilayered subsidy system. It didn't help matters when Chairman Combest resigned soon after the farm bill passed and opened his own lobbying firm representing southern commodity interests.[83] The most enduring black eye for the 2002 farm bill, however, was the fact that cotton's big victory set the stage for a far more damaging loss. The federal windfall for cotton farmers created an easily foreseeable problem that threatened the entire farm support system. In the dust that settled from the 2002 farm bill is written a cautionary tale about the pitfalls of policy overreach.

A Review of the WTO and the Renewable Fuels Standard

Corn's and cotton's fortunes diverged soon after 2002, reopening long-standing tensions in farm bill politics. Between the 2002 and 2008 farm bills, two major developments recalibrated the American farm policy debate. The first was a decision by the WTO against U.S. supports for upland cotton. The second was congressional action that created the renewable fuels standard (RFS). These two issues combined to create the stark divergence between farm policy's traditionally most-combative commodities. The Brazil dispute and the RFS helped shape the 2008 and 2014 farm bills.

Begun in 1988, the Uruguay Round of the GATT negotiations created the WTO in 1994.[84] The negotiations included an agreement on agriculture (the Uruguay Round Agreement on Agriculture [URAA]) that restricted the use of export and domestic subsidies for commodities and provided greater market access among member nations.[85] They also produced both a code of conduct that member nations were to follow with agricultural commodity policy and the specific causes of action that member nations could pursue before the WTO against any other member who did not adhere to the agreement.[86] The WTO process works on a self-enforcement basis under which member nations bring disputes before WTO panels in order to enforce the agreement on agriculture's provisions against other member nations.[87] With WTO approval in Congress, the United States agreed to abide by the agreement on agricultural commodity subsidies.

The WTO also had a profound impact on the American and world cotton and textile industries.[88] U.S. textile production and milling underwent a substantial decline during the 1990s, moving production offshore to capture gains from cheaper labor costs. This decline, in turn, reduced usage of U.S. cotton and increased exports significantly. The U.S. is the second largest producer of cotton in the world but the world's largest exporter.[89]

The 2002 farm bill became law on May 13, 2002; Brazil initiated a formal dispute against U.S. cotton subsidies before the WTO on September 27, 2002.[90] Brazil was increasing its role as a competitor in the world market for cotton. The dispute represented a major shift in the political dynamics for farm policy. The U.S. provided one of the highest levels of farm support in the world, and cotton support equaled nearly 80 percent of the total crop value.[91] For upland cotton the marketing year average price was below the target price every year and below the loan rate in 2002, 2004, 2005, and 2006. In those years American cotton farmers were getting direct payments, the maximum countercyclical payments, and either LDPs or loan gains.

The U.S. also provided prohibited export subsidies for cotton known as the Step Two program. The Step Two program had been created by the 1990 farm bill to help make cotton exports competitive—especially in European markets. The program provided cash payments or marketing certificates to domestic users or exporters of upland cotton. Congress terminated the program in 1997 but reinstated it in the 1999 disaster bill. The program was continued by the 2002 farm bill.[92]

Nineteen months after Brazil initiated the dispute, the WTO panel issued its report, on September 8, 2004.[93] Brazil won overwhelmingly. The panel found that U.S. cotton supports ran afoul of WTO commitments because they were trade distorting and shielded American farmers from market signals to the detriment of Brazilian cotton farmers.[94] The WTO panel also concluded that

Step Two and export credit programs violated U.S. commitments, while target prices and loan rates were too far above market prices. The subsidies provided an incentive to produce more cotton than the market demanded, which lowered prices for cotton farmers around the world and continued high levels of American production that would otherwise be unprofitable.

In a potentially far-reaching part of the decision, the WTO panel concluded that the decoupled payment programs caused serious prejudice to Brazil because they were triggered by low market prices. The subsidies were found to have suppressed prices regardless of decoupling.[95] This conclusion meant that decoupling payments was not enough to protect them from WTO commitments. In the WTO's view, price-based assistance could protect farmers from market signals and permit them to continue feeding an oversupplied, depressed-price market. Moreover, the U.S. was simply too large of a presence in the world cotton market, and its cotton subsidies had an influence on world market prices. The U.S. position was also hurt by the sheer size of the payments to cotton farmers, including direct payments.

The U.S. was told to end Step Two export subsidies and revise domestic support programs. This would remedy the problem underlying the dispute by removing the adverse effects of the supports.[96] Congress and USDA attempted to bring the export credit programs into compliance, including by repealing the Step Two subsidy.[97] Changes to domestic cotton support programs, however, would have to wait until the 2002 farm bill was reauthorized by Congress. That reauthorization was scheduled for 2007. It would take time for Brazil's victory to change U.S. cotton policy, but it would have an impact. For one, the dispute hung over the next two farm bill debates.

If the Brazil cotton dispute presented a threat to American farm policy, renewable energy policy seemed poised to permanently alter the farm coalition's internal dynamics. Put simply, wars in the Middle East and spiking oil prices prompted Americans and their elected representatives to become increasingly concerned about the economic and geopolitical implications of the nation's heavy reliance on imported oil for transportation fuel. Attention turned to renewable fuel alternatives, such as ethanol made from corn, and drove the development of renewable fuel policy. The push for renewable fuels was not new, but the extraordinary legislation produced in 2005 and 2007 certainly was.

A major catalyst for plant-based transportation fuel alternatives came during the energy crises in the 1970s, when the Organization of the Petroleum Exporting Countries (OPEC) cut supplies and gas prices increased. The crisis was exacerbated by the Iranian Revolution in 1978–79. Congress first responded with a tax subsidy for using ethanol in gasoline and then passed a federal ethanol tax incentive in 1980. Subsequently, the Clean Air Act amendments of 1990 added to ethanol's position by mandating minimum oxygen percentages for gasoline. In 1992 Congress added ethanol to the definition of alternative fuels and began mandating that federal vehicle fleets include the purchase of vehicles that could burn alternative fuels.[98] Spiking oil prices and reliance on supplies from the Middle East, in the age of the war on terror and active wars in Iraq and Afghanistan, pushed renewable fuel policy to the next level.

In the 2005 energy bill, Congress created the RFS, which mandated that specified percentages of renewable fuel had to be blended into the domestic gasoline supply.[99] Specifically, this new law required blending 4.0 billion gallons of renewable fuel into the domestic fuel market in 2006, increasing each year to reach 7.5 billion gallons by 2012.[100] Almost all renewable fuel at the time was ethanol produced from corn. Two years later, a Congress returned to Democratic control surpassed the 2005 RFS with the Energy Independence and Security Act (EISA) of 2007. Among its many provisions was an expansion of the RFS, increasing the mandate from 9 billion gallons in 2008 to 36 billion gallons by 2022.[101] It capped corn-based ethanol at 15 billion gallons per year and added cellulosic ethanol, advanced biofuels and bio-based diesel in a nested mandate.

The RFS has had a profound impact on farmers, rural communities, and farm policy. While the mandate represented a small portion of the domestic gasoline market, it quickly became a major player in the farm commodities markets. Because most renewable fuel was ethanol produced from corn, the mandate created substantial demand for corn in the domestic market and altered some fundamental aspects of that market. The RFS drove ethanol production and investment in production facilities, especially in rural communities throughout the Midwest. The massive domestic demand driver in the corn market also increased crop prices and American corn production. The RFS has therefore had a substantial impact on farm programs, crop insurance, and conservation policy, with consequences that go directly to the regional and commodity fault lines. The RFS would also stir up a debate about the use of corn for fuel as compared to using it for feed or food. Its impacts would first be felt during the debate for what became the 2008 farm bill.

Farm Policy under the Influence: The Food, Conservation, and Energy Act of 2008

Reauthorization was scheduled for 2007. The Brazil cotton dispute and the RFS had altered the political landscape during the brief intervening five years. The WTO dispute represented the latest intrusion by the larger world on domestic farm policy through trade negotiations and markets.[102] The RFS and ethanol production, however, were having a far more significant impact on American agriculture and farm policy. In a Congress that was riding a political bandwagon over renewable fuels, the 2007 energy bill that would increase the RFS was being debated at the same time as the farm bill.[103] By 2007 ethanol production was demanding over three billion bushels of corn each year, pushing corn and wheat prices above the 2002 farm bill's target prices. This put corn and wheat in a different position with respect to the countercyclical program as compared to cotton. The larger crops were unlikely to receive payments and budget estimates would begin to reflect this impact from higher prices, leaving fewer funds in the baseline for the new farm bill.[104]

The partisan situation in Congress had also changed. Democrats recaptured Congress in the 2006 midterm elections during the waning years of the George W. Bush administration.[105] Senator Tom Harkin resumed his chairmanship of the Senate Ag Committee. In the House, Representative Collin Peterson (D-MN) took over the House Ag Committee.

Representative Peterson was first elected in 1990 to represent the western district of Minnesota. He had grown up on a farm in the district and become a certified public accountant before he won his seat in Congress.[106] A conservative Democrat, Representative Peterson had been a notable defector on the 1996 farm bill when he voted to report AMTA out of the House Ag Committee. Chairmen Harkin and Peterson represented a change in leadership that was more than just partisan but also regional and ideological. They had taken over for Senator Saxby Chambliss (R-GA) and Representative Bob Goodlatte (R-VA), respectfully. This regional shift was most pronounced in the Senate because Chambliss had long been a staunch supporter of southern commodity interests (cotton, peanuts, and rice) while Harkin was from the heart of corn country and a big proponent of conservation and renewable energy, including ethanol.

A new Congress, WTO and world trade issues, and improving prices under the RFS produced high expectations for significant reforms. Expectations for reform were also helped by a well-timed series of articles in the *Washington Post* that highlighted program loopholes, abuses, and similar political challenges.[107] These expectations would falter on political realities, in large part because the new majority owed much to Democrats who had defeated many Republican reformers in rural districts and needed to deliver on farm policy.[108] In addition, higher prices cut back farm program spending estimates, which complicated efforts to reform those payments. The focus on budget and spending put direct payments in an uncomfortable position, especially as ethanol demand pushed up commodity prices.[109] The warnings Democrats had made in opposition to the payments in 1996 started to look more prescient by 2007.

An extended and difficult process ended up producing little change, however. Farm interests in Congress were able to overcome reformers and a presidential veto to produce a largely status quo farm bill. It took two years to complete the work, but Congress continued direct and countercyclical payments and the marketing loan program. In addition, Congress managed to create new farm programs, including for disaster assistance. Adding programs was notable because the new Democratic majority had reinstated pay-as-you-go (PAYGO) policy under budget rules. New spending had to be offset by cuts to existing outlays or tax increases. After the bonus funds for the 2002 farm bill, the 2007–8 effort returned to the more difficult legislative terrain created by budget disciplines.

In the House, Chairman Peterson opened work on the farm bill feeling the pressures from the federal budget.[110] The path would not be easy, and Chairman Peterson elected to adhere largely to status quo.[111] He navigated competing interests in committee by adding funding for conservation programs and for farmers of specialty crops (fruits and vegetables).[112] To strengthen the coalition, he also added funding for food stamps and renewable energy.[113] The biggest challenge for Chairman Peterson came about because he had to cover the additional spending on food stamps, conservation, and energy. He worked out a deal to eliminate tax breaks for the oil industry to offset the bill's additional outlays. He claimed there was "something in this bill for everybody to like. . . . It's a step in the right direction and has broad support."[114]

Republicans didn't see it that way, however. They raised objections based on the tax offset.[115] Republican opposition to the tax changes was magnified by the fact that the funds were used, in part, to help pay for food stamps. The level of opposition by Republicans resulted in a decision by the Democratic leadership to use a limited rule to protect the farm bill on the floor from damaging amendments. Republicans argued that the rule "poisoned the well" for the farm bill and jeopardized bipartisan support.[116]

The farm bill was also under increasing pressure to reform farm policy, which created the potential for a Republican-Democratic reform coalition that might derail the effort. To head off reforms that could damage the farm coalition with southerners, Chairman Peterson had secured the backing of Speaker Nancy Pelosi (D-CA) and Majority Leader Steny Hoyer (D-MD). Leadership backing was crucial in defeating difficult reform amendments.[117] Reforming various aspects of the farm bill was a bipartisan effort, but so was defending the committee compromise. For example, Representative John Boehner tried to fix an obscure issue with the LDPs that had been discovered when Hurricane Katrina wreaked havoc on the Port of New Orleans in 2005 and corn farmers received a windfall.[118] In addition, Representative Jim Cooper (D-TN) offered an amendment to reduce spending on the increasingly popular crop insurance program by reducing premium subsidy and spending the savings on conservation programs.[119]

The real tests for the farm bill were Republican opposition to the tax provisions and a bipartisan effort for major reform. The latter was more concerning to farm interests because Representative Ron Kind, who had come close to winning his reforms in 2001, was taking another shot.[120] Farm program supporters argued that the bill contained significant reforms and framed the Kind amendment as a threat to the "delicate balance achieved in the committee bill" and a "threat to producers, consumers and rural America."[121] Chairman Peterson had included some payment reforms in the committee bill largely as a defensive mechanism, and, coupled with Democratic leadership's pragmatic needs to preserve its majority, they were enough to defeat the Kind amendment by a large margin.[122] When Republicans mounted one final challenge to the tax provisions, they were defeated and the House passed the farm bill.[123] That passage, however, had to be secured by further deal making from the Speaker to add Democratic votes after Republicans walked away from supporting the bill.[124]

In the Senate, Chairman Harkin was much slower to move, and his effort stumbled over multiple challenges; his challenges would begin in committee.[125] At the center of the committee's struggle was the budget. It placed limits on what could be spent that did not align with what Senators wanted in the bill.[126] Senator Kent Conrad (D-ND), a high-ranking member of the Senate Ag Committee who also chaired the Senate Budget Committee, had been instrumental in providing authority that would allow the Ag Committee to spend above the farm bill baseline, but the additional spending would still have to be offset. This proved to be less helpful than it was likely intended. Budget limits and regional differences over farm programs would play out in a lengthy political drama.

Chairman Harkin wanted the additional spending used to protect and improve the CSP he had created in the 2002 farm bill.[127] Senator Conrad wanted it used to include standing or permanent disaster assistance policy for livestock farmers, revenue-based assistance for crop farmers that would be a supplement to the existing direct and countercyclical payments, and marketing assistance loan programs. The corn growers complicated matters further when they requested a different revenue-based program that was more favorable to new market realities.

Revenue-based farm programs were an outgrowth of the continued climb of crop prices under the RFS and the popularity of revenue-based crop insurance. Chairman Harkin's attempt to add corn's revenue-based proposal ran into resistance from Senator Conrad, complicating the relationship between these two powerful members of the committee.[128] Southern senators, led by ranking member Senator Chambliss, had general objections to revenue policy. At the same time, crop insurance interests viewed the revenue program as a direct threat to their business.[129] The complications for farm policy increased quickly, and mutual opposition to the chairman helped cement an alliance between Senators Conrad and Chambliss.[130]

Compounding the chairman's troubles, Senator Max Baucus (D-MT)—chair of the powerful Senate Finance Committee and a high-ranking member of the Ag Committee—stepped forward in support of Senator Conrad. What followed was an unexpected battle over revenue programs that pitted the Conrad-Baucus program against Chairman Harkin's program.[131] Senator Baucus added to the political drama because he was willing to have the Finance Committee offset the cost of Senator Conrad's program. Finance Committee assistance would free up funds for the farm bill, but it put Chairman Harkin at the mercy of Senators Conrad and Baucus.

Both senators had also aligned with Ranking Member Chambliss to re-create a northern plains–southeastern alliance reminiscent of Senators Richard Russell (D-GA) and Milton Young (R-ND) in the fifties and sixties.[132] Chairman Harkin needed to offset additional spending for conservation and his revenue program. When he proposed to phase down direct payments to do so, he greatly aggravated southerners, which further strengthened the alliance between Senators Conrad and Chambliss.[133] Infighting among these various power centers on the committee slowed things down significantly, and the farm bill effort dragged into October. Chairman Harkin was defending against two fronts that had allied against him, and as time passed, he lost ground.[134]

With Chairman Harkin moving in their direction, Senators Baucus and Conrad pushed a $5 billion trust fund for farm disaster assistance through the Finance Committee so that it could be added to the farm bill on the Senate floor.[135] Chairman Harkin remained short on funds to cover all the demands for the bill, and with pressure growing, he had to cut a deal with Senator Chambliss and southern senators. Southerners were on the defensive over the WTO dispute with Brazil and sought to protect cotton.[136] Chairman Harkin traded farm program reforms opposed by southerners in return for their agreement to additional funding for conservation and nutrition, as well as an optional revenue program.[137]

That deal ended the impasse. The committee proceeded with marking up the farm bill but stumbled again over the revenue programs and yet another intra-agricultural dispute. Senator Roberts threw up a roadblock on behalf of the crop insurance industry. The crop insurance industry was concerned that revenue-based farm program payments would be too similar to revenue-based crop insurance policies and that as a result corn farmers would purchase lower coverage levels or less insurance. The Corn Belt was the most lucrative part of the country for crop insurance. Resolving the insurance question required intense negotiations in the hearing room and the committee's back room.[138] With an assist from Senator Conrad, Senator Roberts won concessions on the revenue program for the crop insurance industry, and the committee finally reported its bill. CBO's estimate that the revenue program would save funds helped.[139] It had been a five-month undertaking to produce general status quo for commodity support and conservation programs. The committee added revenue-based program options and increased spending on food stamps.

Left unsettled by the committee was the issue of farm program payment reform, a fight that loomed on the Senate floor.[140] Everyone expected the floor debate to be difficult as Chairman Harkin explained that budget restrictions had complicated negotiations. The committee had bogged down in an intra-agricultural dispute that pitted the Midwest against the South and Great Plains. Despite this, Chairman Harkin claimed that the committee had produced a "grand compromise" on a "forward-looking bill to make historic investments in energy, conservation, nutrition, rural development and promoting better diets and health for all Americans."[141]

The chairman saw far-reaching changes embedded in the conservation title. He had worked to make certain the bill represented a further policy shift in the direction of working lands conservation assistance embodied by the Conservation Stewardship Program (CSP).[142] He argued that the pressures on natural resources from expanded corn production under the RFS could not be addressed with traditional acreage retirement programs such as the CRP. He also looked to CSP as a model for the future of farm policy—supporting farmers while adhering to WTO commitments to avoid future disputes like that with Brazil over cotton supports. Conservation had long been a more politically acceptable method for helping farmers, dating to the Dust Bowl and the 1936 act.

The chairman's vision for farm policy contrasted with that of his ranking member, Senator Chambliss. Senator Chambliss warned senators that the farm bill was a "delicate compromise" and that "further efforts to take funds from the farm safety net could stall this bill."[143] His comments hinted at the regional dispute with the Midwest underlying the bill, in part owing to the push for renewable energy investments. To southerners like Senator Chambliss, renewable energy was too much of a Corn Belt issue, and they were frustrated with the RFS. For example, Chambliss made sure to emphasize that "100 percent of the ethanol manufactured in this country today comes from corn" mostly from the Midwest.[144] The RFS had put pressure on the traditional regional-commodity competition even as crop prices and farm incomes were rising. Southerners were unhappy with the RFS, blaming it for expanded corn acres, some of which were coming out of cotton, rice, and peanuts. They were also expressing concerns from their farmers about the increased cost of feed for the livestock industry, particularly broiler chickens in states like Arkansas and Georgia and hogs in North Carolina. Southern cotton farmers had to compete with corn for acres, and southern livestock producers had to compete with ethanol producers for corn, which was becoming more expensive. Added to southern frustrations, Congress was in the process of expanding the RFS further. The RFS affected farm policy in a familiar manner, applying pressure to the traditional fault line running between corn and cotton.[145]

The regional tensions had been smoothed over in committee and with extra funding from the Finance Committee. The 2007 floor debate featured a rerun of the payment reform dispute and the partisan battle over floor procedures and amendments—lingering problems that had complicated the 2002 debate. The partisan procedural fight, coming as it did on the eve of a presidential election year, initially brought the debate to a standstill.[146] Negotiations dragged past the Thanksgiving holiday. Agreement was finally reached in December. That agreement broke the partisan gridlock that had stalled the bill, but it also set up the final showdown over reform.[147]

Senators Byron Dorgan and Chuck Grassley pushed for further reforms to farm program payments.[148] The RFS and an optional revenue program were mere irritants to southerners compared with farm program reforms. Southerners viewed the reform amendments as unfairly targeting their farmers, who grew more expensive crops and were not benefitting from the RFS.[149] But it remained a tough issue. The *Washington Post* had effectively highlighted continuing payment abuses. Added to this, farmers continued to receive $5 billion each year in spite of high prices and record income.[150] The farm bill had to produce some level of reform, if for no other reason than as a defense on the floor. While this was well understood, the trick was finding reforms that could be accepted politically by the southerners.

Deal making in committee had attempted to end some egregious practices exposed in the press.[151] The committee lowered the income threshold for payment eligibility and required that all payments be attributed to an actual person—the latter an attempt to end the practice of using multiple entities to bypass payment caps.[152] These changes helped clean up farm policy but were particularly helpful as a defensive strategy on the Senate floor. They provided evidence that the committee bill contained reforms that could be used to block more drastic payment limits and eligibility standards. For example, Senator Blanche Lincoln (D-AR) argued that the committee bill contained "the most significant reform in the history of our farm program" and further reforms would "probably have some dire unintended consequences," especially for southern farmers.[153]

Southerners demonstrated opposition by threatening to filibuster the bill over reform amendments.[154] And that proved to be their best defense against further reforms. They were able to take advantage of the larger impasse over amendments by negotiating sixty-vote thresholds for reform amendments. With procedural protections, they were able to protect their farmers from tighter payment limits and income requirements.[155] The ramparts having held, senators had little left to fight about. The calendar was also moving quickly into the winter holidays.[156] The Senate agreed to additional funds for conservation and food stamps and then finally passed the farm bill on December 14, 2007.[157]

In the New Year, conference negotiations dragged through April. Conferees fought over spending and policy. They were also under pressure from President Bush and his threat to veto the bill.[158] Negotiators struggled over how to pay for the increased spending on food stamps because the only real offsets available required changes to tax policy, which involved the House Ways and Means Committee and the Senate Finance Committee. Using tax law changes to pay for food stamps inflamed Republican opposition. It was particularly problematic because any compromise would need to bring enough Republican support to override what was expected to be an almost certain veto.[159] That deal did not emerge until late April, although conferees were not able to satisfy the president's demands.[160]

The final bill was largely a continuation of the 2002 farm bill policies, including for commodity support. Conference negotiations managed to expand the farm support system by adding a suite of standing disaster assistance programs. They also added a crop revenue program as an option for farmers who elected out of the countercyclical program (target prices). Conferees found additional funding for conservation, energy, and food stamps; the latter program was renamed the Supplemental Nutrition Assistance Program (SNAP) by the final bill.[161]

It had been a long road, but the Food, Conservation, and Energy Act of 2008 passed both the House and Senate with veto-proof majorities.[162] That road had a few bumps left in its final twists and turns. President Bush vetoed the farm bill. As Congress prepared to vote to override the veto, members discovered a technical glitch. In short, the enrolling clerk of the House had inadvertently left out the Trade title of the farm bill, raising various procedural questions and problems.[163] To fix it, Congress had to undertake an unusual process to simultaneously override the veto and repass the complete bill. President Bush vetoed the bill a second time, which required Congress to once again override his veto.[164] Once technical problems and presidential vetoes were dealt with, the 2008 farm bill became law in June.[165] The journey had started in May 2007; the bill had to be passed twice. In the end farm policy had survived the first impacts of the Brazil WTO challenge, danced around a difficult reform effort, and (twice) rebuffed a presidential veto in an election year.

Concluding Thoughts on the Modern Era's Opening Stages

The 1996 farm bill launched the modern era of farm policy by decoupling federal commodity assistance from both production decisions and market prices. It constituted a near-complete break from the target price era and served to reset the status quo. The baseline captured in the decoupled annual payments locked in a distribution of that baseline among the covered commodities. Events and policy developments subsequently carved divergent paths for corn and cotton, the two traditional combatants in the farm coalition. And that, the different trajectories for these two commodity interests, writes the story that was unfolding for the modern era. Importantly, this divergence created the potential for recycling the internecine battles of the past but in a very different political landscape.

With low prices Congress parlayed an anomalous budget situation at the turn of the century to improve crop insurance in 2000 and then restore price-based countercycle payments in addition to direct payments and marketing assistance loans. At first glance this could appear to be a balanced outcome. Crop insurance was favored by corn and target prices favored by cotton, with wheat benefitting from both but to a lesser degree. They are very different policies, however, and that complicates any sense of balance. Crop insurance is partially purchased by farmers and is actuarially sound with prices that are reestablished each year. It is viewed as a policy that helps manage the production and price risks within the growing season for the crop. Target prices are designed to protect farmers against the market, offsetting low prices in a year or over multiple years with federal payments. Target prices are without cost to the farm and are decoupled, meaning farmers do not have to plant the crop to receive the payment for the crop if payments are triggered.

Expansion of price-based payments came with consequences because the rules for support systems had changed under the WTO. Grabbing for the additional benefits of an anomalous budget was not all that unusual for any commodity with power balanced in its favor. With the WTO, it proved to be shortsighted. The South was abetted by farm state Democrats in a reaction to 1996 and the price collapse. Southerners went nearly as far in the opposite direction as the 1996 bill had gone in response to the price increase and congressional reform agenda. Again, the rules had changed, and while southerners won against midwestern Republican opposition, the WTO dispute with Brazil validated the opposition.

To the extent there was any balance coming out of 2002, the WTO dispute completely upended it. The overwhelming loss put cotton supports in immediate and serious jeopardy. They had to eliminate a special export subsidy and were under pressure to revise the support system they had just secured. The WTO dispute also put cotton in jeopardy within the farm coalition because it placed the entire farm policy system and all commodities at risk. The WTO decision was an indictment of the entire 2002 policy apparatus; it just happened to focus on cotton. Moreover, the potential for retaliation against other sectors of the economy magnified the risks to cotton and the farm coalition.

At this juncture the RFS came into play with the potential to further disrupt the balance. It created an unprecedented new domestic demand, but one concentrated on corn. It magnified corn's advantages even as it drove up prices for all crops. From NAFTA to the WTO, not only had the rules changed, but the market situation had as well. The domestic textile industry shifted overseas, leaving cotton more dependent on exports and the world market; paths diverged further.

The RFS is arguably the most successful farm policy to date, even though it is technically energy and environmental policy. It has driven crop prices for all commodities to record highs, surpassing historical impacts from exports and war. That success has also had consequences beyond the comparison between corn and cotton. For one, it rendered the target price policy nearly obsolete or, at the least, irrelevant because it drove crop prices above the target price levels and seemed to promise to keep them there. The history of target prices demonstrated the extreme difficulty involved with increasing them, particularly if budget disciplines come into play. The 2002 target prices even for cotton were not historically high compared to 1990 levels; they were simply well above market levels at the time. Cotton prices would also adjust upward in the RFS era to settle above target prices, albeit slower than corn or wheat.

The RFS resulted in a recycling of one component from the fifties and sixties in that it sent corn looking for different policy options. That search landed on policies modeled off the other farm policy success in crop insurance and, in particular, revenue-based insurance. Senators from the northern plains and Corn Belt pushed for revenue-based payments, likely in part because target prices were no longer relevant for crops benefitting from ethanol demand. Revenue-based farm program payments would supplement crop insurance or compete with it or, possibly, a little of both. The policy also held the potential of further codifying the divergent political fortunes of corn and cotton by incorporating recent market price increases more effectively and cost-consciously than target prices could. Coming in the wake of the WTO decision and the RFS, this policy added pressure to the situation between corn and cotton. The 2007–8 debate in the Senate began to expose this situation, but direct payments largely contained the competitive pressures and the policy debate was held to a relatively minor skirmish.

Crop insurance and revenue programs were not the only policy developments of note during this first dozen years of the modern era. Conservation policy underwent significant developments, but they were relatively quiet in comparison to the others. Some of the 2002 bonus funding was spent on conservation programs. Congress increased the CRP cap to 39.2 million acres and increased spending for EQIP. At the direction of Chairman Harkin, it also created an entirely new conservation program designed to provide conservation-based income support or green payments. This was the Conservation Security Program, which was revised and renamed in the 2008 farm bill as the Conservation Stewardship Program. These were quiet developments because they were relatively noncontroversial but they also created divergent paths that held significant implications post-RFS.

The reserve or retirement programs such as CRP take acres out of production while the working lands programs (EQIP and CSP) assist with conservation on acres that remain in production. By increasing commodity prices, the RFS also increased the value of land in production, which put pressure on policies that held land out of production. In some circumstances it became more valuable to have land in crops, even if marginally productive, than under a CRP rental contract. As CRP contracts expired, some of the acres went back into production, and in the 2008 farm bill, Congress lowered the acreage cap from 39.2 million acres to 32 million acres. That bill also revised CSP and included a direction to USDA that it add almost 13 million acres each year, increasing the program and its baseline. Therefore, with the RFS, high crop prices, and programmatic changes, conservation policy began a noticeable development trend in the direction of working lands programs and away from reserve programs.

On the surface the 2008 farm bill offered a return to the 1996 status quo; strong prices under the RFS had reduced the baseline back to the distribution from direct payments. Below the surface, however, fundamental changes were underway. At the time the RFS was pushing farm policy in new directions with impacts from commodity payments to crop insurance to conservation. For the farm coalition, corn was consolidating gains while cotton was in preservation mode with the WTO bearing down. The bill was set to expire in 2012, but upheaval was once again on the horizon.

8

Old Fights Plague the Agricultural Act of 2014, 2011–2014

Introduction

Significant upheaval in the political landscape followed close on the heels of the 2008 Farm Bill. In 2008 Senator Barack Obama (D-IL) won a resounding, historic victory against Senator John McCain (R-AZ) in the presidential election.[1] The race featured concerns over the continuing wars in Iraq and Afghanistan but was consumed by the housing crisis, the financial industry collapse, and the resulting economic recession.[2] Economy-wide financial problems that first appeared in August 2007 overwhelmed the national and global economies by September 2008. Much of the blame for the crisis was placed at the feet of the U.S. mortgage industry, where relaxed credit standards in the early 2000s brought about high rates of delinquency and foreclosures that shocked the financial industry. The economic downturn in 2008 was the worst since the Great Depression, and it had a profoundly negative impact on the economy, jobs, incomes, and savings.[3]

From a farm bill perspective, the biggest impact was on food stamps. The 2008 farm bill had renamed the food stamp program SNAP and eased the application process by allowing those eligible for other low-income federal assistance to automatically qualify. Participation in countercyclical programs such as SNAP and unemployment insurance increased drastically in the recession, which spiked federal expenditures.[4] Predictably, this drove concerns about federal spending, the deficit, and the national debt.[5] President-elect Obama prioritized economic stimulus during his first days in office, but he notably singled out farm subsidies for millionaires as a "prime example of waste that I intend to end as President."[6] The 2008 farm bill had been on the books less than six months.

Two years after the historic election of Barack Obama, a political perfect storm brought about a wave election in the midterms that threw Democrats out of power in the House of Representatives.[7] Much credit was given to a conservative political movement known as the Tea Party and convulsions in the body politic over the size and reach of the federal government. The Tea Party movement featured frustration with the economic recovery and a brutal partisan fight over Obama legislative initiatives such as health care reform and efforts to combat climate change.

The House Ag Committee gavel passed from Collin Peterson (D-MN) to Frank Lucas (R-OK), a fifth-generation farmer and rancher from western Oklahoma. Representative Lucas had been elected to the U.S. House in a special election in 1994 to fill the seat of Glenn English.[8] He had served on the staff of U.S. Senator Don Nickles (R-OK) and been a member of the Oklahoma State House of Representatives (1988–94).

While control of the Senate did not change partisan hands, a major change took place in the Senate Ag Committee. In 2009 Senator Edward Kennedy (D-MA) died, and Senator Tom Harkin (D-IA) stepped down as chair of the Senate Ag Committee to replace Kennedy as chair of the Committee on Health, Education, Labor, and Pensions (Senate HELP Committee). Senator Blanche Lincoln (D-AR) assumed the chair of the Senate Ag Committee.[9] Senator Lincoln, however, was defeated in her reelection by John Boozman in the 2010 midterms, and the chair was open again.[10]

As the new Congress settled in, Senator Debbie Stabenow (D-MI) became chair and Senator Pat Roberts (R-KS) the ranking member.[11] Senator Stabenow, born and raised in a small Michigan town, had worked in public schools before she began a political career with her home county board of commissioners.[12] She served twelve years in the Michigan House of Representatives and four years in the Michigan Senate. She was elected to the U.S. House in 1996 and the U.S. Senate in 2000, becoming the first woman elected to the U.S. Senate from Michigan.

From Harkin to Lincoln to Stabenow, Senate Ag Committee

leadership had passed briefly through southern hands but landed back in the Midwest just as work began on reauthorization. In the House, Chairman Lucas was expected to have a lot of new members on his committee with little or no farm policy experience but strong ideological leanings. The farm bill's outlook was complicated because consideration would be led by these two very different Ag Committees in the middle of a presidential election year. The 2008 farm bill had been developed and passed by a Democratic Congress and vetoed by a Republican president. Since that time the presidency had gone to a Democrat, and the House of Representatives had flipped to Republican control. For the farm bill coalition, the political upheaval was regional, partisan, and philosophical. It produced no small amount of uncertainty and anxiety.

The farm economy weathered the Great Recession quite well, thanks in part to the demand for corn from the RFS. Farmers experienced record incomes during 2010 to 2013, topping $120 billion and then $130 billion in net cash income from farming.[13] Farmers also continued to receive roughly $5 billion each year in direct payments, as well as subsidized crop insurance and protection against downside price risks.

Farm policy had not escaped the Brazil cotton dispute, however. Brazil sought to retaliate against U.S. exports because the U.S. had failed to fully comply with the panel's ruling.[14] In 2009 the panel found that Brazil could implement cross retaliation against the U.S., including against intellectual property. When Brazil announced retaliation, the U.S quickly negotiated a deal. Brazil would forgo retaliation in exchange for $147 million in annual payments and an agreement that the U.S. would address cotton subsidies in the next farm bill, scheduled for reauthorization in 2012.[15]

Reauthorization played out under these pressures. The Tea Party–infused Republican majority in the House promised disruption with its focus on cutting federal spending and reducing the federal government's role and impact in the national economy. The new House majority was also highly antagonistic and combative toward President Obama and the Democratic Party. These political and ideological forces would hammer the farm bill coalition and greatly affect the debate. The new chairs would have to navigate a difficult process, and the farm bill would have to survive unprecedented political tests.

Budget Battles and the Secret Farm Bill

History has provided two fundamental lessons about farm bill debates. At least since 1965, farm bills have needed a coalition between farm interests and nonfarm interests to pass Congress. Since 1974 budget discipline has exerted incredible stress on that coalition. Those lessons provide insights for the problems that were to quickly consume work on a new farm bill. The reauthorization process began under dark clouds gathering from the federal budget debate and extraordinary partisan conflict.[16] Farm program spending was a target for significant cuts, and the $5 billion per year ($50 billion over ten years) in direct payments were an obvious place to start.[17]

The Tea Party strategy for confronting President Obama was the federal budget.[18] Tea Party members demanded substantial reductions to federal spending, which culminated in the threat that the House would not vote to increase the federal debt ceiling.[19] Defaulting on the national debt was considered catastrophic in general but especially for the still-struggling economic recovery. Congress barely reached an agreement to avoid the catastrophe of a federal default, passing the Budget Control Act of 2011 at the last possible moment.[20] That bill contained unique provisions to tackle the debt and the deficit and to reduce spending.

Congress created and empowered a special select committee on deficit reduction, commonly referred to as the Super Committee. It was provided unprecedented authority to cut $1.2 trillion in federal spending across all jurisdictions.[21] If the Super Committee could reach a universal agreement, it would produce legislation that would pass Congress under expedited procedures similar to budget reconciliation.[22] If the Super Committee failed to come up with legislation, however, across-the-board reductions in spending, known as sequestration, would automatically be implemented.[23]

The Super Committee may well have embodied the farm coalition's worst nightmare: a group of (mostly) non–Ag Committee members given full authority to eliminate farm programs with little input and very few options to stop it. If the Super Committee could wield the budget axe that it had been given, direct payments would certainly be a victim.[24] At roughly $50 billion in the ten-year budget window, direct payments made up nearly the entire baseline for commodity programs. Eliminating them would essentially wipe out that baseline, leaving behind little with which to write a new bill. All of the interests that formed the farm bill coalition, including crop insurance, conservation, and food assistance, were aware of this risk. With record farming incomes in a recession and a government focus on reducing outlays, many also grasped that the payments had become indefensible and the program a major political liability.

House and Senate Ag Committee leaders responded to the threat by entering into desperate negotiations to reduce spending and head off the Super Committee.[25] In September the Ag Committee leaders agreed to a $23 billion reduction over ten years that was spread across farm programs, conservation, and SNAP; the bulk of the savings would come from eliminating direct payments.[26] The negotiations resulted in a closely held effort to rewrite the farm bill to achieve targeted spending reductions while delivering on the regional and commodity demands for farm policy.[27] Because the deal involved eliminating direct payments, the effort caused a scramble among the regional commodity interests over the design of policy to replace them.[28] That scramble took place within the post-1996 regional competition over distribution of the federal expenditure baseline. It was complicated by the intense partisan dynamic due to the Republican Party's dominance in the South and the House.

Southern interests (cotton, rice, and peanuts) demanded a fixed-price program similar to traditional target price and deficiency payments, but the WTO dispute and potential for retaliation complicated cotton's situation. Midwestern (corn) and northern Great Plains (wheat) interests wanted a revenue-based program that would supplement crop insurance by making payments when revenues (prices multiplied by yields) fell below a moving, historical benchmark. Budget cuts left little room to bridge the differences.[29] At its core the dispute was about more than whether farmers were paid using a statutory price or calculated revenues. The program details and parameters can be esoteric but produce real consequences because they dictate how payments are to be triggered. Cotton had demonstrated one way in which this worked in the 2002 farm bill, reaping a windfall from a high target price.

The cost of a fixed-price policy depends on the price levels established in the bill. Congress lists the reference price for each crop (e.g., corn at $3.70 per bushel; peanuts at $535 per ton), and it does not change. Effectively, this constitutes a price floor for the commodity, albeit one decoupled from planting the crop. A higher statutory price relative to market prices the more likely it will trigger payments. In CBO's estimations this adds costs to the baseline for the policy. It also creates a relative value for base acres of the covered crops.

Under CBO estimates this has additional political implications because triggering payments on small crops like peanuts (two million base acres) has a far smaller impact on the baseline than does triggering payments on large crops like corn (ninety-seven million base acres). Even without the history behind target prices, this acreage footprint explains much of the disagreement among the commodities. In negotiations under the vise of budget discipline, smaller crops have leverage to demand higher target prices because they do not score as much in the CBO estimates. Fixed-price policy favors the southern crops.

By comparison the payments from, and CBO costs of, revenue programs depend on both price and yield variables. An important political issue with revenue policies is that they were designed using rolling five-year Olympic averages of market prices instead of a statutorily fixed price. An Olympic average drops the highest and lowest in the five years and averages the remaining three years as a mechanism to smooth the average. This incorporates some degree of market price risk in the calculations for assistance. For example, if prices fell for multiple years within the five-years used in the Olympic average, the price component in the benchmark revenue calculation effectively declines as well. Politically, using a percentage of market average prices makes it difficult for commodity interests to secure more favorable treatment because all crops would presumably use the same percentage. It would be obvious if one crop received a higher percentage than others, but it is less obvious when using statutory prices fixed to different units of production (e.g., dollars per bushel versus dollars per ton).

Additionally, use of yields as part of the calculation for triggering payments alters the price-protecting aspects of the policy. A crop year with high yields may produce low prices, but the yields may offset the price decline and hold revenue up. Consider an overly simplistic example: a 100-bushel crop at $2.00 per bushel produces $200 per acre in revenue; a 200-bushel crop at $1.00 per bushel also produces $200 per acre in revenue. Yields may also vary substantially across areas with potentially different impacts in the main production regions.[30]

The revenue policies are also related to crop insurance, which the South has traditionally favored less than the Midwest. One aspect of the similarities with crop insurance is concern about overlapping payments with insurance coverage. Applying a cap to revenue programs can help avoid the potential for overlap, but it will also limit the amount of potential payments. If one goal of a policy is to maximize payments to a commodity, then such a cap would not be acceptable.

In short, while they contain similar features and cover the same commodities, fixed-price and revenue policies involve differences that are important to farmers and their respective interest groups. Upon a closer look, the dispute appears to recycle some of the same regional and political dynamics that existed with flexible loan rates versus 90 percent of parity. The revenue program was a relatively newer method with mechanics that ran counter to traditional price-based farm policy. Yields can prevent payments even with low prices. More fundamentally the declining benchmark is contrary to the desire for a floor on prices, which has been the longest running dispute in farm policy.

Over the course of three months in the fall of 2011, the Ag Committee leaders attempted to find a deal among these competing regional policy demands. They were operating under intense budget pressures and faced an institutional burden because they were working to rewrite farm policy without the benefit of an open, regular process in which disagreements could be sorted out through debate and votes.[31] Regional disagreements over specific policy designs were complicated. The South and the House wanted a fixed-price policy similar to the countercyclical program created in 2002. The Midwest and the northern Great Plains wanted revenue-based policy but complicated matters because they differed on which yields to use in the revenue calculation. The Midwest preferred county average yields, but the northern plains preferred individual farm level yields.

With great difficulty and lingering disagreements, Chairwoman Stabenow and Chairman Lucas were able to negotiate an agreement, but their efforts were for naught. The Super Committee failed to complete its legislative task by the statutory deadline and was disbanded.[32] The expedited, unprecedented process that Congress put in place to bridge the self-inflicted debt-ceiling crisis had unraveled; the goal of cutting federal spending crashed on the hard political realities of actually cutting federal spending. The Ag Committees were the exception. They reached a deal to cut spending in their jurisdiction. The effort for the failed Super Committee turned out to be merely the first stage of a long, difficult farm bill negotiation.

House Divided and Farm Bill Delayed in 2012

The new year presented the prospect of producing a farm bill in the regular-order process that presented many challenges. For one the 2012 campaign season promised to increase the political temperature owing to the intense battle over President Obama's reelection along with control of the House and Senate. The politically charged atmosphere, if it amplified the budget battle, heightened farm community concerns that it would produce more severe outcomes for them.[33] Those risks, however, did not cool the regional disagreement.

The most pointed example of the increasingly tense regional conflict can be found in reported comments from the cotton industry that "some grains and oil seeds are trying to take your money . . . the money that's in the baseline for rice and peanuts and cotton in order to enrich their revenue programs."[34] Taken at face value, these comments exposed that the real dispute was over interest group preferences for the distribution of estimated federal expenditures rather than policy or farmer needs. That this claim was made in the thick of partisan budget-cutting demands raised difficult questions for the farm bill. To the extent that these comments represented the southern perspective, they exposed troubling aspects of the intra-farm coalition dispute and highlighted the challenges of writing a farm bill. The comments also risked elevating bad optics in a partisan political environment obsessed with federal spending.

The Senate and House Ag Committee efforts went their separate ways in 2012, differing on policy as well as process. Chairwoman Stabenow wanted her committee to move aggressively; Chairman Lucas waited for directions from the House Budget Committee.[35] The budget resolution produced by House Budget chairman Paul Ryan (R-WI), however, added challenges for Chairman Lucas. It required major reductions in spending from the farm bill across the commodity, crop insurance, conservation, and nutrition (SNAP) titles.[36] When Chairman Ryan dropped the House budget, it fell as a partisan weight that threatened to break open the previous year's compromise. The effort could have taken the farm bill as one of its casualties, but much depended on how Chairman Lucas responded.[37]

The House budget directed six House committees to produce savings that would be used to replace the across-the-board cuts known as sequestration, which had resulted from the Super Committee failure. At $33 billion the House Ag Committee was tasked with the biggest contribution.[38] Of all the options for reducing spending, the committee chose to achieve the entire $33 billion in savings by cutting SNAP. This could not have been more inflammatory for the farm-SNAP coalition, particularly because SNAP was exempt from sequestration.[39] It was also a highly partisan move that happened to coincide with the Senate Ag Committee's efforts to begin marking up its bill.[40] It was an ominous sign for the farm bill's chances in the House. The Super Committee agreement between the Senate and House Ag Committees had been for $23 billion, including $4 billion from SNAP and $15 billion from farm programs. The first move by House Ag in 2012 was to cut more ($33 billion) and to take it all from SNAP.

In their budget blueprint, House Republicans were making a very partisan, ideological argument. They were attacking what they saw as government programs that had gone from a "safety net" to a "hammock that lulls able-bodied citizens into lives of complacency and dependency."[41] This line of argument opened farm policy to uncomfortable comparisons and political risk. Farmers were receiving nearly $5 billion per year at a time of record incomes. With a cap at $40,000 per individual farmer ($80,000 for a married couple), farm program recipients received far more in federal payments than any individual SNAP recipient, with an average yearly benefit of $1,606.[42]

This gulf between the treatment of the beneficiaries of the two programs echoed painful views from a troubled past.[43] It also held the potential for driving a wedge through the coalition of rural and urban interests brought together in 1973. The loss to Brazil in the WTO dispute over cotton payments alone should have counseled efforts to strengthen that coalition. The U.S. was paying $147 million each year to Brazil to avoid retaliation. The dissonance is difficult to reconcile. And yet the House proceeded on that troubling course.

The Senate Ag Committee largely avoided the kind of partisan debate over SNAP that was taking shape in the House.[44] The committee bill included a comparatively minor $4 billion reduction to SNAP spending that changed how benefits were calculated for those households also receiving federal energy assistance.[45] The changes to SNAP did not affect participant eligibility. They would not result in people becoming ineligible for SNAP benefits, which was a key principle for Democrats. The Senate revisions to SNAP would reduce benefits for some families, not remove them; it affected participants in a limited number of states, most of them in the Northeast. In committee Senator Kirsten Gillibrand (D-NY) strongly opposed the cuts to SNAP but was unable to find the votes to defeat the provision.

Almost all of the intense politics in the closely divided committee were concentrated on commodity support programs. Overall, the chairwoman held a one-vote majority (eleven to ten) with strong support from six Democrats, but she lost Senator Gillibrand over SNAP. Senator Roberts could count on five midwestern Republican votes for a working majority. Key swing votes and interests belonged to Senator Leahy (D-VT) and dairy, as well as Senator Conrad (D-ND), Senator Baucus (D-MT), and Senator John Hoeven (R-ND), who were interested in a different revenue program design. The substantive difference between the revenue programs was whether county average yields or actual farm level yields were used to calculate revenue.[46] In addition, the committee included four southern Republicans. Southern commodity interests and senators had two program demands for the farm bill.[47]

First, southerners wanted the traditional fixed-price program for all crops with increased target prices (renamed "reference prices"), particularly for rice and peanuts. Second, they wanted a specific cotton crop insurance policy known as "STAX," which stood for stacked income protection insurance. This program permitted cotton farmers to purchase two insurance policies on their crop: standard yield or revenue coverage supplemented by subsidized county-level yield or revenue coverage. The STAX policy was intended to finally settle the Brazil WTO dispute as agreed to by the U.S. in the temporary settlement to head off retaliation. STAX was proposed as that fix. It replaced commodity payments for cotton, although not the marketing assistance loan program. In place of payments from fixed prices or revenue, cotton farmers would be able to purchase the new insurance policy to cover losses at the county level, in addition to any coverage for losses on the individual farm.

Although STAX was not a Title I commodity program, it was inextricably linked to commodity policy. Cotton demanded the inclusion of a reference price in the insurance policy, contributing further to the regional conflict over fixed prices. The demand kept cotton aligned with rice and peanuts politically but came with specific problems. Operationally, including a fixed reference price would be an unworkable component of an actuarially rated insurance program. For example, if actual market prices were below the reference price at the time farmers purchased their insurance policies, they would be purchasing insurance with the knowledge that it would pay an indemnity. Reference prices also would have been unacceptable for Brazil because they would have merely shifted the countercyclical program from 2002 into the insurance setting with a reference price recoupled to cotton-planted acres and production.[48] The STAX policy with reference prices would have been worse than the decoupled countercyclical program for which Brazil had won its WTO dispute.

The battle lines were clearly drawn. Senator Roberts and his midwestern Republican allies adamantly opposed the southern fixed-price program, as did the chair and some of the Democrats on the committee.[49] They also opposed including a fixed reference price in STAX. Southerners strongly opposed the revenue program design favored by the chair and ranking member. Northern plains senators Conrad, Baucus, and Hoeven sought a modified revenue program that was expensive and opposed by much of the rest of the committee interests. The odds and votes favored the chair, but not by much.[50]

Chairwoman Stabenow and Ranking Member Roberts were both contending with split views about farm policy within their ranks and had to be concerned about an alliance between these interests. It was not an unfounded concern given that it involved powerful senators, particularly Senators Conrad and Baucus, who chaired other committees and had experienced success in 2008 against then-chairman Harkin. Going into the farm bill markup, the united front from committee leadership appeared to have the upper hand but was facing strenuous opposition.

The actual farm bill markup would prove anticlimactic. Even with a last-minute effort to derail the markup, it was clear that the South had lost the policy fight.[51] The southerners were unable to get their preferred farm program included. Moreover, the bill included farm program payment reforms championed by Senator Chuck Grassley (R-IA) but vigorously opposed by southerners. Senator Grassley had pushed for two basic reforms. First, he wanted a hard cap on payments, limiting them to $50,000 per individual. Second, he wanted to close a loophole in the requirement that individuals be actively engaged in farming to receive farm program payments. These policy and political losses magnified opposition but also highlighted the significant limits on southern power to exert its traditional influence.[52]

The markup documents provide further clues about the political power dynamic at the committee level.[53] The final negotiations were confined to the revenue program design. Versions of the bill altered that design multiple times but did not add a fixed-price program or reverse payment reforms. Last-minute changes to the committee bill indicate that Senators Conrad and Baucus pressed their case to the brink of markup but that southern senators had little leverage to force major changes. The commodity program debate was settled by an amendment from Senator Baucus to revise the revenue program, and it was accepted by a voice vote with little controversy.[54] The actual markup was a relatively brief, five-hour meeting. At its conclusion the Senate Ag Committee voted to report the bill to the full Senate over unified southern opposition; the bill had been little changed at the committee table.[55]

After the bill had cleared this first hurdle, there was much anticipation in the farm interest community about how southerners would respond.[56] Chairman Lucas and Ranking Member Peterson were critical of the bill's farm programs. They were concerned that it would not provide sufficient assistance if prices collapsed.[57] Their response indicated alignment with southern commodity interests and policy demands that were reminders of farm policy fights from the parity era. Then the House Ag Committee remained the strongest bulwark on 90 percent of parity loan rates as the Senate Ag Committee pushed for flexibility in the program. In 2012 the debate involved the House prices fixed in statute versus the Senate revenue calculation.

On the Senate floor, Chairwoman Stabenow and Ranking Member Roberts opened the argument by explaining the bill's bipartisan agreement to reduce spending through reform; bipartisan legislation that reformed policy to reduce outlays made it the only bill of its kind.[58] The chairwoman emphasized that the bill eliminated the $5-billion-per-year direct payment program and closed long-protected loopholes in farm program eligibility.[59] She explained the fundamental shift in policy, moving away from payments that protected against market forces that were anachronistic in the modern farm economy. In place of outdated policies, the bill incorporated risk management concepts to "supplement crop insurance ... [with] a simple market-oriented and risk-based program" that represented "significant and historic reform in agriculture policy."[60] Stabenow added that "for years, Congress has struggled to balance the needs of different commodities, different programs." The revenue program "solved" this problem by using "the market as a guide" and treating "every commodity the same."[61] The chairwoman's arguments amounted to an effort to frame opposition to the bill as less than principled; opponents were objecting because the policies had not been designed to favor their commodities. Senator Roberts, who as House Ag chairman had written and pushed through the 1996 farm bill that created direct payments, added that it was a "true reform bill."[62]

These were not only selling points for a closely divided Senate in an election year; they were also defensive mechanisms against opposition to both farm programs and the revision of SNAP. This defensive strategy was different from that deployed in 2002 or 2008. It reflected the challenging political environment for the bill—legislators were under pressure to cut spending while dealing with altered farm coalition dynamics. Gone were claims of a balanced bill tilted in favor of the southern commodities but necessary to protect food assistance. Instead, the chair and ranking member were employing southern opposition as proof that the committee's work constituted real reforms to controversial programs. This was a calculated risk that took advantage of budget politics and the divide within the Republican caucus.

Senator Baucus and other farm state Democrats also embraced the market-oriented shift in farm policy. This represented a notable change from many long-held positions.[63] Concerns about multiyear depressed crop prices lurked below the surface, however. These senators remained uncertain that the new policies would be adequate when the inevitable happened.[64] The southern counterattack on the floor started there, seeking to exploit concerns about depressed prices. It was a tough sell with market prices increasing and farm incomes breaking records.[65]

The southerners' main objection was that they viewed the bill as unfair to southern farmers. For example, Senator Saxby Chambliss's (R-GA) "greatest concern with this bill is that the commodity title redistributes resources from one region to another not based on market forces or cropping decisions." Senator Chambliss also attacked the revenue program as "filled with inequities" and "unbalanced" because it "seeks to place a one-size-fits-all policy on every region of the country" that would not work for rice and peanuts.[66] His argument was that of the cotton industry, which based the measure of fairness on CBO baseline distribution.

This argument, however, missed the mark. Senator Roberts made the point clearly: "Money is shifting among commodities because farmers are farming differently . . . not because we in Washington are intentionally picking winners and losers."[67] The bigger problem for the opposition was the overarching emphasis on budgetary matters at a time of strong prices and incomes. Thus, 2012 was not the year to fight over the fairness or amount of commodity payments. The political environment was not at all conducive to baseline competition among crop interests. Demands that one region be given what it deemed an acceptable share of CBO expenditure estimates were politically misplaced. The reform message packed too much political power.

Amendments on the Senate floor further proved that point. Efforts to take reforms even further than the committee was willing to go were overwhelmingly successful. Seventy-five senators agreed with Senator Grassley's amendment to apply a strict payment limit to the marketing assistance loan program, including loan gains and LDPs.[68] Even more notable were amendments to crop insurance. Sixty-six senators voted to apply adjusted gross income (AGI) standards to crop insurance premium assistance.[69] The amendment would have reduced premium subsidies for farmers making over $750,000, and its passage sent shockwaves through the agriculture community.[70] That amendment followed on the heels of another successful amendment to crop insurance by Senator Chambliss that reattached conservation compliance.[71] In fact, only the sugar program was able to escape the farm bill debate without being subjected to reforms, but it was forced to survive two contentious votes.[72]

In the Democratic-controlled Senate, minor reductions to the SNAP program and its popularity in a postrecession economy provided strong political defense against Republican attacks.[73] The floor debate delivered emphatic statements that the $4 billion committee reduction was a sufficient contribution from the program. Republican senators tried to reduce the program's spending by reducing the number of people who received SNAP benefits but were defeated.[74]

The committee deal also provided protection from Democratic senators concerned about any reductions to SNAP.[75] Senator Gillibrand, for example, pushed an amendment to reduce spending on crop insurance and reverse the cuts to SNAP in the bill.[76] Senators defeated this effort with little difficulty.[77] Senators appeared to understand that SNAP would have to take some reductions in the process, but there was little appetite to cut the program nearly as deep as the House had proposed. Senators were particularly opposed to changes that would reduce participation or eliminate eligibility.

Where the debate struggled most was the partisan fight over the amendment process. On the Senate floor, a mountain of amendments stood between the thousand-page farm bill and Senate passage. This procedural fight had plagued the last two farm bills, but an equally partisan budget debate added a large shadow over the process.[78] Republicans pushed back against the tactics used by Majority Leader Harry Reid (D-NV) to control the process and block problematic amendments that were not relevant to the farm bill.[79]

In a politically charged atmosphere, some senators demanded votes on amendments that were not related to the bill but that carried difficult political implications. To avoid an unlimited amount of mischief, Majority Leader Reid made use of procedural rules to block amendments from being considered without unanimous consent. Senator Reid's move can be viewed as an escalation of previous procedural battles, including those that bogged down the 2002 and 2008 farm bills.

Chairwoman Stabenow, Ranking Member Roberts, and Senate leadership conducted a week of intense negotiations to resolve the impasse. The chairwoman's bipartisan negotiating persistence was particularly notable, and it paid off.[80] The Senate was able to reach an agreement on a list of seventy-three amendments—out of more than three hundred filed—that would be considered in an expedited process known as a vote-a-rama.[81] This is a series of stacked votes on a finite list of amendments; each amendment is provided two minutes of debate before it is voted on. Vote-a-rama is most often used for budget resolutions. For the farm bill, it provided senators an open, but controlled, amendment process that blocked overtly partisan amendments that had no bearing on the underlying bill.

Senators debated and considered changes to the bill, but Chairwoman Stabenow and Senator Roberts combined forces to ensure that only amendments approved by them were able to get the support needed to pass. The marathon effort demonstrated the strength of the coalition and the power of the limited process. The two managers used both to protect the bill. The ability to consider amendments, even on a limited basis, helped bring the debate to a productive close and allowed the Senate to pass it with a large, bipartisan vote.[82]

The farm bill's success provided a bit of reprieve from some of the partisanship in the Senate, but it was a harbinger of problems within the bill's coalition. Senate passage was an important process accomplishment in its own right; the way in which it was achieved held additional significance. Passage despite near-unanimous southern opposition was a rare result within the farm coalition's history and one that added intensity to the regional conflict.[83] Southerners had not merely lost on policy specifics; they had waged a fierce public battle that proved futile and appeared counterproductive. The loss exposed the breadth of opposition to southerners' preferred policies and the weakness of their vote bloc. They were unable to block reforms, unable to get their programmatic demands, and unable to slow the bill. It was a loss noted in dramatic fashion by one reporter who wrote, "If you would have walked by Senate Ag Committee offices last night, you might have heard the Band's 'The Night They Drove Ol' Dixie Down.'"[84] The Senate result also provided hints of the problems to come.

The Democratic-controlled Senate protected SNAP and midwestern farm policy priorities while the Republican-controlled

House was prepared to make big reductions in SNAP but protect southern farm policy priorities.[85] There appeared to be little negotiating room in those positions should the bill reach conference. Southern interests, moreover, were dependent on the House Ag Committee and Republican leadership. This was a position made incredibly precarious by the Tea Party dynamics on the House floor.

Pressure from the Tea Party to cut spending focused most intensely on SNAP, but not exclusively. Many wanted to end farm program payments and subsidized crop insurance as well; they understood that the coalition helped protect all accounts. Thus, moving too far to the right on SNAP risked a collapse in the center that would put farm spending in jeopardy. It also exposed the tough optics over farm payments that favored commodities whose backers sought deep cuts in food assistance.[86] Committee Republicans opened this seemingly perilous journey by providing target prices for all crops as demanded by southern interests coupled with substantially deeper cuts to SNAP.[87] The House bill reduced spending by $35 billion, $16 billion of it due to reducing participation in SNAP.[88]

Nature added further drama. As Congress debated the farm bill, farmers were struggling with what was to become a record-breaking drought.[89] The House Ag Committee reported its bill on July 10, 2012, but prospects on the House floor were uncertain at best.[90] The politics were tough in spite of a significantly strong bipartisan vote in committee. In particular, Tea Party resistance to both SNAP and farm assistance was substantial, and House leadership was reluctant to give the bill floor time with campaign season approaching.[91] Adding complications, the drought was driving up commodity prices and concerns about food costs.[92]

The drought, coupled with its legislative counterpart, increased political pressure as the election season approached, but House leadership held fast. Before the August recess, the House had passed a short-term extension of the 2008 farm bill that added disaster relief, but it was seen as a political gimmick to help members avoid problems back in their districts. The extension was dead on arrival in the Senate and would be the last word on the farm bill before voters had their say.[93] When Congress left for the August recess, rural House members went home to face constituents, shadowed by comments that the votes were not there for the bill.[94]

Recess did not improve the political situation. The House lacked enough votes from Republicans to pass the new farm bill, mostly because of demands for deeper SNAP cuts. The House also remained unable to move an extension of the 2008 farm bill.[95] Congress allowed the 2008 farm bill to expire and went back into recess for the November elections.[96] House leadership had gambled that no farm bill was better than the pain of debating one on the floor. They were betting that farm district members would not be punished by voters in November or that any punishment would be minor compared with an incredibly divisive and partisan fight over farm programs and SNAP. That calculation carried very uncomfortable implications for the farm coalition.

In November American voters returned status quo to Washington by reelecting President Obama, keeping the Senate in Democratic hands and the House in Republican control.[97] In agricultural circles, the farm bill's uncertain future passed to the lame-duck session of Congress and the year-end efforts to avoid tax increases and other spending problems known as the "fiscal cliff."[98] Efforts to sort out the farm bill, however, failed because differences over farm program design and reductions to SNAP were insurmountable.

In reality there was little chance that a trillion-dollar farm bill could be added to the year-end fiscal cliff legislation. Divided over its fault lines, the farm bill reached a dead end. In the New Year's Eve legislative rush to avoid the fiscal cliff, Congress extended the 2008 farm bill for another year. The extension was largely based on concerns that dairy policy would revert to the 1949 act's provisions on January 1, 2013. Under reversion, USDA would be fixing dairy prices at unreasonably high levels, thus extension was added to avoid falling off the dairy cliff.[99]

At the end of the day, the election year politics of 2012 had proved too much. The farm bill had fought its way through a difficult Senate process only to be put on ice by House leadership worried about its impact at the voting booth. Budget pressures were to blame. Eliminating direct payments had reopened a major regional rift within the farm coalition that damaged the bill's progress. The disagreement over farm programs was minor compared to the partisan budget pressures that were building against the farm program–SNAP coalition. The forty-year coalition between rural and urban House members was under immense pressure. The Senate and House Ag Committee leaders would have to try all over again in the new Congress.[100]

Second Chances and the Final Bill, 2013–14

A minor power play in the Senate Ag Committee made for a rocky start to 2013. To restore southern negotiating leverage in committee, Senator Thad Cochran (R-MS) used his seniority to replace Senator Roberts as ranking member of the Senate Ag Committee.[101] Senator Roberts had long had a contentious relationship with his southern colleagues, dating to the 1995–96 farm bill debate and Freedom to Farm. As ranking Republican on the committee, he had been instrumental in commodity program development in 2012. He was an ardent opponent of the fixed-price policy pushed by the South. After they had lost in the Senate and found unreliable support in the House in 2012, southerners appeared to be taking no chances in 2013 as fighting over the federal budget continued.[102]

Moving Senator Cochran into the ranking member's seat changed the political equation on the Republican side of the committee. It forced Chairwoman Stabenow to alter the previous year's successful strategy; changes that came with real risks.[103] To get out of committee, the revised farm bill would have to move in the South's direction on issues like fixed-price farm policy, payment limits, and eligibility reforms. Each move added risk on a Senate floor where Senators had gone even further than the committee reforms the year before. The moves were also sure to face opposition from Senator Roberts and his midwestern Republican allies.

There were other policy and political implications. Any movement in Senator Cochran's direction would put the Senate farm bill closer to the expected House position. This would complicate the chairwoman's negotiating position in conference and could weaken her ability to avoid SNAP cuts objectionable to her Senate colleagues and the president. In other words, what would appear to be even minor changes in committee could result in major problems for the bill once it left committee.

By spring both committees moved nearly in unison. The complications were readily apparent. The Senate moved toward the House on commodity policy, as the House moved further away from the Senate on SNAP.[104]

Chairwoman Stabenow had agreed to Ranking Member Cochran's demand that the bill include a fixed-price farm program.[105] It was included in addition to the revenue programs passed the year before, but amendments at markup again provided clues about the intensity of the fight. For example, a bloc of amendments revised the reference price for all commodities except peanuts and rice, moving from a fixed price to a moving average. The bill was again the product of intense negotiations.[106] Although it remained less than southerners had demanded, it was enough to allow the committee to report it.

These relatively minor revisions, however, highlighted the depths of the regional problems. Adding back a price program for Senator Cochran and the South provoked strong objections from midwestern Republican senators aligned with Senator Roberts.[107] They had backed the 2012 version of the farm bill but walked away from the 2013 version. In their view the price program was "a step backwards to an old, outdated policy."[108] Having Congress fix prices in the statute was a deal breaker, and their votes were replaced by southern Republican votes, a move that added uncertainty for the floor.

The House Ag Committee moved its version of the farm bill the next day.[109] A strong bipartisan vote masked escalating problems regarding SNAP. In addition to the controversial changes it had included the year before, the House Ag Committee reinstated an asset test for SNAP that was expected to remove another two million current beneficiaries from the program.[110] It was an escalation of the fight a year after House leadership had buried the farm bill because of concerns about a floor fight over the program. Although SNAP was the most significant problem, the dairy provisions also presented political challenges owing to strong opposition by Speaker of the House John Boehner (R-OH).[111]

Senate debate began quickly in 2013 in a chamber once again consumed by the partisan budget feud.[112] Senators were being dragged through a tough farm bill debate for a second consecutive year with no assurances the House could pass it. Chairwoman Stabenow reiterated the substantial policy reforms in the bill, emphasizing the "new historic agreement between conservation groups and commodity groups around conservation and crop insurance."[113]

This "historic agreement" had been the only other controversial matter in the 2013 farm bill markup. The conservation community had demanded that Senator Chambliss's successful amendment from 2012 be included in 2013 bill. Committee negotiations worked with a coalition of farm and conservation interest groups to revise the policy, which became part of a compromise effort to protect crop insurance. In particular, the top goal of the farm groups was to remove income-based reductions to premium subsidy. They also wanted help to defend against expected amendments on the floor.

Farm groups wanted to protect crop insurance, and conservation groups wanted to ensure that farmers were complying with wetlands and highly erodible lands provisions after direct payments were eliminated.[114] Yet the deal was not strong enough to prevent the Senate from again adding an AGI test to premium subsidies.[115] This reform to crop insurance had succeeded previously on the Senate floor, and the coalition's ability to mount a successful defense would have required switching a large number of votes. This would have been difficult in any debate; difficulty increased at a time of intense pressure to revise and reform crop insurance.

In the budget-obsessed environment, crop insurance stood out because the bill added policies to the insurance program. The bill included the STAX program for cotton farmers as well as a similar program to provide supplemental policies for farmers to purchase. The details may have been obscure to most senators, but the increased spending was clear. Crop insurance expenditures were also estimated to increase on the heels of the record-setting drought in 2012. The drought had required billions in crop insurance indemnity payments.[116] Those payments hit insurance company bottom lines and would affect future farmer premiums, but they helped Congress avoid calls for substantial disaster assistance. The program, however, was a growing political target.

The regional dispute over commodity support policy remained the biggest problem for the Senate bill, and minor changes to obscure commodity policies garnered outsized attention. Southern senators argued that adding a price-based program was "a step in the right direction."[117] For midwestern Republicans, adding the price-based farm program was "a step backward" from the market-oriented reforms passed the year before.[118] Senator John Thune (R-SD) pointed out that Congress was not "capable of setting accurate fixed prices for the next 5 years."[119] The southern price policy, he argued, risked a return to the situation in which farmers would make planting decisions based on government programs and payments. It was an "outdated concept from past farm bills" that had been left out of the 2012 market-oriented reforms.[120] Opposition was bipartisan but regional. Senator Sherrod Brown (D-OH) raised concerns about having "farm programs in one part of the country become more market-oriented while others do not."[121] Midwestern Republicans were trying arguments that might have resonated in a Senate interested in reform.

Upper hand in the debate might have provided a moral victory, but it held no further measure of success. The fixed-price program was essential for southern support of the bill, and with Senator Cochran managing the bill, southern support was necessary. An obvious reversal from 2012, this single policy disagreement had pushed intraparty differences among Republican senators to a counterproductive point. It managed to bog down the bill in the Senate.[122]

Senator Cochran appeared to be less than confident about his chances against attacks on the price program. Instead of fighting the amendments, he blocked them from being considered. In retaliation, senators objected to each other's amendments. With the Senate unable to reach an agreement on amendments, the process halted once again.[123] While this situation had existed in 2012, it was more complex in 2013 because the most problematic amendments were ones relevant to the bill and not easily dismissed.[124] Republicans blocked each other's amendments and refused to negotiate, a situation that bordered on absurdity. Chairwoman Stabenow was left to work out the intra-Republican dispute in order to move the farm bill forward.

Demonstrating the unique level of intransigence produced by the price program was the treatment given other difficult program priorities such as SNAP. Senator Roberts sought to cut SNAP spending further. His amendment was in line with the House Ag Committee's position, and the amendment was given consideration on the floor along with a vote. The Senate rejected it. The vote should also have signaled to the House the obstacles for large SNAP changes and reductions.[125] Similarly, the Senate conducted a repeat performance of the debate to revise the sugar program. That amendment was also defeated.[126]

It is fairly remarkable that the price program was the only aspect of the farm bill that was protected on the floor. But that program stood between success and failure. For that reason Senator Cochran's tactics succeeded, and the institutional pressure to complete the bill worked in his favor. The Senate passed the farm bill on June 10, 2013, without voting on any revisions to the price program.[127] Southern senators voted for the bill, but many midwestern Republicans did not. Inside the Republican caucus, the 2013 vote was nearly the mirror opposite of 2012. What remained unchanged was the farm bill's fate. It rested with the House for a second time.

The debate in the House was mired in partisan problems from the start.[128] House Agriculture Committee chairman Frank Lucas acknowledged the difficult path it had already traveled before reaching the House floor. His efforts to stress bipartisan support for the farm bill were undercut by the partisan atmosphere in the chamber. He argued that it was the "most reform-minded bill in decades" and that it saved $40 billion in federal spending from farm programs, conservation, and SNAP. The bill eliminated direct payments, he added, in favor of "a more market-oriented approach," but farm programs had to "work for all crops in all regions of the country."[129]

House Ag Committee ranking member Collin Peterson (D-MN) began the debate supporting the chairman. He noted that there were problems over SNAP but focused on farm programs and his frustration with the direction taken by the Senate. He explained that his "father almost got bankrupted by Ezra Taft Benson and some of the nonsense that went on during that period of time. . . . So the chairman is right": the House bill reflected "a compromise between commodities and regions," made "major reforms to farm programs," and ensured that farmers wouldn't receive "a government subsidy for doing nothing."[130]

To understand fully the politics of what took place in the House, it is important to note that the chairman and his allies were able to protect most farm priorities on the floor. They protected both the sugar program and crop insurance from further reforms, including Democratic amendments to reduce crop insurance to offset the $21 billion in SNAP cuts.[131] Chairman Lucas was also able to convince midwestern Republicans to stand down over the price-based program.[132] This was as close as the House came to the regional dispute that had so troubled the Senate debate for two consecutive years. The paucity of debate on the fixed-price issue demonstrated the strength that southern members held in the Republican caucus.

In fact, there were only two exceptions to the chairman's defense of farm programs. Representative Jeff Fortenberry (R-NE) led a successful amendment to lower payment limits and reform farm program eligibility so that individuals who were not actively engaged in farming did not receive payments.[133] The amendment's success was yet another indication of the politics for farm payments among budget pressures plus record prices and incomes. The only other breach was an amendment by Representative Virginia Foxx (R-NC) to place a cap on the overall spending on farm programs on the basis of CBO projections.[134] The amendment was considered unworkable and probably passed in part because everyone understood that it would not survive conference.

By comparison, Chairman Lucas was not able to protect the farm bill from a highly partisan and controversial amendment to SNAP. Representative Steve Southerland (R-FL) offered an amendment that added provisions to encourage states to add new work requirements to the SNAP program.[135] The amendment provoked substantial anger from Democrats. They argued that it would provide "perverse" incentives to states to remove people from the program.[136] It was viewed as designed to redirect funds from low-income recipients into state coffers for those states that implemented work requirements to reduce the number of SNAP beneficiaries. The amendment carried heavy partisan and even racial implications; it was seen as harming poor minority citizens in the South, especially. It also looked as if it were taking from Democratic constituents to help offset state funds reduced by Republican-controlled governments.

It should not have been surprising that the amendment produced a wave of angry opposition. Ranking Member Peterson cautioned the House that the amendment would break "the deal we had and [was] offensive" in how it would treat SNAP recipients.[137] Adding further partisan complications, House Majority Leader Eric Cantor (R-VA) broke his silence on the farm bill to offer his support to the amendment.[138] This was taken as a sign that House leadership, and especially the Tea Party faction, stood behind the amendment. It may also have served as a warning to wavering Republicans that voting against the amendment would have political repercussions. Leader Cantor's support was credited with helping the amendment pass.[139]

The Southerland amendment proved to be the proverbial straw that broke the back of the farm bill coalition. Democrats who had supported the bill pulled their support over opposition to the amendment and House leadership tactics. Many Republicans, however, also withheld their support because they wanted to cut spending further. A near-complete destruction of the coalition meant there were simply not enough votes left. As a result, the House defeated the farm bill on June 20, 2013.[140] It was the first farm bill defeated on the House floor since 1962.[141]

The rural-urban, bipartisan coalition disintegrated almost forty years to the day after the House had agreed to combine the two policies in the 1973 farm bill.[142] All the more troubling, the defeat followed House leadership refusal to bring the bill to the floor in 2012—a decision that did not appear to have affected the election. The election produced no noticeable consequences on the farm bill's behalf. The Tea Party faction in the House had increased its numbers in 2012, at the expense of some rural district Democrats.[143] A total of sixty-two Republicans voted against the farm bill, including a number of committee chairs and those who had won amendments on the floor; all of them had voted for the Southerland amendment.[144] The breakdown in the forty-year coalition appeared to be complete.[145]

The path forward was not clear. House leadership probably had only two alternatives. Either they would have to back down on SNAP cuts and try to reestablish the traditional coalition with Democrats, or they would have to double down with the Tea Party. The latter would entail splitting the bill and muscling through otherwise unacceptable SNAP cuts with pressure from leadership to persuade enough Republicans to vote for the bill to get it to conference. This was the path they chose.[146]

House leadership cobbled a bill out of the wreckage that included all amendments agreed to on the floor but removed the entire nutrition title that contained SNAP.[147] House leaders claimed that the defeat was because of "the inclusion of a nutrition policy in the agriculture bill."[148] They were walking a fine line between the Tea Party demands for more cuts and reform (including to farm programs) and the potential political implications of splitting farm and food assistance policies.[149] Chairman Lucas claimed the split was merely a "step towards getting a 5-year farm bill on the books this year" but admitted that he would have preferred to pass the whole bill in June.[150] He added his "personal pledge" that the Ag Committee would follow up with a nutrition-only bill.[151]

Democrats were not placated with vague promises on SNAP and were concerned about what it took to get the sixty-two Republicans to change their votes.[152] They unleashed a furious attack on the decision to remove the nutrition programs and employed the limited tactics available to the minority party. As Democrats saw it, Republicans were enmeshed in divisive, destructive partisan tactics rather than constructive efforts at legislating.[153] The Democrats also did not shy away from highlighting the most troublesome aspects of the move. For example, they argued that Republicans were "taking food out of the mouths of [their] own poor constituents . . . taking food out of the mouths of babies."[154] Finally, they noted the futility of the entire effort because it would not pass the Senate and President Obama would veto it.

The Republican effort was widely opposed and criticized, including by farm interests.[155] The move carried significant risks for the chairman. It sacrificed the traditional bipartisanship behind farm bills in order to appease conservatives who opposed nearly everything in the bill. The split bill had no Democratic backing, and the chairman had lost the support of Representative Peterson.[156] The veteran lawmaker was characteristically blunt: "I don't see a clear path forward from here . . . no assurance from the Republican leadership that passing this bill will allow us to begin conference with the Senate in a timely manner."[157] Democrats lacked the votes, and House leadership muscle carried the day. The House passed the altered bill on a straight party-line vote on July 11, 2013.[158]

Passing a bill without SNAP did not settle things.[159] Majority Leader Cantor followed through with a separate nutrition-only bill that cut SNAP spending further.[160] Political and public pressure on the House increased but had little impact on the effort. The stand-alone bill was controversial.[161] It reduced SNAP by $39 billion and had been written largely by Majority Leader Cantor in consultation with conservatives. It did not go through the Ag Committee markup process, and it had no Democratic input or support.[162] If partisanship could have been made more intense at that point, the Cantor SNAP bill did it.

There was little question that the bill was mostly an effort to appease the Tea Party faction while rescuing House leadership from the June debacle. Rooted in the ideological perspective of the Republican budget blueprints, the bill included the controversial work requirements for benefit recipients.[163] Representative Southerland, the author of the controversial amendment blamed for killing the farm bill in June, reached for the Bible. He stated, "God created Adam and placed him in the garden to work it."[164] Majority Leader Cantor added his views about the "dignity in work" and the bill's contribution to helping "put people on the path to self-sufficiency and independence."[165]

These were echoes from the budget blueprint's call for reforming programs "to ensure that America's safety net does not become a hammock."[166] The SNAP bill was one of the few examples of how the ideological perspectives in the budget were to be put into operation in legislation. It thus provided a visceral demonstration of how controversial budget cutting can be, and how difficult. Lost to supporters of the effort was the troubling distinction between treatment of farm programs and of food assistance.

Democrats were not amused. Representative Marcia Fudge (D-OH) said, "Today's exercise is nothing more than a waste of time and an insult to every American in need." Given that Cantor's bill contained the "same toxic amendments that derailed the farm bill's passage the first time," she added, the "fact that we are considering this legislation makes me question whether the Republican leadership even wants a farm bill to pass."[167] House Democrats were powerless to stop the bill, and it passed without a single one of their votes.[168]

Subsequently, House leadership put the farm bill back together for the long-awaited conference. The procedural maneuvers cut a confusing trail through a partisan thicket. In the course of less than four months, the House had killed the farm bill, split it in two, passed both with Republican-only majorities, and then joined them back together. The last step was seemingly anathema to the very Republicans blamed for much of the bill's struggles. Adding to the confusion, the House moved the farm bill forward to conference at the same time it tore things apart over the budget. Negotiating the budget featured an even larger partisan fight that resulted in a federal government shutdown on October 1, 2013.[169] Farm bill negotiations began during the sixteen-day shutdown but struggled to bridge the partisan and regional policy divides between the House and Senate.[170]

The most notable outcome of the long-awaited conference was SNAP. After nearly destroying the entire farm bill effort in a drive to reduce the program by as much as $40 billion, House negotiators settled for a minor change estimated to save approximately $8 billion.[171] Revisions were largely limited to the same provisions acceptable to the Senate that reduced benefits for some participants but did not change eligibility requirements.[172] A blunt assessment would be that Senate Democrats won on SNAP. That is not to say that SNAP was an easy negotiation, but it was the clearest win in conference; most of the negotiations were consumed by farm policy disagreements.[173] There were six key farm policy issues in conference.

First were the southern priorities for a fixed-price program and the new STAX cotton insurance policy intended to settle the Brazilian WTO dispute. Cotton had demanded and the House Ag Committee had originally included a reference price in the insurance policy. This contributed further to the regional conflict over fixed-price policy. As discussed previously, this was an unworkable feature for an insurance program and unlikely to resolve the dispute with Brazil.[174]

The second issue was whether to make payments using the actual acres planted to the supported crop or to continue to use the base-acre, decoupled payment system that was the key feature of the 1996 farm bill.[175] This issue was also linked to the fixed-price program. Moving to planted acres meant farmers would more likely be making planting decisions on the basis of expected government payments if market prices were low. It would have reversed the most enduring reform for farm programs and one that launched the modern era.

Third, the House bill had eliminated the 1949 act as permanent law. Doing so would jeopardize future farm bills because reversion acted as a motivator for Congress to reauthorize or extend farm programs. The fight over dairy policy at the 2012 fiscal cliff was an important reminder. It was also a repeat of efforts by Republicans in 1995 and 1996, when Democrats in the Senate fought to keep reversion as protection for future efforts after the drastic decoupling reforms of Freedom to Farm. This lesson had been learned, particularly given the price collapse of the late nineties and the return of target prices in 2002.

These three issues were effectively combined. The House might have been using 1949 act and planted acres as leverage for the price program—matters to be traded away in negotiations. Neither provided much trade value, however, because farm interests opposed the House position. Farmers preferred the flexibility of decoupled programs, and the political risks of paying farmers for crops they didn't plant was minor compared to planting for payments. The reform had held through two farm bills, and the 2014 bill preserved this key legacy of the 1996 farm bill. Conference, however, provided farmers the option to reallocate their existing base acres in a manner that might better reflect recent planting decisions but did not permit them to add base acres. Eliminating the 1949 act was not acceptable to any of the farm bill interests and could not have been a real sticking point.[176]

If the intention was to use these issues as leverage and negotiating concessions, it was successful. Conference agreed to include the House's price program, known as price loss coverage (PLC), with benefits that would flow in a noticeably southern direction.[177] But the farm program design fight between the Senate and House and among the Midwest, Great Plains, and South was a draw. Congress could not settle on a specific policy for the commodities and thus handed the decision to farmers. The final bill provided farmers with a onetime election for their base acres between PLC and the agriculture risk coverage (ARC) program that provided revenue-based assistance as designed in the Senate. This left the debate about the future direction for farm policy unsettled (as discussed in the next section).

Fourth, the Senate had reattached conservation compliance to crop insurance and included the AGI restrictions for crop insurance premium assistance. Chairwoman Stabenow prioritized conservation compliance but had worked to defeat the AGI amendment. The AGI amendment was far more controversial to the farm coalition. Conservation compliance was a high priority for the conservation members of the larger coalition. It was easier to accept and likely made for a good trade in return for dropping the AGI provision. Trading AGI for compliance and revising STAX settled most of the crop insurance issues.

Fifth was the dairy program. It had been most controversial in the House. It involved a dispute between dairy producers and the dairy-processing industry. In short, dairy farmers wanted protection against declining margins, or the difference between milk prices and feed costs. This support policy, however, provided production incentives. To address this problem, dairy producer interests added a provision that sought to limit supplies in times of low prices or oversupplied milk. Limiting supplies to increase prices has long been controversial in farm policy; it remained so in 2013 for the dairy program. The processing interests fought it throughout the debate, and they were backed by House Speaker Boehner. Dairy farmer interests were represented by Ranking Member Peterson, and the two appeared to be playing the legislative equivalent of a game of chicken. Failure to resolve the dairy issue could have blocked the entire bill, but the increasingly real potential for reversion to 1949 act policies for dairy forced an agreement.[178]

The sixth and final challenge for negotiators was farm program reform. Reforming farm programs represented the end stage of the regional fight.[179] The total package of reforms included payment limits for the price, revenue, and marketing loan programs. It also included the provision to close a loophole in the payment eligibility requirement.[180] Both the House and Senate had agreed to changes designed to prevent nonfarmers from qualifying for farm payments. This issue also had a long history, dating to the 1987 efforts to close down the circumvention of payment limits through passive investors in farm entities. That reform had included a loophole: a provision that permitted a person or entity to qualify as actively engaged by claiming that he or she was a manager in the farm operation.

The eligibility loophole reform was the most difficult. In conference Senator Cochran, Chairman Lucas, and Representative Peterson all opposed provisions that had passed both the House and the Senate.[181] Removing provisions that had passed both chambers was problematic procedurally because it could be subject to points of order on the floor. It was also politically risky given the strong support for reforms in both chambers. There was no easy compromise, however. The eligibility loophole reform was the final item standing between passage of the farm bill or an unknown future. Negotiators, unable to reach agreement, handed the issue to USDA. Instead of the reform language included in the House and Senate bills, the conference language required the secretary of agriculture to sort it out in rule making.[182] Conference concluded in January 2014.[183]

When the House took up the conference report, Chairman Lucas noted, "This has been a long and seemingly epic journey."[184]

Ranking Member Peterson called the effort "a challenging and, at times, frustrating process.... We did what we have always done ... worked together."[185] Consideration of the bill in the House was shockingly subdued given all that had occurred over the course of more than two years, but it appeared that reality had taken hold and Republicans supported the bill overwhelmingly.[186] Substantial numbers of House Democrats, however, still opposed the bill because of the reductions to SNAP and likely because of the partisan nature of the effort. Opponents argued that the minor revisions "will make hunger worse in America, not better," but the House passed the farm bill on January 29, 2014.[187]

In the Senate, Chairwoman Stabenow proclaimed that this was "not your father's farm bill" because it was "focused on the future ... to make sure that policies worked for every region of the country, for all of the different kinds of agricultural production."[188] She added that the bill contained "the greatest reforms to agricultural programs in decades" because it "ended direct payment subsidies ... [and] shift[ed] to a responsible, risk management approach that only gives farmers assistance when they experience a loss."[189] Senator Chuck Grassley, who had led the payment reform effort, opposed the conference report because the reforms he fought for were not included while the House's fixed-price program was included.[190] The regional dispute smoldered, but the Senate passed the farm bill conference report on February 4, 2014.[191] President Obama signed the farm bill in East Lansing, Michigan, with Chairwoman Stabenow and other senators on February 7, 2014.[192] The president's pen brought an end to a congressional debate that had been initiated in February 2011.

Concluding Thoughts on the Troubled 2014 Farm Bill

Old fights plagued the 2014 farm bill, but the edges were sharper, more partisan. When President Obama signed the Agricultural Act of 2014 into law not a single congressional Republican was present. After a bruising effort, maybe the partisan and regional wounds were too fresh for celebratory reconciliation; perhaps there was little interest in traveling to Michigan in February. The bill had been three difficult years in the making, and the absences were notable regardless of the reason. The farm bill concluded on a very partisan note that only added to questions raised by the process itself.

Because there is always the risk of overlearning the lessons of the most recent policy battles, what follows in this concluding section is an attempt to evaluate the 2014 farm bill through the larger lens of policymaking and history. This section also serves as a bridge between the long history discussed in the preceding chapters and the concluding discussions in chapter 9. Budget discipline has proved to exert immense pressure on legislating coalitions, concentrated at the fault lines. But the history carved in those fault lines determines how the coalition reacts to the pressure and how those reactions shape both the political and policy outcomes. The 2014 farm bill debate provided an excellent case study.

The 2014 farm bill consisted of two different policy and political debates in terms of topics and outcomes: a victory for the Senate on SNAP and a draw between the two chambers (and regional interests) on farm programs. The Senate passed a farm bill twice on the back of large bipartisan votes. The House could barely pass a bill, doing so only in the wake of shocking defeat and on strict party-line votes. Before 2013 only the feed grains–wheat bill in 1962 had been rejected in a similar, dramatic fashion on the House floor. That bill did not include all commodities. More important, it predated the inclusion of food stamps and was not subject to the impacts of budget discipline.

The South and Midwest have fought many battles over farm programs for a long time. Food assistance has been subjected to ideological opposition and worse. Congress has consistently wrestled over the size, scope, and purpose of the federal government through debates defined by ideology and partisanship. The institution of budget disciplines has affected each farm bill since 1981; the 2002 farm bill thus far stands as the only exception. The 2014 farm bill effort combined all into a single, messy undertaking, but the differences between the House and Senate debates hold the keys to understanding these matters and to gleaning more general lessons. The farm bill debate in the House was ideological and partisan, focused on drastically curtailing or ending benefits for one part of the larger coalition. The farm bill debate in the Senate was a competition for the distribution of benefits, limited to the farm coalition. That distinction makes all the difference.

The House provided a difficult political environment for the farm bill, and SNAP consumed almost all the energy in the debate. The debate was partisan and ideological, driven by long-running animosity to the program rooted in many difficult issues. This partisanship might have been the product of institutional tendencies. House members are elected in small districts for only two years, and decisions are made by a simple majority vote; these political dynamics have been exacerbated by gerrymandering. The debate was far more destructive to the larger coalition and seemingly dismissive of the vote power it possessed. These features in the House combined to produce a dysfunctional and counterproductive debate—a lesson in how rigid ideological aspirations are ill-suited to complex legislative realities.

Opposition to SNAP in the House was exclusively partisan fueled by ideological animosity concentrated on the right of the political spectrum—home to the Tea Party faction. The roots of this dynamic in the House debate can be found by examining the party's budget blueprint. For example, the blueprint expressed concern that "a government that promotes dependency and undermines institutions of faith and family will inevitably weaken the nation's greatest strength: the exceptional character of its entrepreneurial, self-reliant, and hard-working citizens."[193] Therefore, providing SNAP benefits could be viewed as requiring taxes that stifle the freedom of one constituency to provide benefits to another, benefits that promote dangerous behaviors such as dependency and weakness. If true, SNAP would represent a rather dire threat to the nation.

Some of this may be mere political rhetoric, campaign themes masquerading as governing principles. To the extent this rhetoric calcifies within the complexities of legislating, however, it becomes an impediment to achieving any outcomes that require compromise. What could be entirely legitimate points for debate about the appropriateness of providing benefits and the way to pay for them becomes stealing from one constituency to give to another. The rhetoric frames the debate as about something other than the policy. Instead, it subsumes the debate in terms of moral failings, weighted with religious connotations and grave implications for the future of the republic. For those members who campaigned on this vision and voted for the budget blueprints, drastic policy responses became the necessary and only outcome; spending was an "evil" that had to be defeated.

This perspective does not leave a lot of room to negotiate legislative solutions with those who do not see the issue in the same dire terms. To them, the dire terms ignore some important realities, such as the participants who do work but still need assistance to purchase food.[194] Worse still, the rhetoric paints the opposition and the beneficiaries of the programs in a very harsh light. This is complicated for assistance to purchase food that encompasses children, the elderly, and those with debilitating conditions. These complications are magnified by troubling historical baggage. From sharecroppers to civil rights to striking workers, attacks on this federal assistance have evolved in a narrow and partisan direction that should counsel caution.

Almost all federal assistance in the farm bill belongs to the same family of New Deal responses to the Great Depression. Opposition to food assistance was initially bipartisan—coming from Republicans and southern Democrats—but was compromised by southern support for farm assistance and the increasing damage from racism and segregation. Opponents shifted to concentrate on organized labor in the seventies, and then President Reagan repackaged opposition to food stamps as a matter of budgetary and spending concerns. Through the seventies and eighties, this opposition continued to be bipartisan. When Republicans reclaimed congressional majorities in 1994, however, most of their political victories came at the expense of southern Democrats. Efforts to drastically reduce food stamps continued to be framed in terms of budgets and spending but were now concentrated in the Republican Party and began to affect farm assistance more directly. Those efforts produced limited change at significant political cost. The House position in 2012 and 2013 recycled these strands of a difficult history but offered little new or different.

A simple explanation is that taking benefits away from constituencies makes for tough politics, especially when the constituencies are large and the politics are burdened by difficult historical baggage. Taking away benefits is generally not conducive to achieving legislative outcomes through compromise and bargaining. It limits the ability to form coalitions. Coalitions bring votes necessary for success. For opponents of food assistance specifically, the coalition narrowed to a single party, concentrated regionally. In a process decided by consensus, these developments operate to create an ideological trap.

A narrow ideological goal to eliminate program benefits establishes a very high bar for success that can be difficult to achieve. The legislative process remains an exercise in counting votes. Taking benefits away creates a motivated bloc in opposition. To succeed requires a larger, more motivated bloc to counter that opposition. The various SNAP efforts failed this test.

Consider that spending associated with SNAP was estimated at roughly $700 billion over ten years and that beneficiaries of the program exceeded 40 million people. House Republicans' aspirational blueprint might have reduced SNAP spending by $127 billion over the ten-year budget window, in part by removing millions of people from the program.[195] The House Ag Committee's various legislative efforts would have reduced SNAP by $33.7 billion or $16.1 billion or $20.5 billion.[196] The highly controversial House SNAP bill in September 2013 would've reduced spending from the program by $39 billion over ten years.[197]

Republicans produced a confusing array of directions that were counterproductive to an outcome, and none of them eliminated the spending entirely. On one side, millions of citizen-constituents lose assistance to purchase food. The other side fought for what could be perceived as half measures. If the program were indeed a threat to the republic, the gulf between that threat and the response was obvious and substantial. Therefore, whatever the perceived benefits of cutting SNAP, they were insufficient to offset the pains, both political and real, of taking benefits away from a large constituency.

The ideological attack on SNAP had other shortcomings. Paeans to limited government and the need to prevent safety nets from becoming hammocks can strike discordant notes in a farm bill debate. The farm bill distributes taxpayer-funded benefits to many constituencies, the most prominent being low-income individuals and farmers. Arguments made for limited government during the farm bill process can appear selective and risk being hypocritical, holding different programs and beneficiaries to different standards.[198] Worse still, they can invite uncomfortable comparisons between beneficiaries and relative benefits—a debate unlikely to favor farm payments.

The Tea Party members were somewhat consistent in that they wanted to end both the farm and food assistance programs; splitting the bill was their strategy for eliminating each in turn. They simply lacked the votes for taking benefits away from both sets of constituents. Functionally, the same majority that voted to cut SNAP had previously voted to continue benefits to farmers. The House farm bill debate became ensnared in this trap and foundered. If there was any political benefit, it served as a distraction from farm policy. As the primary target for the Tea Party faction, SNAP certainly absorbed ideological fire that may have otherwise been directed at farm assistance. If targeting SNAP was part of a negotiation strategy, however, it was risky and ahistorical. Food stamps had to rescue farm programs from rock bottom in 1964. Farm programs simply do not have the vote strength to pass Congress without the broader coalition, which has been apparent since at least 1973.

House Republicans ultimately failed to achieve any of their big-picture ideological aspirations. After the June loss on the floor, a majority did vote to eliminate nearly $40 billion in SNAP spending, but the result was short-lived. It was also meaningless when taken out of context and disconnected from the entire process. The vote on Cantor's SNAP bill came only after an embarrassing loss. The bill itself was unable to survive conference negotiations with the Senate, and the president said he would have vetoed it. Southern demands for farm assistance survived, much to the chagrin of the Tea Party faction. This too recycled history. President Reagan also wanted to end federal assistance to both groups but was limited in doing so by the need for deals with the Boll Weevils and by the farm crisis. The new Republican majority during the Clinton presidency also failed to end farm programs, opting instead to capture baseline in fixed annual payments.

By comparison, the Senate debate was a classic farm bill contest, featuring a more traditional competition among the South, Midwest, and Great Plains—cotton, corn, and wheat—over farm programs and the payments. More regional than partisan, much of the dispute was contained within the farm coalition and the minority Republican Party. Because the Senate effectively avoided the fight over SNAP, the budget pressures came to bear almost exclusively on this primordial fault line. Budget pressure increased tension that had been building, especially between the coalition's traditionally most committed combatants, over divergent fortunes under the RFS and WTO.

Institutional differences provide some explanation. Senators are elected statewide and serve a more diverse constituency for which things like gerrymandering are irrelevant. With six-year terms they are somewhat shielded from the near-constant electoral pressures of the House. This can temper the ideological intensity, although not always. Because the commodities are regionally divided, the commodity disagreements can have more resonance in the Senate. Senate rules are also generally designed to better force consensus through super-majority procedural mechanisms such as the filibuster. Institutional dynamics, however, probably better explain the more politically sober route that avoided a partisan fight over SNAP than the regional contest over commodity programs.

The specific policy reasons for the dispute are more difficult to decipher but can generally be understood as a fight over the distribution of benefits. This is an important factor of the Senate policy debate that differentiates it from that which took place in the House. The changes to farm programs were not eliminating benefits to a group of constituents, although, to some extent, cotton was facing changes that came close to an elimination of benefits. To the extent that the legislation altered the distribution of benefits, it did so on a limited basis. Farm programs possess a much smaller constituency than SNAP does, and this constituency was divided among itself. Finally, the changes in benefits were cloaked in the rubric of reform, which was particularly well suited to the political climate.

Out of context, a disagreement about whether payments are triggered by fixed prices or revenue makes little sense. The long record of this dispute and the environment for reform provide the context. Recall the discussion in the concluding section of chapter 6 regarding the 1996 farm bill reforms. On one end is status quo, and it does not constitute reform; the other end is complete elimination, which might be total reform. In between are various components such as phaseouts and buyouts. Parallel to that continuum of reform is also a continuum of political difficulty; status quo typically is the least difficult and total reform the most difficult because it is taking benefits away from a constituency.

The 2014 farm bill eliminated direct payments entirely, although cotton farmers were able to receive partial direct payments as a transition to STAX. Direct payments had had an eighteen-year run in which farmers received federal payments each year regardless of prices, production, or economic circumstances. For this program, it was total reform. The 2014 farm bill also removed cotton from the list of commodities that could receive income-support payments. This was significant reform but not total because cotton support was not completely eliminated. The bill did not remove marketing assistance loan or other commodity-specific benefits, and it provided a new supplemental crop insurance policy for cotton farmers. These two reforms were the direct result of political pressures: the partisan budget pressure to reduce spending and that created by the WTO dispute with Brazil.

Congress also wrestled with program payment reforms that included limits on benefits, restricting eligibility by closing the loophole for the requirement to be actively engaged in farming and extending these reforms to crop insurance. The result fell short of the amendments agreed to in either the Senate or the House floor debates but arguably made incremental progress in line with the 2002 and 2008 farm bills. This issue has a long history dating to the New Deal but increased intensity in light of direct payments and improved transparency provided by the digital age. It is an issue that also splits along traditional regional fault lines, again with the South and Midwest the most committed adversaries.

Finally, the 2014 farm bill did not eliminate all farm program benefits, providing neither buyout nor phaseout of the support system. Instead, it transferred direct payment baseline into crop insurance and contingent farm programs that triggered payments on relatively lower prices or revenues. To the extent this was reform, it is best understood as a rebalancing of the benefits among the commodity interests that aligned squarely with the history of farm policy; history was a weight on the debate regardless of whether the actual participants fully understood or acknowledged it. In fact, the Senate debate can be understood as having directly recycled the regional dispute over high fixed loan rates (90 percent of parity) and flexible loan rates that dominated the debates of the 1950s.

The 1938 system of flexible loan rates proved most favorable to corn, and a return to some version of that flexible system was part of the reform corn interests, including the Eisenhower administration, sought after World War II. Similarly, the revenue-based program makes use of a more flexible assistance trigger than the fixed-price program: the five-year Olympic moving average benchmark prices and the use of yields compared to a reference price. Within the coalition, shifting policy to a revenue basis was considered more favorable to corn, secure in strong domestic demand, followed by wheat but least favorable to the southern commodities with larger export risk exposure. The southern perspective certainly seemed to be that corn was seeking to consolidate gains under the RFS. Southerners viewed revenue programs as a clever way to boost corn farmers' payments while cutting others, effectively grabbing baseline away from cotton, rice, and peanuts. With cotton on the ropes, midwestern interests may have seen the chance to make real progress on program reforms they had been fighting to achieve for a long time.

The fight over 90 percent of parity was best understood as a more fundamental dispute over acres and the domestic feed market; high loan rates underwrote the acreage allotments that diverted cotton and wheat acres to feed grains. The fight over revenue is better understood as competition over shares of baseline for decoupled benefits but with more fundamental real-world consequences. Where decoupled programs benefit some crops more than others on a per-base-acre basis, they can create a form of cross subsidy made possible by the CBO baseline.[199] In short, the relative support levels have consequences in a decoupled system because farmers are free to plant other commodities on the supported acres.

First, the CBO baseline and scoring system can be used to the benefit of smaller acreage crops because those with fewer base acres have a much smaller impact on the CBO baseline. Increasing fixed-price levels for smaller crops does not score as much as increasing them on crops with large base acreage. The same increases for the larger crops have a much bigger score and far greater impact on the baseline. Under the 2014 farm bill, corn has almost ninety-seven million base acres but peanuts only two million base acres.[200] To illustrate, consider an overly simplified scoring example in which policy changes increase expected payments for both crops by $10 per base acre per year. The cost in the baseline for peanuts would be $20 million per year or $200 million in the ten-year budget window. For corn, however, the cost would be $970 million per year or $9.7 billion in the ten-year budget baseline. In a budget-constrained situation, it is much easier to make policy changes on small acreage crops, and over time this can create a payment system that is inequitable on a per-acre basis. While this inequity might be a political irritant in isolation, it can be more problematic in the context of a decoupled support system because it can operate as a cross subsidy.

Second, under the decoupled system, a farmer is free to plant any covered commodity on the base acres but will receive payments only for the covered commodity that is the source of the base acre. In other words, a farmer can plant soybeans on corn base acres but receive a corn payment on those base acres if the program triggers it. This has regional and commodity significance. Corn, soybeans, and wheat are planted throughout the country, but cotton, rice, and peanuts are isolated regionally in the South. In other words, southern farmers can plant corn, soybeans, or wheat on cotton, rice, or peanut base acres and receive the benefits of cotton, rice, or peanut program payments while planting the same crops as farmers in other regions. Farmers outside the South are not capable of planting cotton, rice, or peanuts. All farmers are competitors in the market for these basic commodities.

Generally speaking, if rice base is worth $100 in the program and corn base is worth $20, it might not be fair but is probably not worth the fight. If, however, farmers with $100 rice base plant corn on their acres, they receive the higher payment while planting the same crop for the same market as the corn farmer receiving $20 per acre and planting corn. Both are planting the same crop, but the federal payments provide much more to one farmer of that crop—a competitive advantage in the marketplace. In a very real sense then, this recycles the acreage diversion fight from the parity era but updated for payments and decoupling.

As can be expected, midwestern farmers do not appreciate cross subsidizing a group of marketplace competitors. For southerners, this cross subsidy may well be necessary to balance out the more favorable conditions available to midwestern farmers. The production environment (weather and soils) tends to be better, yields are generally higher, crop insurance is cheaper, and biofuel production spurred by the RFS is regionally concentrated in the Midwest. Additionally, livestock production that has expanded in the South now competes with biofuels for corn and soybeans.

As designed by the Senate Ag Committee in 2012, the revenue program rebalanced the distribution of benefits among the major commodity interests. All commodities were to receive the same percentage of a benchmark revenue that was calculated using the five-year Olympic moving average of the national average prices received by farmers in the marketplace. Conceptually, these features combined to make the program more market-oriented, equitable, and transparent. In practice, these features are more favorable to crops with better security anchored in the domestic market, such as corn with feed and ethanol demand. Politically, the southern crops benefited more from the 2002 farm bill system, and thus, the market-oriented rebalancing of benefits could be framed as a form of baseline theft. And that was the fight in the Senate in 2012 and 2013, and with the House in conference.

In the Senate the farm bill fight came down to political power in terms of vote counting at the committee and floor levels. That vote counting took place within the prevailing political mood and climate: the WTO dispute, the RFS, high prices, strong farm incomes, economic recession, and budget deficits. Those factors tilted the playing field in favor of reform and significantly limited southern power and influence; southern ability to control this issue had waned, and it showed. This was especially true given the Republican Party's dominance in the South built on an ideology of smaller, limited government and reduced federal spending. Reforms carried the day; from revenue to payment limits and eligibility, they encroached further with changes to crop insurance.

Historically, Presidents Eisenhower, Nixon, and Reagan had been proponents of farm program reform. It was a focus of Republicans after they had taken control of Congress in the 1994 midterm elections. With southern Democrats few in number and with no votes in the Senate Ag Committee, the 2012–14 regional feud was almost completely between southern and midwestern power centers of the Republican caucus. The farm policy fight was no longer midwestern Republicans against southern Democrats, with the divide largely confined to the Democratic caucus. The results of this iteration of the regional farm policy dispute were relatively unprecedented in modern times and align with the somewhat modest midwestern Senate victories during the post–World War II debates. As always, however, the final outcome depended on tough negotiations with the House in conference.

The limited power of southern Senators on farm policy also limited the ability of Senate Republicans to achieve partisan objectives on SNAP. It also increased the pressure on the House to deliver on regional farm policy, which, in turn, further limited that chamber's ability to accomplish partisan SNAP goals. Those limits were somewhat minor compared to the limits the House had caused itself, however. The Tea Party faction was nearly nonexistent in conference and its influence constrained by the institutional nature of the process. While Tea Party members could be disruptive in the House floor debate, they could be less so with a conference product; hard-line ideological opponents lacked the votes to defeat the bill and lacked the influence over the Senate and the president. The House SNAP position was only moderately effective as negotiating leverage for southern commodity program demands. The House had gone so far afield on the program that its negotiators had no realistic way back to a deal, and the situation was ill-suited for extracting major concessions on farm payments. Ultimately, the Senate position won out on SNAP, and the farm policy fight ended in a draw. Where Congress could not decide, it pushed the decision to farmers or, in the case of eligibility, USDA.

The primary lessons for future farm bills can be found in the points discussed previously, especially in the comparison between the House and Senate debates. While there is political power in the push for policy reforms, there is an important difference between taking benefits away from a large constituency and making modest or moderate revisions to the distribution of benefits among a smaller constituency. To put a finer point on it: ideologically rigid and overly partisan demands that harm one group more than they help any other are less likely to succeed; where they leave little room to compromise, they are most likely to fail.

Thus, the outlook for any future farm bill necessarily begins with the political climate for SNAP in the House, the Senate, and the White House. Barring unforeseen political developments that change the vote-counting or budgetary dynamics in Congress, the worse that climate the worse the odds are for a farm bill. For the farm coalition itself, the conventional wisdom is that a fight over SNAP leaves farm programs and crop insurance vastly more vulnerable, likely to the same forces attacking SNAP. For the pending farm bill debate in particular, improved conditions in the national economy have lowered SNAP participation and spending, and both are expected to fall further in future years.[201]

For farm policy and farm bill supporters, the primary lesson is to strengthen the coalition and avoid changes to SNAP that are partisan and ideological. Farm policy does not have the votes to make it through Congress on its own. This has been true since at least 1964, and continued demographic trends only reinforce this reality. The politics of attacking food assistance, beginning with comparisons over the amount of benefits each receives, are bad all around for farm interests. And the coalition with food stamp interests is more than smart politics. Food stamps help people purchase food and benefit farmers; except for cotton and biofuels, food is the ultimate result of all the farm commodity production being supported by farm programs.[202] Drastically changing a program with such widespread benefits and beneficiary-constituents is a difficult political undertaking. It would require offsetting benefits that were at least as equal and widespread. Given the history of SNAP and farm programs, it might require even more.

While the political situation for SNAP provides the clearest and most straightforward outlook for any farm bill in the foreseeable future, the outlook for farm policy, from commodity programs to crop insurance and conservation, is less clear. The historical perspective raises the possibility that the farm coalition debate is entering a recycled round of competition reminiscent of the fifties and possibly even the early sixties. This perspective would place 2014 somewhere in the historical vicinity of 1954 but with fairly significant differences.

A final, important farm issue for the current debate is as old as farm policy itself: crop prices. Having peaked in 2011 and 2012, crop prices fell drastically after the 2014 farm bill was debated and signed into law. This is consistent with history: prices increase and policy changes only to be tested by the subsequent price environment. That is likely to be the case for the immediate farm bill future; low prices are challenging the farm policy revisions in the 2014 farm bill, but the politics for future farm bills are far more uncertain. Adding uncertainty are a few specific issues within farm policy.

The 2014 farm bill furthered a trend that began in 2000: it placed crop insurance as the top priority in farm policy, leaving it with the biggest baseline and adding new policies.[203] The larger baseline makes crop insurance a larger political target in a budget-stressed political environment, and reform measures made real inroads on crop insurance in the 2012 and 2013 Senate floor debates. In addition, the return of relatively lower prices will challenge the trend toward crop insurance. The prices used for crop insurance reset each crop year, and multiple years of relatively lower prices add pressure for farm payment programs that respond to such a scenario. In a budget-constrained debate, this may increase pressure to cut from crop insurance to offset the costs of bolstering farm programs.

For farm programs specifically, relatively lower prices will challenge both the cotton and revenue reforms in the 2014 farm bill.

That bill provided farmers a choice between the House price program and the Senate revenue program with farmers generally breaking along the commodity-regional divide. Midwestern corn and soybean farmers elected the revenue program, southern peanut and rice farmers overwhelmingly elected the price program, and wheat split about evenly between the two.[204] Cotton, however, was removed from the farm payment programs, forced by the WTO dispute to settle for a crop insurance policy that has not been well received by cotton farmers. Cotton interests have demanded a policy response that returns their base acres to the payment programs. Congress acquiesced, adding seed cotton as a new covered commodity on former cotton base acres in the Bipartisan Budget Act of 2018.[205] This unprecedented move to revise farm programs for a single commodity, accomplished outside a farm bill, likely alters the political dynamic for the upcoming farm bill, but the exact implications are unknown as of this writing.

Finally, the conservation title was subject to little controversy in an otherwise controversy-filled effort but is entering the current debate burdened by potentially difficult issues as well. The 2014 farm bill reduced spending in the title, much of it from the CRP through a lowering of the acreage cap. Lower prices are exposing a real divide in that community in the wake of an increasing trend toward working lands conservation. Some conservation interests and some farm interests seek to increase the CRP acreage cap, pulling acres back out of production and into the program. Doing so will come with significant baseline costs, however, and require offsets from other programs. This can be seen as a direct threat to the working lands programs, and it pits them directly against CRP interests in a potential fight over baseline. History has proved time and again that removing acres does little to control production or improve prices. Reserving acres for ten to fifteen years is a rather blunt policy instrument, while shifting baseline might lock away resources needed for pressing issues such as for combatting nutrient losses in the Corn Belt—a regional dimension that could magnify the political dispute.

The challenges for the 2014 farm bill were strongly rooted in a long and often difficult history, and the difficulties for that bill had many wondering about the future—was this a potential end to the farm bill? Farm policy and the farm bill have proved incredibly resilient over the course of eight decades. This resiliency should counsel against jumping to any conclusions about its imminent demise. The 2014 farm bill, anchored in and informed by history, does provide plenty of lessons for future legislative efforts. That debate also offers reason for concern.

The fault lines in the farm bill—the regional ones in the farm coalition and the larger fault line with food assistance interests— have been the most important force in shaping the various policies in the bills over decades. These fault lines are likely to continue to play the key role in any future farm bill debate, driving policy developments for as long as the programs are on the books. While specific outcomes cannot be known in advance, tracing the fault lines and the pressures on them offers the best method for understanding what the process produces. Finally, a concluding perspective is that the manner and method in which the fault lines were exposed in the 2014 debate place the farm bill on arguably the most politically precarious footing since at least the early 1960s.

9

Trying to Reason with the Fault Lines

In Search of Answers and Finding More Questions

This book stems from a quest for answers to questions about farm policy and politics—an attempt to understand better what I was working on, from how programs are designed to what the authors in Congress were seeking to accomplish. Additionally, I wanted a better understanding of the politics that swirl around congressional attempts to rewrite, authorize, and reform the farm bill. As mentioned in the introduction, history can provide valuable lessons, and some of those lessons have future applicability or relevance. As of this writing, Congress is working to complete the process for reauthorizing the programs of the farm bill as it works toward the 2018 sunset dates in the 2014 bill. The history herein may provide insights into that debate and the numerous debates presumably to follow.

Researching this history and the many iterations of farm bills over time raised questions that went beyond farm policy and concerned legislating and policymaking more generally. There is within this history eight decades of insight on the dynamics of legislating. The farm policy story provides ample opportunity to explore matters such as vote counting and coalition building at the various stages of the legislative process. Reviewing and analyzing the developments for the various policies—what drove minor evolutions and major revisions—were instructive about farm policy and politics but also about policymaking itself. Trying to reason with the process led to questions about the U.S. system of self-governance. The history of farm policy notably threads through major milestones in American history and government.

Farm programs sprang from the same New Deal efforts by Franklin Roosevelt to combat the Great Depression that also produced Social Security and other components of the modern administrative state and social safety net. Farm programs have traversed civil rights debates and been caught up in the ideological movements of the Reagan administration and the Gingrich-led Republican takeover of Congress. The policy has been subsumed within debates about the role, size, and scope of the federal government in our society. The Tea Party push to cut SNAP in 2014 was one aspect of a larger conservative effort to redefine the federal role as it was shaped in the New Deal. The history of farm policy is entangled in the history of our national debate about providing for the general welfare of our citizenry. These matters will be addressed in turn in this final chapter.

Fault Lines Carved by Crop Prices

The history of farm policy travels a path etched by crop prices through time; with each bill Congress reacts to ever-changing economic and political conditions. The productive capability of American agriculture has repeatedly demonstrated that farmers can supply more of the bulk commodities than demand can absorb, absent extraordinary demand drivers such as war. For example, Willard W. Cochrane has said, "As I read the historical record of the United States, I see surplus stocks accumulating in agriculture and those stocks pressing against population needs in every peacetime period since the Civil War except 1900–1914."[1] The balance between supply and demand is a challenge because the consumer's food purchases change little in the aggregate and generally do not keep pace with supply.[2] While demand changes little (outside war), supply has expanded rapidly when acres are added (closing of the frontier) or returned to production (the seventies). But supply is also fueled by technological advances rapidly adopted by farmers in their quest for efficiency and competitive advantages. This was most evident in the post–World War II years and has reappeared in recent years with genetic engineering and precision technology. Professor Cochrane memorably framed the supply issue as "a race between population growth and farm technological advance" in which the "average farmer is on a treadmill with respect to technological advance," constantly seeking to reduce costs in a market where the farmer does not control input costs or crop prices.[3]

This history began with depressed crop prices in the twenties, the catalyst for forming the farm bloc in Congress and the initial McNary-Haugen policy responses. The founding interests—corn, cotton, and wheat—were a natural coalition but not monolithic. They represent the main row-crop production regions in the nation and in Congress and continue to occupy leading roles in farm policy debates. Each brings different elements to the coalition; whereas crop prices are the defining issue, each commodity possesses variations on it. The introduction to this book presented the basics of farm risk: risk to yields or production and risk to prices. Farmers sink costs into the soil and depend on the weather to produce a crop. Any crop produced is subject to often-volatile markets for its value or ability to cover costs and provide enough profit to stay in business. In good years farming can be lucrative; in bad years it can be bankrupting.

From hogs to ethanol, corn, anchored by strong domestic consumption and demand, has consistently held the most secure market position among the major commodities. Corn is the largest crop in the United States, which is the world's largest producer of corn. Security from domestic consumption has historically placed corn on the policy vanguard. Corn farmers, interests, and congressional supporters have held the most market-oriented policy perspectives and consistently pushed for payment reforms. Corn was the first to get price-supporting loans and benefitted from the highest loan rates under flexible loan rate calculations. Corn was also traditionally less affected by acreage allotments or marketing quotas. In recent times corn farmers have had the highest adoption of crop insurance and have led the way on revenue-based insurance and farm programs.

By comparison, cotton has always been far more dependent on export markets and world prices, a position exacerbated in modern times by the near-total offshoring of the textile industry. Cotton lacks the security of domestic markets, and cotton farmers are relatively insecure in a world where foreign nations can be highly protective of their own cotton production, generally subsidizing it heavily. It stands to reason that cotton interests, from the fields to the halls of Congress, would demand the strongest policy in response to market risks. They have been the most resistant to changes in programs, a consistent bulwark of traditionalist, protectionist policies. Southern cotton interests fought the hardest and longest for 90 percent of parity loan rates and were the last of the three major commodities to exit the parity system. They resisted Freedom to Farm and decoupling until the politics and benefits were overwhelming; they led the return of target prices. Cotton has also been the strongest opponent of farm payment reforms. In fact, the acreage reduction policies that caused so much trouble in the parity era are the only issues that cotton's market situation does not adequately explain.

Wheat has resided between these two polar opposites, historically splitting its production between domestic and export markets. It lacks the domestic security of corn and the world market insecurity of cotton, but wheat farmers are subject to both. There are strong foreign competitors in wheat, and it is a fundamentally important crop for human food needs. Farm policy started in wheat country and expanded wheat acres have underwritten many farm economic problems. The closing of the frontier took place in the Great Plains, and much of that ground was planted to wheat for World War I. The market collapse was felt in wheat first and most forcefully. Wheat acres were also at the heart of the Dust Bowl calamity. Wheat was slower to exit the parity system than corn, but its shifts affected the politics enough to weaken cotton's positions in Congress. The Soviet wheat deal in the seventies helped drive export market enthusiasm in the Nixon administration. Wheat led with target price policy to establish a federal partner for the risks of expanded production, and the consequences of target prices were once again strongest in wheat country. In recent years wheat farmers have adopted crop insurance at a pace slower than corn farmers but faster than cotton farmers, and wheat's leaders from North Dakota and Montana were proponents of revenue-based farm programs.

In purely political terms, these market dynamics form the primordial fault line in the farm coalition and for farm policy, with corn and cotton the most committed partisans. The fault lines are sharpest and deepest between corn and cotton; disagreements are more intense and tend to drive much of the policy's development. Wheat is the swing voter in the farm coalition but has traditionally been more aligned with cotton. When wheat drifts away from cotton's traditionalist position, it tips the balance to corn, but wheat and corn have rarely been strong allies. Politically, wheat is in a complicated position, not only because it is the swing voter in the coalition but because it is largely split along northern and southern regional lines in the Great Plains. Wheat is burdened by the Dust Bowl legacy in the southern plains and a strong populist political streak, particularly from Nebraska north. The climate of the Great Plains is less conducive to farming than that of the Corn Belt; wheat yields have far higher variability and far more weather-based vulnerability.

In summary, these are the basic political dynamics of the farm coalition understood as fault lines that run throughout its history and that remain relevant today. The coalition is strongest in the depths of crisis; it is weakest when afforded the luxury of internal competition and conflict. Crop prices and market risks are the essential elements, but they are rarely exclusive. Far more pressures come to bear on the coalition and the policies; development happens in response to the combined weight as it is managed by the individual legislators and the times in which they work.

Concluding Perspective on Farm Policy Development along the Fault Lines

Stepping back from the hand-to-hand legislative and political combat of each bill, an evolutionary pattern emerges. The largest movements in market prices produce the most drastic changes in policy. At the pattern's most basic, Congress has the most room to revise policy when crop prices are increasing; evolutionary processes accelerate but may push changes too far, too fast. Increasing crop prices are inevitably followed by decreasing prices, if not an outright collapse. What generally follows is largely a corrective evolutionary process that adapts the new policies to the updated situation. The adaptation can appear to be a reversion but rarely is, although it may contain reversionary elements.

Another way to look at it is to grasp that policymaking is a long game played out over many iterations. Policy and political victories are not permanent; initial gains can be consolidated and built on, but they are subject to intense competition that may reverse any gains. Sometimes the larger the win the more tenuous it is because it can lead to an equally powerful reaction. Often the most lasting gains and those easiest to build on are broad-based victories that benefit all partners in a coalition. The most difficult and precarious situations take place under the intense pressure of a zero-sum game in which one partner's gains come at the expense of other partners. Arguably, policymaking is at its best when making incremental changes over time as new information and circumstances, informed by a deep well of history and experience, are absorbed. Farm bill history provides plenty of examples.

The Great Depression was the biggest and most consequential event. It was singular in how it upended the existing status quo. Congress had been developing farm policy for more than a decade, making slow and steady progress on the substance of legislation. The Great Depression interrupted the evolutionary process and propelled it forward with the 1933 and 1936 emergency responses to Depression and Dust Bowl catastrophes. The Depression pushed policy into much more extreme or radical territory; the new policy was unlike the evolving concepts that had preceded it, all of which had been unable to alter the status quo. The domestic allotment plan attacked low-price and oversupply problems at the field level. It involved heavy land-use planning and limits on what a farmer could plant based on what USDA estimated the market needed. A Supreme Court decision and the Dust Bowl continued this policy under the guise of conservation. These bills changed forever the evolutionary path for farm policy. They established a new status quo that all changes and evolutions would have to work against.

The 1938 act took the multiple emergency responses and codified them into a single complex, regimented scheme. It sought to control supplies through acreage allotments, limiting what could be planted to individual commodities—a policy with many fatal flaws that became painfully obvious. Price-supporting loans were the principle benefit to the farmer and an enticement to comply with controls. Loan rates calculated on a flexible scale determined by supply were an attempt to dampen the loans' production incentive. In practice the loan policy had two problems. The first was the ability to forfeit oversupplied commodities in times of low prices, putting crops in government storage. World War II delayed exposure of this problem, but the postwar technological revolution brought it about quickly and forcefully. The second problem was that the flexible schedule produced different loan rates for the major commodities. Viewed through the fault lines, the program benefitted corn farmers the most and cotton farmers the least.

Because domestic demand from livestock helped keep corn out of oversupply territory, it received the highest loan rates and was less affected by acreage allotments or reductions. Corn farmers were also less affected by the marketing quotas; program operation was much less intrusive for a market not typically oversupplied. By contrast, chronically oversupplied cotton and wheat—especially cotton—received the lowest loan rates. Moreover, cotton and wheat had the strictest acreage allotments. For cotton especially, this was by demand and design with potentially darker purposes. Wheat farmers also had their acres consistently reduced, and both crops were frequently subjected to marketing quotas. The 1938 act deepened the fault lines and codified the differences within the coalition. When farm policy emerged from the depression and dust of the thirties, cotton and wheat fought against a return to that policy—a return pushed by corn.

Seeking to correct the loan-rate equity problem, Congress pushed all loan rates to 90 percent of parity. The problem was that Congress increased loan rates and eased the heavy regimentation of the 1938 act under the auspices of war and wartime demand. Higher prices returned stronger farm incomes, while 90 percent of parity loan rates provided an equitable (across all commodities) and comfortable federal backstop. Codified in the 1949 act, high loan rates established a status quo that would prove difficult to revise even as it became painfully misguided policy in the postwar economy with significant consequences. What followed 1949 was a long and difficult corrective process.

A Republican Congress backed down from 90 percent of parity in 1954 and suffered in the election. The new Congress, prodded by President Eisenhower, attempted to ease acreage allotment problems with the Soil Bank in 1956. The 1958 act permitted corn farmers to vote themselves out of the parity system. All were hard-fought policy changes, requiring as much political pressure as the Eisenhower administration and Corn Belt could muster against the southern stronghold. They were assisted by the complete breakdown of the parity system. This process continued into the 1960s and through 1970; Congress made incremental progress. The 1961, 1962, and 1965 acts pushed further into compensatory payment policy, especially for corn, including the payment-in-kind system. Wheat and then cotton experimented with versions of the two-price system, struggling to find an effective alternative to parity. By 1970 Congress had taken up the Nixon administration's set-aside policy to replace the acreage allotment system in the last important evolutionary step before another major disruption.

The early seventies price spike was a substantial disruption to the evolutionary process. It produced drastic policy changes and a reset of the status quo. Congress demoted price-support policy beneath the new target price policy and deficiency payments. Target prices were designed not to increase market prices but rather to offset low prices in the farmer's income. Target price policy was coupled with a concerted push by the Nixon administration to get farmers to expand production and capture export markets in Secretary Butz's "promised land." These drastic changes in policy added fuel to the high price environment and went too far. Farmers expanded and took on debt and then slid into economic crisis and another process of correction.

Farm policy evolved on a corrective path through the 1977, 1981, 1985, and 1990 farm bills. Congress chased inflation with target prices, but the crisis worsened. The Reagan administration fought for reform with ideological opposition that was a magnified version of that which had driven efforts by Presidents Eisenhower and Nixon. It faltered on the realities of economic crisis but managed to advance corrections to the 1973 status quo. Target prices, coupled to production, incentivized farmers perched on the edge of financial ruin to produce into oversupplied, depressed markets.

Congressional farm interests negotiated against the Reagan reform pressures to increase target prices, resulting in slow changes that checked increases in 1981 and then halted them in 1985.

Federal outlays were exaggerated by the state of crop prices and the farm economy, but Congress was correcting the policy. It was slowly moving away from production incentives and the challenges of acreage expansion in the seventies through the PIK and set-aside policies. By 1985 Congress had frozen base acreage and reinstituted the CRP to retire unproductive and environmentally sensitive lands. It also added conservation compliance, combined with CRP, to initiate modern conservation policy.

Additionally, Congress was working toward making the policy more market oriented, beginning with the loan program. The 1985 farm bill introduced the marketing loan feature, which permitted repayment at a lower rate when prices were below the loan rates. This helped avoid the forfeiture problems that had long-plagued the loan policy. It also permitted loan deficiency payments in lieu of an actual loan, pushing this program in the direction of income support and away from price support. Finally, Congress provided export subsidies and foreign market development in order to improve the demand side of the equation.

The 1985 farm bill recycled previous policies but did not repeat them, demonstrating the evolutionary nature of policy development. The marketing loan feature was reminiscent of Eisenhower's efforts to push Congress off 90 percent of parity, most notably in its use of a moving Olympic average price calculation—market-based flexibility in the setting of the loan rate. Similarly, CRP, set-asides, and compliance recycled elements of the Soil Bank. Export subsidies and foreign market development were rooted in the 1954 act known as Food for Peace (or P.L. 480), which sought to use surplus for geopolitical purposes.

Evolutionary corrections advanced further in 1990 with the advent of triple base and extension of the marketing loan features to all commodities. Triple base moved policy further in the direction of acreage decoupling and dampening the production incentive. This process was interrupted by strong crop prices and political upheaval with the Gingrich-led Republican revolution in the 1994 midterms. From that upheaval and spiking prices, Congress leaped far ahead in the corrections to 1973, fully decoupling farm commodity assistance from both acres and market prices. The 1996 farm bill once again reset the status quo.

Adding market price decoupling to acreage decoupling in 1996 can be viewed as too drastic, an interruption of the evolutionary process. This view is supported by subsequent history; when prices inevitably collapsed, congressional reaction was almost as equally drastic in the opposite direction. That reaction came in the form of market loss assistance payments in the late 1990s and codified in 2002 with the return of target prices. Here then was another multilayered, complex assistance scheme: countercyclical payments, decoupled from plantings, when prices were below target prices; annual direct payments decoupled from prices and plantings; marketing loans coupled to production as the final backstop when prices were below fixed loan rates.

From this perspective of history, the 2008 and 2014 farm bills appear to be parts of yet another process of correction from the combined interruption of 1996 and overreaction of 2002. The corrective process is being driven at least in part by the RFS, which has driven up prices, rendering target prices and loan rates effectively irrelevant to corn, soybeans, and to a lesser extent, wheat. Introducing revenue-based commodity policy may well constitute the early evolutionary steps because this policy combines both price and yields in the commodity support calculations and the price calculation uses a rolling five-year Olympic average of market prices. This makes commodity support more equitable across commodities, transparent, and market oriented, but it also risks recycling the commodity feuds previously around flexible or fixed parity loan rates.

Whether this evolutionary correction continues or is disrupted by relatively lower prices remains to be seen. One takeaway from this history might be to counsel caution with respect to making more sudden changes in policy in reaction to 2014 and the lower price environment. The process of evolutionary correction might make for better policy in the long run if it's allowed to run its course. That cotton was removed from the commodity programs in 2014 only to be returned in 2018 outside a farm bill may complicate matters further.

Evaluating the Policies Carved by the Fault Lines

Paradox permeates farm policy. Cursed by abundance and stuck on a technological treadmill, farmers are traditionally independent minded and ideologically conservative despite the shelter provided by federal assistance. Federal assistance has proved remarkably resilient, stretching across more than eight decades of American history and nearly two dozen legislative efforts. It has, however, produced an almost unbroken chain of failures, lurching from disasters to surplus to boom to crisis and back again to boom times. An ever-shrinking base of constituents compete for federal benefits that are blamed for decreasing their numbers, increasing environmental problems, and harming farmers around the world. The programs have passionate advocates and critics, a political base that some view as inordinately powerful but that is so insecure it has long required a broad-based coalition reaching far from the farm fields. Setting aside much of the heated rhetoric of the world of political advocacy, this history can be used to evaluate what has and has not worked, while offering lessons for future debates.

The historical record is fairly clear as to what has not worked. The parity system of price-support loans and production controls failed. The reasons underlying this system's failure are important. Price-support loans are production incentives, and the potential for commodity forfeiture creates substantial liabilities. Failure was clearly demonstrated by the surplus problems of the fifties and sixties as well as by the current peanut program.[4]

Furthermore, acreage-based production controls are ineffective and often counterproductive. Bulk commodity supplies cannot be controlled by limiting the number of acres planted, whether through allotments or set-asides. At least there do not appear to be reasonable acreage reductions that could be politically acceptable from farm areas to consumer interests. Total production is little affected by such policies. It often continues to increase because farmers tend to remove the least productive acres and increase intensity on the remaining acres, underwritten by federal assistance. Farmers still need to derive income from the land. They will divert acres to other commodities and spread surplus problems. Reducing American acres only encourages farmers around the world to replace those acres, especially if reducing U.S. acres lifts world prices. This has been especially true for export crops like cotton.

Parity policy was built from crisis and the need for emergency responses; many of its problems came about once the emergency faded, but the policy didn't adjust. More important, the idea of production control was something borrowed from industry and proved ill-suited for farming. A manufacturing company can reduce production, lay off workers, and take other similar measures when demand drops to preserve its bottom line and satiate the demands of shareholders. It is much more difficult to similarly adjust farm production, given the unknowns of weather and the realities of thousands of independent actors in competition with each other, all trying to maximize profits on individual operations. Whatever merits production control may have had in an emergency and ignoring the impact on those considered surplus labor, those merits dissipated after the emergency faded. High, fixed loan rates that encouraged production and forfeitures only made matters worse.

Income-support payments such as target price policies can also create production incentives if they are coupled to production decisions and if target prices are set too far above market prices. In times of low prices in an oversupplied market, the farmer needing to derive income will logically plant the crop that will generate the best return, a decision that necessarily factors in the federal payments. In the aggregate this incentive increases production and further decreases prices, which, in turn, increases federal payments. The cycle can be self-feeding, increasing federal outlays and harming markets while also potentially driving farmers to produce on land that is environmentally sensitive or marginally less productive. Coupled target prices contributed to the problems of the seventies and eighties.

Farm programs are also blamed for reducing the number of farmers and driving concentration, but the record is less than conclusive.[5] Technological advances and tough competition, particularly in times of economic or financial stress (or crisis), are likely bigger factors. Policies have contributed, however, most notably in the South when policies to reduce acres drove small sharecroppers off the land and out of cotton farming. Federal assistance also went into helping farmers adopt technology to replace labor and consolidate lands while acreage allotments became capitalized in farm lands and concentrated in larger operations. Recent iterations of farm payments have also contributed by underwriting technological adoption and consolidation, while being capitalized in base acres and land rents. It remains difficult, however, to separate the impact of federal assistance from the larger economic, technological, and sociological forces.

Failures and problems in a troubled history are not the complete picture. There have been relative successes from which can be gleaned a better understanding of policy development for future application. The most successful policies have been crop insurance and conservation. Acreage decoupling has also proved to be a relatively successful development. Potentially, aspects of revenue-based assistance, such as rolling average calculations, may constitute development in a positive direction.

Since the Agriculture Risk Protection Act (ARPA) of 2000, crop insurance has helped farmers better manage both price and yield risk. It has grown in terms of crops, acres, and liability, as well as popularity with farmers. Inclusion of price risk in the form of revenue policies coupled with increased premium subsidy has contributed substantially to the program's growth. Other key aspects of the program include the close relevance to actual farm risk and the requirement for actuarial soundness. Insurance covers the farm's production using actual production histories and futures prices. Farmers pay a portion of the premium, the cost of which depends on the insurance ratings for their fields and the level of insurance purchased. Farmers can purchase insurance coverage at levels ranging from 50 percent to 85 percent, with a range of options regarding fields and prices.[6] Farmers can tailor their insurance according to their assessment of risk needs and costs. A farmer receives an indemnity only in the case of an approved loss, and the coverage level determines how much of a loss the farmer bears before the indemnity (i.e., deductible). Ultimately, farmers pay for the risk coverage, which may influence decision making and risk taking to some degree, at least in comparison to payment programs.

Crop insurance is limited to those risks within a single crop year. The farmer purchases policies each year, and those policies include updated yields as well as the actual futures prices for that year. In that way crop insurance does not provide assistance for multiple years of low prices, a feature that has continued the demand for farm programs. Getting premium subsidies and ratings correct is also a big challenge, and some sectors are less satisfied with the program. The South, for example, has been the most reluctant. Farmers pay a portion of the premium; some areas and crops have higher rated costs. The program also provides a relatively high level of premium support. Nationally, almost 62 percent of farmer premiums are covered by the federal government.[7] Premium subsidies account for the bulk of federal outlays for crop insurance. As participation and prices increase, some argue that it has become too expensive. This has focused criticism and opposition on the program in recent debates, something that is expected to continue into the foreseeable future.

Since 1985 conservation policies have provided real successes on the ground and for the taxpayer dollar. The 1985 deal to include compliance and CRP has proved successful in reducing erosion.[8] Assessments by USDA have found that the voluntary, incentive-based conservation approaches in the farm bill are achieving results and have made significant progress for natural resource issues such as sediment, nutrient, and pesticide losses.[9] The CRP program has removed millions of environmentally sensitive acres from intensive row-crop farming, which has helped reduce soil erosion and reestablish wildlife habitat.[10] The advent of working lands programs has also arguably helped make conservation more relevant to row-crop production, providing cost-share and other assistance for practices that are a part of farming.

Designing effective conservation policies can be difficult, and budget disciplines provide challenges to appropriate funding levels. Conservation policy, historically ineffective in controlling production to improve prices, has removed acres from production that arguably should not be farmed because of environmental sensitivity. But acreage reserve or retirement programs are a very limited, blunt tool; they can be inadequate for addressing the large variety of farming's natural resource challenges. Notably, Congress has increased or decreased program acreage coinciding with low or high crop prices (respectively). This back-and-forth reflects a difficult reality for land retirement programs: pressure to reduce acres in the program when prices and, presumably, land rents are strong but reverse pressure to return acres to the programs when they are weak.[11] This also exposes one of the conflicts inherent in the policy between earning a return from the land and protecting natural resources.

These challenges are part of the reason for the growth of working lands conservation policies. Notably, these policies took off with decoupling in 1996 and creation of EQIP, followed by the complex, complicated CSP created in 2002. Unlike the back-and-forth of reserve programs, Congress has consistently innovated with working lands programs and expanded them in terms of funding and acres. This is most evident in CSP, to which Congress has repeatedly emphasized adding millions of acres each year. Innovation is also evident in the creation in 2014 of the Regional Conservation Partnership Program (RCPP), which not only consolidated various conservation policies into a single authorization but also sought coordination across multiple programs and on a regional scale. Congress explained that it wanted to push innovation and advance efforts to integrate practices and approaches across multiple programs on a regional scale.[12]

The problem with conservation policy is the scale, scope, and complexity of the challenges compared to the relative paucity of funding. Crop insurance and farm programs cover far more acres. Conservation also suffers from bureaucratic challenges, including a lack of flexibility for the farmer. Conservation risks being a burden to farmers, particularly when prices are low and incomes challenged. Conservation policies may well "operate on the outdated premises that conservation and farming are mutually exclusive and that cropland that is not explicitly identified for conservation will not protect natural resources or ecological services."[13] Compliance provisions also indicate a one-sided focus on how to make commodity programs or crop insurance more conservation oriented. They can be too rigid given the nature of farm production and risk.

Reviewing the growth and expansion of working lands policies in recent farm bills coupled with reductions in the reserve programs gives the appearance of a trend in policy preferences toward working lands. Conservation operates through individual programs on individual farms, often to address isolated practices. Congress recently has looked at a coordinated, regional approach across multiple farms, which may well reflect some frustration that the existing patchwork of programs, policies, and practices is not achieving a desired level of conservation.[14] A contrary view might be that this patchwork reflects policy preferences when prices are trending upward and that it could be reversed with lower prices.

Acreage decoupling has proved successful in giving the farmer flexibility to make market-based planting decisions. It has also dampened the production incentives of the payment programs. Decoupling's issues have been mostly political. The optics of farmers getting payments for crops they do not plant can be a problem, although it hasn't proved to be a big one. Acreage decoupling has also created the odd competition for baseline distribution. All in all, acreage decoupling has been more beneficial than detrimental to both farmers and the federal taxpayer, which qualifies it as a success in the history of farm policy.

Finally, there is the question of revenue-based payment policies such as the ARC program in the 2014 farm bill. The policy incorporates yields, which deals with the negative correlation between yields and prices (to some degree).[15] Arguably more important is that it also uses moving averages of actual yields and prices to trigger payments. This permits the program to adjust over time and not fall too far out of line with market realities, while providing some assistance to farmers during the adjustment period. If use of the rolling average price calculations proves successful, it might be able to fill in where the 1996 price decoupling failed. It will continue to incorporate price risk by adjusting over time.

As farmers work their way through a run of relatively lower crop prices on the back of high yields, it remains to be seen how they will view revenue and moving average policies when the program adjusts. Farmers will need to see corresponding adjustments to input costs (fertilizer, seed, and fuel) to match, so as to avoid the cost-price squeeze. Most important are adjustments in land rents. They will not be easy adjustments, not the least because they may be affected by the view that farmers are receiving federal assistance.

While not technically farm policy, the RFS has also been a key a factor. It has succeeded in creating demand for commodities unheard of outside war. The RFS is not in the jurisdiction of the Ag Committees nor USDA but has created a new demand for bulk commodities, mostly a boon for corn, soybeans, and midwestern farmers. This demand is artificial, coming from a federal mandate, which might well create its own problematic consequences. The demand the RFS has generated, however, has pushed crop prices to record levels. At this point it has been far more successful as commodity policy than as energy policy. It remains to be seen whether these successes will be maintained or whether adjustments in the markets lead to a new round of problems for farmers.

From the perspective of farm policy history, the RFS can be viewed as in line with the abundance viewpoint that emerged from post–World War II planning at USDA. The abundance view was based on conclusions and lessons from war; farm price and income problems were more of a demand issue than oversupply. Previous abundance policy concepts included domestic and foreign food assistance, but they were insufficient to overcome production increases from the technological revolution. Today, acres have expanded under the RFS while good crop years have produced more than even the artificially enhanced market can absorb, so prices have fallen. The age-old challenge remains: balancing supply and demand on acres that expand easier than they adjust with production driven by rapid adoption of technology.

Food for Thought from the Fault Lines

While the focus of this book and research has been on farm policy and the fault lines within the legislating coalition, the study also brought to the surface many questions about the American system of self-governance, legislating, and policymaking. Farm policy has traveled a long path through a notable swath of American history and government, particularly as it was reconstituted by the New Deal. While an adequate treatment of these matters is beyond the scope of this book, they are relevant to farm bills and farm bills may be relevant to them. Somewhere between ignoring them completely and delving into them fully is the following concluding discussion; acknowledged to be a cursory, inadequate treatment, it is included nonetheless to stir thinking and debate but also to add context to farm bill history.

Legislating is an exercise by a very large, diverse group to reach often difficult, complex decisions concerning significant issues or problems in society.[16] This work would be next to impossible if legislators did not adhere to a formal process for discussing and debating issues; agnostic as to the outcome, this process channels the effort toward a decision point within the realm of politics.[17] The process is often chaotic and choked with competition among interested groups and legislators.[18] It also creates real challenges for new policies and for changes to existing policies because they must overcome significant hurdles in order to become law. This, in turn, should temper and moderate proposals through a deliberative undertaking that serves a "larger instrumental purpose of improving public policy."[19]

Additionally, we govern ourselves through a system of proxies. We vote for officials to represent us in Congress and the administration. They serve as our proxies in a system that channels disagreements and demands that are ultimately decided by votes. Interest groups and other institutional actors serve as additional proxies that seek to influence debates and outcomes. The system itself works through proxies: committees, leadership, and caucuses in the legislative process. The bottom line, however, remains voting—a clear, transparent (and relatively peaceful) method for setting disputes.

The ultimate arbiter of the decision in this process is voting because it provides representatives an equal voice as to amendments and the final legislative products of the debate. Counting votes to reach a decision permits disagreement by providing a straightforward, transparent, and equitable method for settling it; the view with a majority of votes wins. It does not stifle disagreement but rather productively channels it into the debate, rendering a clear verdict transparently tallied in votes.[20] This, in turn, permits those who disagree with the majority's decision a clear measure of where they stand and what is needed to change the outcome. Moreover, opponents have multiple steps in the process to work toward changes, along with the opportunity to push for the election of more like-minded representatives. In short, the "tacit theory of the authority of legislation is a sense that discussion and validation by a large assembly of representatives is indispensable to the recognition of a general measure of principle or policy as law."[21] Legislation that becomes law therefore "deserves respect because of the achievement it represents in the circumstances of politics—action-in-concert in the face of disagreement."[22]

The long history of farm bills provides excellent case studies as discussed throughout this book and summarized herein. Farm bills have repeatedly demonstrated the process in action. Writing legislation is concentrated at the committee level, which allows for specialization and development of expertise among members of the committee.[23] The committee system also magnifies the influence of interest groups in the process, which often raises concerns.[24] It is important to consider, however, that interest groups are generally considered more successful at blocking legislation than at passing it and are "more influential when the issue has low public visibility and their objectives are narrow and technical."[25]

A key component of the legislative process remains that the committee's product must still be accepted at multiple stages that represent different constituencies.[26] Representatives in the House represent the smallest, most homogeneous constituency; senators represent entire states; and the president the entire nation. Thus, interest group influence may be moderated or limited—if not checked—by the need to pass the bill on the House and Senate floor, where the views of all voters and citizens are presumed included and counted. The process is neither perfect nor capable of producing perfect outcomes, but it does provide valuable deliberation that includes the ability to change the legislation at each stage.[27]

The dynamics within the legislative process are also behind the need for coalitions and logrolling. Coalition building is an essential ingredient of policymaking owing to the need to accumulate enough votes to pass a bill.[28] Rare is the interest or issue that has the political power and vote strength to survive the legislative gauntlet on its own. Therefore, the process acts as a forcing mechanism to bring together disparate interests. It forces negotiation and, ultimately, compromise to achieve success.[29] In this way, the process also serves as an ongoing system for checking and balancing interests, forcing them not only to make deals but also to justify their interests and legislative demands.[30] No interest gets all that it wants, nor does it ever achieve its policy goals on its own. At some point every interest has to convince enough others to support it; the more support it can gather the stronger it is.

Building a coalition, however, is only part of the effort because coalitions must be maintained and sometimes expanded—both of which create the fault lines that have been the focal point of this history. While the coalition is a source of strength, its weakest points are found at the fault lines, and stresses focused on those lines can pose significant risk to the coalition and the legislation. These pressures can involve competition among the coalition members over the resources and outcomes provided by the policies. Competition can create conflict about how the policy should work and for whom, especially when one interest overreaches on behalf of itself or when other members of the coalition view the outcomes as inequitable. Pressures on a coalition's fault lines can also be external, forcing all the interests first to come together and then to react, defend, make changes, or adapt to survive.

The farm bills of the Eisenhower and Kennedy administrations provide strong examples. Failure of the parity system was evident in the growing stock of surplus commodities in federal storage and the increasingly intense conflict within the farm commodity coalition. President Eisenhower and Secretary Benson provided external pressure for reform. In 1954 farm interests succeeded in both committees but were unable to sustain those victories on the House and Senate floors of a Republican Congress aligned with the popular president. A compromise amendment in both chambers garnered sufficient votes to force an initial step toward flexibility; it survived conference, but Democrats were restored to power in Congress and southerners in the Ag Committees.

Internal pressures from acreage diversion weakened and damaged the coalition. In 1956 the House sought to maintain 90 percent of parity, but that policy lost on the Senate floor. Southern Democrats restored it in conference, but the bill was vetoed by President Eisenhower. The veto, sustained in Congress and coupled with administrative actions, forced supporters of the policy to concede defeat. A subsequent veto in 1958 forced further compromise that permitted corn to vote itself out of the parity system. By 1962 support for acreage policy had narrowed to southern interests that pushed the coalition and their preferred policy too far in committee. With civil rights matters also working against them, southerners sacrificed votes they could not afford to lose. A coalition of interests opposed to them formed and defeated the bill on the House floor. Disagreement, channeled through the process, allowed internal and external pressures to combine to bring about incremental changes in policy.

Other examples appear in the historical record. Reforming farm program payments has proved more politically powerful on the House and Senate floors than in committee, especially in the age of decoupled payments. By comparison, specific policy changes to farm programs are generally near impossible to achieve on the floor; witness midwestern efforts to revise the price programs in the House and Senate in 2013. Payment limits and reforms are much easier for the nonspecialists of the two chambers to understand than are the esoteric program specifics. In fact, the potential for problems on the floor have often led the Ag Committees to take proactive and defensive measures on issues such as payment reforms and conservation.

The rise of food stamps featured the competing interests on behalf of consumers, particularly low-income constituents of an increasing number of members of Congress from urban districts. Southerners on the House Ag Committee initially resisted food stamp legislation, but changing demographics upended traditional representation and vote counts in Congress. As food stamp votes strengthened, so did their ability to push changes on their priority. This culminated in the logrolling efforts in 1964 for wheat-cotton policy and the Food Stamp Act. As their strength grew in Congress, consumer and food stamp interests forced the farm coalition to settle internal disputes and expand the coalition, ultimately combining farm policy and food stamps in 1973.

Congressional reforms in 1974 raised new legislation and policymaking matters. Budget procedures, especially with the revised usage of budget reconciliation beginning in 1981, altered the process. Farm bills bear witness to the overriding importance of the budget process on legislation.[31] It created a powerful external pressure on the farm bill coalition and has forced significant policy changes that might not have occurred in the normal process, most notably in 1996 and 2014. The CBO scoring procedure, especially when combined with acreage decoupling from 1996, has resulted in a zero-sum negotiation for farm bills. Budget reconciliation has further empowered this element of process, elevating it to a primary role in policy development. Triple base in 1990 and decoupling in 1996 stand as prominent examples.

Intentionally or not, the addition of budget discipline and procedures to the legislative process has fundamentally altered it. Republicans in the Reagan era turned budget reconciliation into a powerful procedural and political weapon. Instead of the agnostic process discussed previously, budget disciplining of legislative outcomes may have folded substantive outcomes into the legislative process, privileging one view (budget) above others.[32] Arguably this budget discipline elevates traditionally more conservative principles about federal spending and may restrict disagreement. It may also upend policymaking dynamics and result in odd outcomes. For example, budget discipline appears to disadvantage small benefits for large constituencies much more than large benefits for small constituencies. Reconciliation also specifically limits the ability of minority interests in the Senate to block legislation or force super-majority procedural requirements of the filibuster. These aspects of budget reconciliation, in turn, may be feeding the partisan polarization trends in Congress. They are certainly increasing conflict within coalitions, such as those necessary to pass a farm bill. Budget disciplines force prioritization, which pushes partisans to target priorities they disagree with while protecting their own priorities.[33] Notably, the brunt of Republican-led budget reconciliation efforts in 1981 and again in 1995–96 was felt by food stamps and not farm programs.

The recent farm bill effort (2011–14) was disrupted by a threat of budget cutting that pushed committee leaders to make reductions in estimated expenditures. Eliminating direct payments under budget pressure unleashed the regional dispute over farm programs. Largely confined to the Republican Party, the dispute between midwesterners and southerners lacked some of the more partisan edges that were present for the SNAP debates, particularly in the House. Process features did moderate the partisan, ideological demands for drastic changes unacceptable to Democrats, much of the Senate, and the president.

By comparison Tea Party interests in the House went too far and in such spectacular fashion that they eviscerated their own negotiating leverage when the bill went to conference. The obvious result of such an uncompromising position for all parties interested in the bill was damage to everyone's priorities but the Tea Party's. The Tea Party's demands in the House effectively left only two options in conference: the Senate's position on SNAP or no bill. And it highlighted the most fundamental reality of legislating: that no single interest or faction can get everything it wants and that compromise is necessary.[34] At the same time, it does raise further questions about how folding in substantive budget outcomes has altered the process and policy.

Again, the many iterations over eighty years gives farm policy its value to these questions. The farm bills have brought about evolution in policies, programs, and coalitions corresponding to changes in society and in Congress, learning from experience and adapting to changing landscapes. What began as a very heavy-handed intervention in the agricultural sector to force adjustment has incrementally evolved to be less market distorting with a decreasing level of interference. This history cautions against drastic changes and unreasonable, ideologically driven demands.[35] It favors tempering or moderating proposals that improve policies in a deliberative manner.[36]

Farm bills also expose questions about legislating and America's experiment in self-governance.[37] The New Deal, of which farm policy was a key component, began the rise of the modern administrative state in response to the emergencies of the Great Depression. At the heart of partisan disputes over the legacies of the New Deal is the debate over the proper role and scope of the federal government in American society—nothing less than the very parameters of how we govern ourselves as a nation.

Critics and opponents of the modern system appear fixated on spending as a proxy for the proper role and scope of government. Republicans have particularly focused on spending in terms of taxation and damage to the general welfare; expenditures raise concerns over the level of taxes necessary to pay for it all.[38] The narrow lens of spending, itself, produces winners and losers who are themselves constituents, voters, and members of interest groups. Thus, a core question is whether the constitutional charge to provide for the general welfare includes public policy responses to balance out the winners and losers.

Agriculture has been a leading sector on these issues as society has moved from agrarian to industrial and service economies. Before the Great Depression, many farm interests viewed industry as using federal policies—mostly tariff policies that increased the cost to the farmer and limited market opportunities—to its benefit and the farmer's detriment. Agriculture had to adjust to modern national and world economies; in fact, adjustment was the key word in New Deal farm policy. This policy had as its centerpiece the concept of federal production controls, an attempt to mimic industry's ability to fit supply to demand. People left farming in droves, often migrating to cities and suburbs where their general welfare needs also changed. They were replaced by technological advances on consolidated lands, trends that World War II greatly accelerated and magnified. Farm policy was ill-suited for the postwar world and failed, but it was not eliminated. Instead, it was permitted an incremental evolution with great difficulty punctuated by drastic and abrupt changes connected to brief periods of high prices.

While not unique to cotton, there is something stark about the thread within this history of cotton labor in the South that could make it particularly instructive. At the time of the New Deal, cotton production was considered "woefully backward" in terms of its near-total reliance on hand labor instead of mechanization and technology.[39] It was neither reasonable nor fair to expect cotton production to remain so. It had to modernize to keep pace with other sectors within agriculture and the economy at large. The historical baggage made problematic the methods used to force modernization, namely, federal acreage reduction policies—problems compounded by the Depression and the Jim Crow South.

What should have been done? This lingering question may be particularly relevant given current debates about jobs, blue-collar workers, issues of class, and the impacts on our national politics and voting.[40] These debates feature prominent criticism of technology, trade, globalization or offshoring jobs, and growing inequality and immigration. They carry echoes of adjustment, modernization, and race. Farm policy was permitted a slow, incremental, and difficult evolution. It also had to be rescued by food stamps in a changing society and Congress. Logrolling the two issues made sense because food stamps are the most obvious and direct link between agriculture and low-income citizens who are struggling viscerally in a changing, modernizing economy. Whether this coalition holds bigger lessons reaching beyond farm bills is a matter for further exploration.

Farmers have certainly benefitted immensely from this coalition, not only in terms of federal benefits but also in terms of their continuing opportunities to correct policies that have failed or fallen out of political favor. Yet as we debate the proper role of the federal government, there is notable cognitive dissonance on the two issues that may highlight how a focus on federal outlays skews the debate.

How should federal policy address matters involving our most fundamental human need for food? Is it proper for federal tax dollars to be spent helping farmers manage the risks of farming? Society benefits from a safe, affordable, and abundant supply of food; we appear to prefer that it be produced by something akin to the family farm and, increasingly, in a manner that accounts for natural resource concerns. Moreover, should tax dollars also help low-income families partake in the fruits of federally assisted farming? Why raise concerns that one form of safety net might become a hammock but not the other?[41] Is it because outlays for one are around $70 billion per year and the other around $17 billion per year, or is it something else? A focus on federal outlays may permit partisans to ignore more fundamental questions about policy, a troubling issue considering the abundant historical baggage.

From out of these multiplying questions, the striking comments about New Deal farm policy by Secretary Wallace appear to echo loudly. He called the emergency actions taken under the 1933 AAA "a shocking commentary on our civilization"; these actions were "not acts of idealism in any sane society" but "made necessary by the almost insane lack of world statesmanship during the period 1920 to 1932."[42] Critics of the federal government's modern role may do well to contemplate those words and the way concentrating on size, scope, and spending might miss or ignore the lessons contained in them. These are challenging matters at the very heart of self-government and the constitutional charge to provide for the general welfare, matters to be sorted out in deliberative processes and political debates. They are, however, beyond the reach of this book.

Concluding Thoughts

As previously noted, the 2014 farm bill is scheduled to expire in 2018. The effort to reauthorize the programs and policies contained in its omnibus legislative language is already underway. Uncertainty and unpredictability are pronounced in the current political environment.

If history holds, current price trends would indicate a largely status quo farm bill with some incremental changes to specific programs. But much depends on SNAP and the viability of the coalition with nonfarm interests. There remains little indication that farm policy can pass Congress without SNAP, but will this deter the most ideologically rigid partisans? Can partisanship and ideology be set aside in favor of seeking incremental reforms and revisions?

If future farm policy discussions seek to learn the lessons from history, they would start with the value of incremental change over drastic revisions (1996) or counterproductive reversions (2002). Such a debate might look for minor revisions that incorporate insurance concepts, conservation concepts, and market-oriented features such as rolling averages. It would continue acreage decoupling and improve transparency and equity among the commodities. In that way the 2014 farm bill would reinforce longer-term trends in market orientation and risk management aligned with crop insurance.

Similarly, future directions with respect to conservation may continue recent policy trends in the direction of working lands over the more traditional retirement or reserve policies. The return of relatively lower prices appears to be calling that trend into question, however, as the current push is for increases in CRP that are likely to be offset out of working lands programs. This would be in keeping with historical responses to episodes of lower prices. It seems out of line, however, with the longer trends in what farmers need and society is demanding. Water quality and drought, coupled with consumer interest in sustainably produced foods, all point to the need for more working lands assistance, barring a return on those matters to the farmer from the marketplace. How this blends with farm programs and crop insurance will be important and likely an increasingly dominate part of the farm policy debate.

In conclusion, the history contained in this book should be important for informing the thinking and planning of future iterations of farm bills and policies. The lessons provided by history are relatively straightforward to understand but difficult to implement. The development of farm policy has been an evolutionary process because changing entitlement policies such as those in the farm bill favors a long game of incremental revisions, with significant effort to hold on to prior progress. Drastic changes generally require a perfect political storm of forces both within the coalition and outside it. Yet much rests on the vital importance of building and maintaining a functioning coalition, without which there is no chance for farm policy or a farm bill because there are simply not enough votes in Congress. For the foreseeable future, maintaining coalitions will be challenged by the predominant focus on spending constraints. Politics will always play a role, whether the politics of delivering for a constituency or of demanding reforms to existing policies. What the fault lines deliver next for farm policy only time will tell.

APPENDIX 1

Graphs and Charts

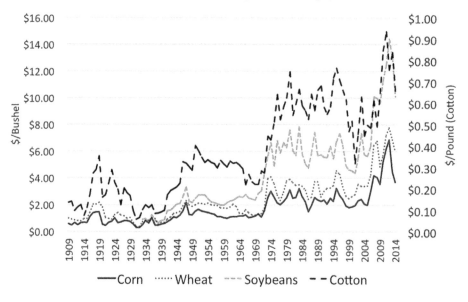

Fig. 1. Prices received by farmers (1909–2016). Compiled by author from U.S. Dept. of Agriculture, National Agricultural Statistics Service, Quick Stats, https:// quickstats.nass.usda.gov/ (select crops and data).

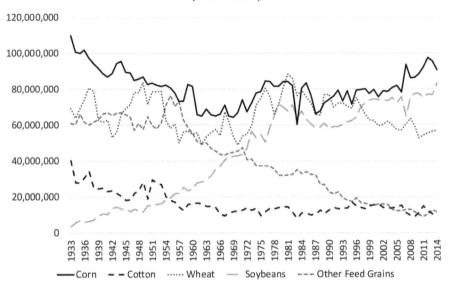

Fig. 2. Acres planted to major commodities (1933–2016). Compiled by author from U.S. Dept. of Agriculture, National Agricultural Statistics Service, Quick Stats, https://quickstats.nass.usda.gov/ (select crops and data).

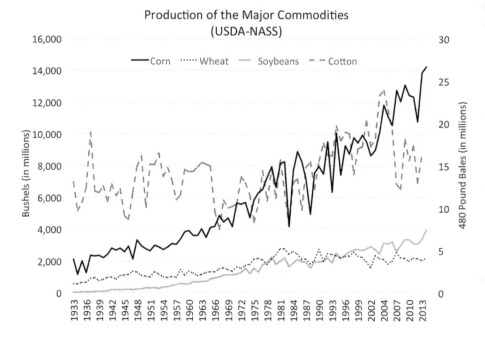

Fig. 3. Production of the major commodities (1933–2016). Compiled by author from U.S. Dept. of Agriculture, National Agricultural Statistics Service, Quick Stats, https://quickstats.nass.usda.gov/ (select crops and data).

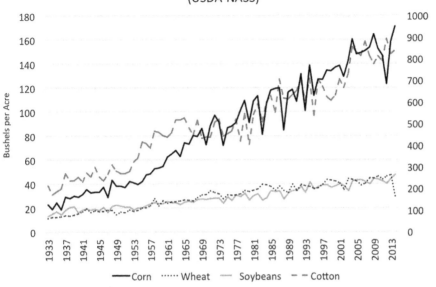

Fig. 4. National average yields of major commodities (1933–2016). Compiled by author from U.S. Dept. of Agriculture, National Agricultural Statistics Service, Quick Stats, https://quickstats.nass.usda.gov/ (select crops and data).

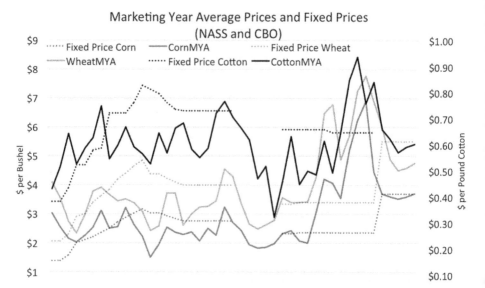

Fig. 5. Marketing year average prices and fixed prices (1974–2018). Compiled by author from U.S. Dept. of Agriculture, National Agricultural Statistics Service, Quick Stats, https://quickstats.nass.usda.gov/ (select crops and data) (1974–2016); projected prices (2017–18) Congressional Budget Office, "CBO's June 2017 Baseline for Farm Programs," (June 29, 2017), https://www.cbo.gov/sites/default/files /recurringdata/51317-2017-06-usda.pdf.

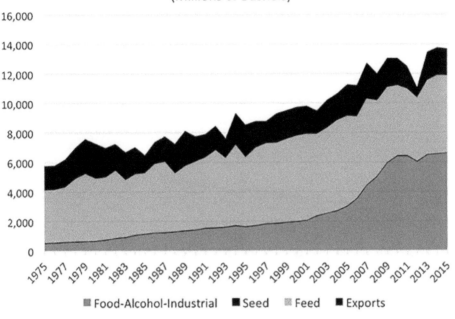

Uses of Corn (USDA-ERS)
(Millions of Bushels)

■ Food-Alcohol-Industrial ■ Seed ▨ Feed ■ Exports

Fig. 6. Uses of corn (1975–2016). The thin black line below feed indicates seed. Compiled by author from U.S. Dept. of Agriculture, Economic Research Service, Data Products, "Feed Grains: Yearbook Tables, Feed Grains Data-All Years" (last updated, November 15, 2017), https://www.ers.usda.gov/webdocs/DataFiles/50048 /Feed%20grains%20yearbook%20tables-All%20years.xls?v=43054.

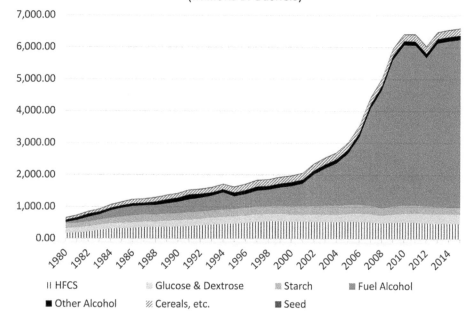

Food, Alcohol and Industrial Uses of Corn
(Millions of Bushels)

Legend: II HFCS | Glucose & Dextrose | Starch | Fuel Alcohol | Other Alcohol | Cereals, etc. | Seed

Fig. 7. Food, alcohol, and industrial uses of corn (1980–2016). Compiled by author from U.S. Dept. of Agriculture, Economic Research Service, Data Products, "Feed Grains: Yearbook Tables, Feed Grains Data-All Years," last updated November 15, 2017, https://www.ers.usda.gov/webdocs/DataFiles/50048/Feed%20grains %20yearbook%20tables-All%20years.xls?v=43054.

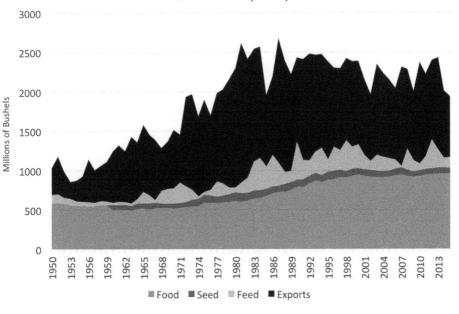

Fig. 8. Uses of wheat (1950–2016). Compiled by author from U.S. Dept. of Agriculture, Economic Research Service, Data Products, "Wheat Data-All Years," last updated November 13, 2017, https://www.ers.usda.gov/webdocs/DataFiles /54282/Wheat%20data-All%20years.xls?v=43052.

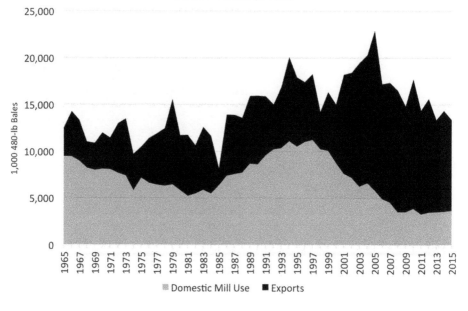

Fig. 9. Uses of cotton (1965–2016). Compiled by author from U.S. Dept. of Agriculture, Economic Research Service, Data Products, "Cotton and Wool Yearbook: U.S. Cotton Supply and Demand, U.S. Supply and Use Estimates and State Production Estimates" last updated November 20, 2017, https://www.ers .usda.gov/webdocs/DataFiles/48516/U.S.CottonSupplyandDemand.xlsx?v=43059.

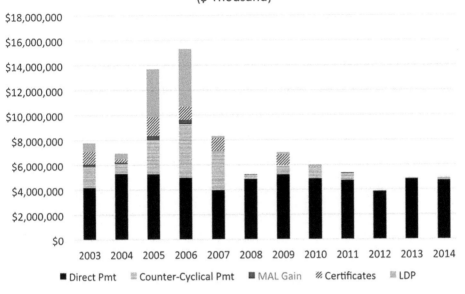

Fig. 10. Commodity program spending (2003–14). Compiled by author from the following sources: For 2003–11: U.S. Dept. of Agriculture, Farm Service Agency, Budget and Performance Management, Budget, ccc Budget Essentials (multiple tables), http://www.fsa.usda.gov/about-fsa/budget-and-performance -management/budget/ccc-budget-essentials/index. For 2012–15: U.S. Dept. of Agriculture, Office of Budget and Program Analysis, "usda Budget Congressional Justifications" (various tables), http://www.obpa.usda.gov/explan _notes.html.

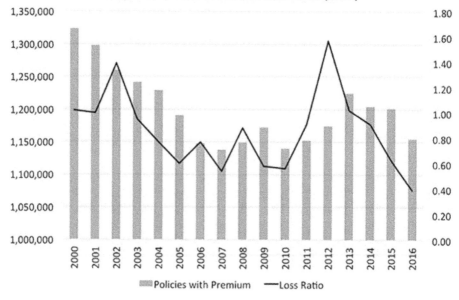

Fig. 11. Crop insurance: policies and loss ratio (2000–2016). Compiled by author from U.S. Dept. of Agriculture, Risk Management Agency, Information Browser, "Summary of Business Reports and Application" (multiple reports), https://www .rma.usda.gov/data/sob.html.

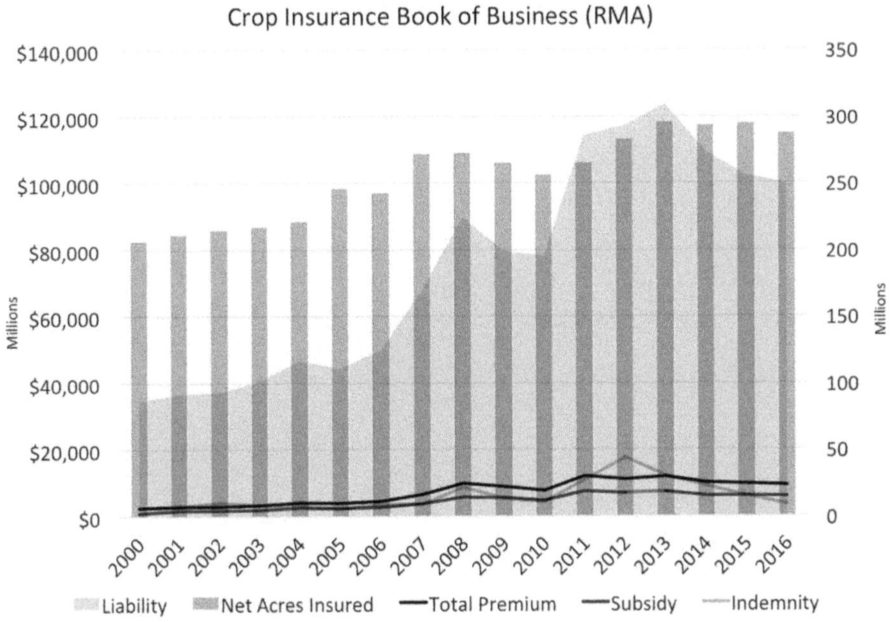

Crop Insurance Book of Business (RMA)

Liability | Net Acres Insured | Total Premium | Subsidy | Indemnity

Fig. 12. Crop insurance: book of business (2000–2016). Compiled by author from U.S. Dept. of Agriculture, Risk Management Agency, Information Browser, "Summary of Business Reports and Application" (multiple reports), https://www .rma.usda.gov/data/sob.html.

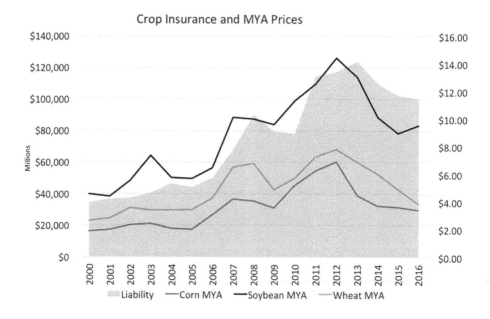

Fig. 13. Crop insurance liability and MYA prices (2000–2016). Compiled by author from U.S. Dept. of Agriculture, Risk Management Agency, Information Browser, "Summary of Business Reports and Application" (multiple reports), https://www .rma.usda.gov/data/sob.html.

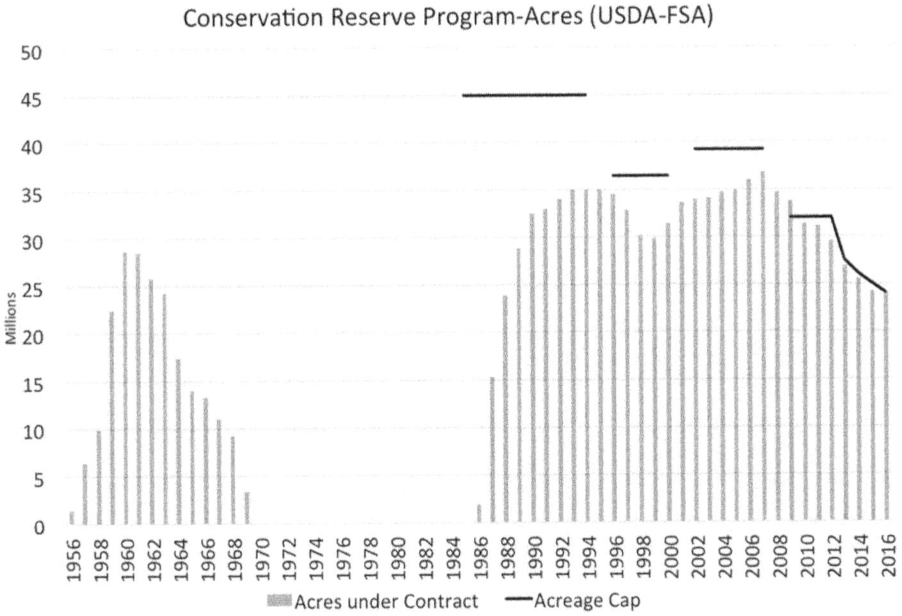

Fig. 14. Conservation Reserve Program acres (1956–2016). Compiled by author from the following: 1956–73: (Soil Bank information), Cochrane and Ryan (1976); 1986–2016: U.S. Dept. of Agriculture, Farm Service Agency, Programs and Services, Conservation Programs, Reports and Statistics, "Conservation Reserve Program Statistics, CRP Ending Enrollment by Fiscal Year, 1986–2016," https:// www.fsa.usda.gov/programs-and-services/conservation-programs/reports-and -statistics/conservation-reserve-program-statistics/index.

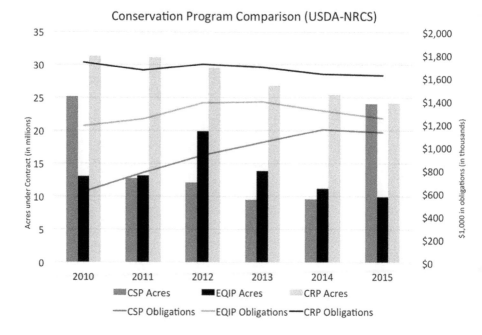

Fig. 15. Conservation program comparison (2010–15). Compiled by author from U.S. Dept. of Agriculture, Natural Resources Conservation Service, RCA Report, Interactive Data Viewer, Program Reports, http://www.nrcs.usda.gov/wps/portal /nrcs/rca/national/technical/nra/rca/ida/.

APPENDIX 2

Bills and Terms

Table 1. Farm Bill Summary

Year	Farm bills	Summary
1933	Agricultural Adjustment Act	Broad emergency authority; increases in purchasing power
1936	Soil Conservation and Domestic Allotment Act	Limits on acres; increases in purchasing power; reduction of soil erosion by planting soil-conserving crops
1938	Agricultural Adjustment Act	Variable loan rates depending on supply; acreage allotments and marketing quotas; crop insurance
1942	Steagall Amendment	90% of parity
1948	Agricultural Act	90% of parity, switching to flexible loan rate schedule
1949	Agricultural Act	90% of parity, then 80–90%, then flexible but at discretion of secretary
1954	Agricultural Act	82.5–90% of parity, then to sliding 75–90%; cross compliance on allotments for conservation payments
1956	Agricultural Act	Flexible loan rates; Soil Bank
1958	Agricultural Act	Mostly cotton and corn (no wheat); corn referendum to move to 90% of 3-year moving average (passed 1959).
1961	Emergency Feed Grains Act	Payments-in-kind out of CCC stocks in return for reduced acres planted to feed grains
1962	Food and Agricultural Act	Mostly corn and wheat (not cotton); elimination of corn allotments; voluntary acreage reductions; fixed loan rate and PIK; wheat referendum on mandatory controls (rejected 1963), compensatory/PIK
1964	Agricultural Act	Wheat and cotton; two-price wheat program; increased cotton price support; addition of ginning subsidy (vote paired with Food Stamp Act in House)
1965	Food and Agricultural Act	All commodities; feed grains with loans and PIK; cotton at 90% of world market price, plus payments; two-price wheat program; Soil Bank (extended to 1970 in 1968)
1970	Agricultural Act	End of production controls; set-aside acres; compensatory payments to supplement loans; $55,000 payment limit

1973	Agriculture and Consumer Protection Act	Target price policy; lowered and fixed loan rates (cotton at 90% world price); $20,000 payment limit; combined with food stamps.
1977	Food and Agriculture Act	Set-aside acres with cross compliance; farmer-owned reserve; increases in target prices; removal of purchase requirement for food stamps
1981	Agricultural and Food Act	Increases in target prices each year; one-year food stamp extension
1985	Food Security Act	Target prices frozen (slight decrease at end); 5-year Olympic average loan rates; cotton and rice marketing loans; set-aside policy; export subsidies; Conservation Reserve Program (45 million) and conservation compliance; food stamp extension
1990	Food, Agriculture, Conservation and Trade (FACT) Act	Triple base; marketing loans for all commodities; CRP (45 million), addition of wetlands easements and compliance; food stamps continued
1996	Federal Agriculture Improvement and Reform (FAIR) Act	Decoupling (acres and prices) with 7-year fixed (AMTA) contract payments; elimination of set-asides; continued marketing loans; addition of EQIP; lower CRP cap (36.4 million); separation of compliance and crop insurance; 2-year extension of food stamps.
2000	Agriculture Risk Protection Act	Increases in crop insurance premium subsidy; premium subsidy for revenue policies; Noninsured Crop Assistance Program (NAP); disaster assistance
2002	Farm Security and Rural Investment	Direct payments; countercyclical (target price) payments; marketing assistance loans / LDPs; addition of CSP; increase in CRP cap (39.2 million); increases in food stamps
2008	Food, Conservation, and Energy Act	Continuation of direct payments, countercyclical payments, marketing assistance loans / LDPs; addition of average crop revenue election (ACRE) program and supplemental disaster assistance; payment reforms (direct attribution); revision of CSP; lowering of CRP cap (32 million); increases in food stamps, renaming of program (SNAP); inclusion of crop insurance

| 2014 | Agricultural Act | Eliminates direct payments; choice between price (PLC) and revenue (ARC) insurance coverage; continuation of marketing assistance loans / LDPs; removal of cotton from ARC/PLC (generic base); lowering of CRP cap (24 million); combines easements (ACEP); revision of CSP; RCPP; reductions in SNAP (heat and eat, $8 billion); crop insurance increased SCO and STAX (cotton) |

Created by the author.

Table 2. Terminology

Term	Definition	Example
COMMODITY PROGRAMS		
Price support loan	Price support policy attempts to increase commodity prices in the market; nonrecourse loans provide a floor on prices; farmer takes out a loan on harvested crop at a predetermined loan rate; if prices are above the loan rate at repayment, the farmer repays the loan (plus interest); if prices are below it, the farmer forfeits the loan commodity and keeps the loan funds; federal government acts as the buyer of last resort at the loan rate.	Loan rate = $2.00/bushel Commodity = 10,000 bushels Loan funds = $20,000 If prices at repayment = $2.10, farmer repays loan If prices at repayment = $1.90, farmer forfeits commodity and keeps loan funds
Two-price policy	Developed initially by McNary-Haugin legislation; splits production into domestic uses and foreign uses; domestic uses are supported at a higher price (loan rate) and export at a lower level to match world prices.	50% of wheat presumed exported and 50% presumed used domestically; farmer producer 20,000 bushels. 10,000 bushels supported at $3.00 bushel loan rate; miller has to purchase certificate at this price to purchase grain. 10,000 bushels supported at world market price, $2.00 per bushel. Total support is $2.50 per bushel.
Payment in kind (PIK)	Payments to farmers in actual bushels held by USDA-CCC in storage or for certificates on those bushels; used to entice farmers to reduce production of the commodity.	Farmer with 100-bushel-per-acre yield on 100 acres for 10,000 bushels of expected production. A 10% reduction in acres planted would be 90 acres and 9,000 bushels. Farmer paid 1,000 bushels from storage (actual or certificates); able to sell on the market.
Income support	Direct subsidy payments to supplement farmer income; entitlement benefit.	See below.

Direct payments	Decoupled payments using a fixed rate and historic yield on base acres for a covered commodity.	Corn: $0.28 per bushel on program yield of 100 bushels per acre; direct payment rate is $28 per base acre; typically paid on 85% of base acres.
Fixed-price policy (also known as target prices or reference prices)	Congress fixes target price in statute, and if average market prices are below it, the farmer receives a deficiency payment on the difference but sells the crop in the market (no forfeiture); examples include countercyclical payments and price loss coverage (PLC).	Target price = $2.00 Market average price = $1.90 Deficiency payment = $0.10/ bushel
Revenue policy	Triggers payments by comparing actual crop year prices and yields to a historical benchmark of prices and yields; payments are generally set against a guarantee that is a percentage of the historical benchmark.	Historical yield = 100 Historical price = $2.00 Benchmark revenue = $200 Actual yield = 100 Actual price = $1.90 Actual revenue = $190 Payment = $10/acre
Olympic moving average	Calculation that drops each of the highest and lowest years in the average (typically five years) and averages the remaining.	~~2009 $3.55~~ 2010 $5.18 2011 $6.22 ~~2012 $6.89~~ 2013 $4.46 Olympic Average = $5.29

CROP INSURANCE

Yield policy	Insurance coverage based on a percentage of the farmer's actual production history (APH), a 10-year average of records for the crop; coverage can be purchased in 5% increments from 50% to 85%; indemnity payments in case of a loss will incorporate market prices.	APH = 100 bushels Coverage = 75% Actual yield = 65 bushels (65%) Market price at harvest = $1.00 Indemnity = 10 bushels × $1.00 = $10 per acre

Revenue policy	Insurance coverage based on a percentage of the farmer's actual crop revenues calculated as the APH multiplied by futures prices for the crop in the spring (or at harvest); coverages can be purchased in 5% increments from 50% to 85%.	APH = 100 Spring price = $2.00 Revenue = $200 Coverage = 80%; $160 Actual yield = 110 Actual price = $1.30 Actual revenue = $143 Indemnity = $17/acre
Premium subsidy	The portion of a farmer's crop insurance premium that is covered by the Federal Crop Insurance Corporation within USDA; percentage depends on coverage level.	Coverage level = 85% Premium subsidy = 38% Rated insurance premium = $30/acre Farmer pays = $18.6/acre
Actuarial soundness	A statutory requirement for the crop insurance program to be operated in terms of the loss ratio for the entire program; premiums paid into the program or risk pool (including the federal share of the premium but not delivery costs) should equal indemnities paid out; measured as a loss ratio of 1.0.	Loss ratio 1.0 = $1 of indemnities matched by $1 of premium Loss ratio 0.5 = $0.50 of indemnities for $1 in premium (underwriting gains) Loss ratio 1.5 = $1.50 indemnity for $1 in premium (underwriting losses)

CONSERVATION POLICIES

Conservation Reserve Program (CRP)	Traditional retirement or reserve policy; federal rental payments on a per acre basis under contract to remove land from production and place it under conserving uses (permanent cover such as grasses or trees); 10-to-15-year contracts; cost-share for establishment of conservation cover.	80 acre field meeting qualifications $50 per acre county average yield $4,000 annual rental payment for 10 to 15 years
Wetlands easements	Permanent (or 30-year) property right in a portion of farm field for restoring and maintaining a functioning wetland; federal government pays for cost of the property right and cost-shares establishment and maintenance	

Grasslands easements	Permanent (or 30-year) property right in field historically grassland to reestablish and maintain as a grassland; grazing permitted; federal government pays cost of property right and cost-shares reestablishment of grass cover
Farmland protection easements	Permanent (or 30-year) easement in land that limits its use to farming purposes, preventing it from being developed into homes or subdivisions; federal government pays for cost of the property right
Environmental Quality Incentives Program (EQIP)	Cost-share assistance for installing or maintaining specific conservation practices on a farm; 60% of total funding reserved for livestock operations
Conservation Stewardship Program (CSP)	Five-year contract for annual payments to farmers meeting threshold conservation on their farms and agreeing to increase conservation on their farms during the contract term; per-acre payments based on environmental benefits, costs, income foregone, etc.

Created by the author.

NOTES

A note on *Congressional Record* sources: Wherever possible, I have cited page numbers from the bound or permanent edition (available from ProQuest Congressional and the Government Printing Office). For some years, I cite the page numbers from the daily edition (which begin with either *H* or *S*, indicating House or Senate).

Introduction

1. This phrase is a nod to one of the earlier works on farm policy. See Tweeten, *Foundations*, 60. The other reference is to Woody Guthrie, "This Land Is Your Land." See http://www.woodyguthrie.org/Lyrics/This_Land.htm.

2. See, e.g., Cochrane, *Curse*, chapter 7.

3. See, e.g., Pollan, *Omnivore's Dilemma*, 10; Michael Pollan, "The Food Movement, Rising," *New York Review of Books*, June 10, 2010, http://www.nybooks.com/articles/2010/06/10/food-movement-rising/; Berry, "Pleasures of Eating," 321.

4. See, e.g., Goodwin and Smith, "2014 Farm Bill"; Babcock, "Welfare Effects"; Marion Nestle, "The Farm Bill Drove Me Insane," *Politico*, March 17, 2016, http://www.politico.com/agenda/story/2016/03/farm-bill-congress-usda-food-policy-000070.

5. The quote is attributed to William Shakespeare. See http://www.bartleby.com/73/1296.html.

6. Santayana, *Life of Reason*.

7. See Graham Allison and Niall Ferguson, "Why the U.S. President Needs a Council of Historians," *The Atlantic*, September 2016, http://www.theatlantic.com/magazine/archive/2016/09/dont-know-much-about-history/492746/; Fredrik Logevall and Kenneth Osgood, "Why Did We Stop Teaching Political History?," *New York Times*, August 29, 2016, http://www.nytimes.com/2016/08/29/opinion/why-did-we-stop-teaching-political-history.html?_r=0.

8. See, e.g., Hunter, "Reason Is Too Large," 1202.

9. Sumner, Alston, and Glauber, "Evolution," 406; Levins and Cochrane, "Treadmill," 550–53.

10. Statistics provided by the U.S. Department of Agriculture's National Agricultural Statistics Service (NASS), Quick Stats, last modified January 29, 2018, http://www.nass.usda.gov/Quick_Stats/.

11. Gary Schnitkey, "Revised 2015 Corn and Soybean Return Estimates," *farmdoc daily* (5):169, Dept. of Agricultural and Consumer Economics, University of Illinois at Urbana-Champaign, September 15, 2015, http://farmdocdaily .illinois.edu/2015/09/revised-2015-corn-and-soybean-return-estimates.html; Gary Schnitkey, "Downward Pressures on 2016 and 2017 Cash Rents," *farmdoc daily* (6):56, Dept. of Agricultural and Consumer Economics, University of Illinois at Urbana-Champaign, March 22, 2016, http://farmdocdaily.illinois .edu/2016/03/downward-pressures-on-2016-2017-cash-rents.html.

12. Agricultural Act of 2014, P.L. No. 113–333 (2014).

13. See, generally, Lawrence, "Profiles in Negotiation"; Hamilton, "2014 Farm Bill," 1ff.

14. Jonathan Coppess and Todd Kuethe, "Mapping the Farm Bill: The Traditional Farm Coalition and Current Production," *farmdoc daily* (6):152, Dept. of Agricultural and Consumer Economics, University of Illinois at Urbana-Champaign, August 11, 2016, http://farmdocdaily.illinois.edu/2016/08 /mapping-the-farm-bill-traditional-farm-coalition-current-production.html.

15. Among those commodities being given less attention, dairy stands out because it has played a difficult but integral role in many farm bills, including the 2014 effort. Sugar and peanuts are also players in farm policy but are given less focus for reasons that will hopefully become evident. For this effort, they will be supporting actors in the larger unfolding discussion, and to those who produce those commodities and for those who represent their interests, I apologize for the less than full attention.

1. The Origins of Farm Policy

1. Davis, "Development of Agriculture Policy," 326; Breimyer, "Agricultural Philosophies," 342.

2. Kirkendall, "New Deal and Agriculture," 83 (attributing quote to George Peek).

3. Hurt, *Problems of Plenty*, 12.

4. Hurt, *Problems of Plenty*, 12; Benedict, *Farm Policies*, chapter 7.

5. Egan, *Worst Hard Time*, 58 (writing that homesteaders broke out nearly twenty million acres of native prairie sod for new farmland by 1925; they plowed another five million acres between 1925 and 1930).

6. Benedict, *Farm Policies*, 153–56, 171. See, generally, Campbell, *Farm Bureau*.

7. Kile, *Farm Bureau*, 24.

8. Benedict, *Farm Policies*, 119, 152; Kile, *Farm Bureau*, 27.

9. Kile, *Farm Bureau*, 28.

10. Kile, *Farm Bureau*, 161–62.

11. Campbell, *Farm Bureau*, 5–6; Benedict, *Farm Policies*, 176.

12. Benedict, *Farm Policies*, 177–78.

13. Benedict, *Farm Policies*, 393–94.

14. Davis, "Development of Agriculture Policy," 298–99; Benedict, *Farm Policies*, 167.

15. Benedict, *Farm Policies*, 169.

16. Benedict, *Farm Policies*, 145–48, 157, 169.

17. Davis, "Development of Agriculture Policy," 298–304.

18. Benedict, *Farm Policies*, 172

19. Benedict, *Farm Policies*, 181–83; Hansen, *Gaining Access*, 27–31. See, generally, Campbell, *Farm Bureau*; Kile, *Farm Bureau*.

20. Benedict, *Farm Policies*, 207–8; Hansen, *Gaining Access*, 37–38.

21. Peek and Johnson, *Equality*, 5–6, 21.

22. Hansen, *Gaining Access*, 40–41.

23. Hansen, *Gaining Access*, 40–45; Benedict, *Farm Policies*, 210–17; Winders, *Politics*, 42.

24. Benedict, *Farm Policies*, 216–19; Winders, *Politics*, 42; Hansen, *Gaining Access*, 41–45.

25. Campbell, *Farm Bureau*, 37–38; Kile, *Farm Bureau* 133–51; Hansen, *Gaining Access*, 50–53; Benedict, *Farm Policies*, 223.

26. Daniel, "Crossroads," 435.

27. Benedict, *Farm Policies*, 223; Davis, "Development of Agriculture Policy," 311; Hansen, *Gaining Access*, 64–65; Winders, *Politics*, 42–44.

28. Winders, *Politics*, 45–46; Benedict, *Farm Policies*, 230–31.

29. Winders, *Politics*, 45–46; Benedict, *Farm Policies*, 239–41.

30. Benedict, *Farm Policies*, 267.

31. See, generally, Finegold, "From Agrarianism."

32. Winders, *Politics*, 47; Benedict, *Farm Policies*, 241–43.

33. Benedict, *Farm Policies*, 247.

34. Breimyer, "Agricultural Philosophies," 334.

35. Benedict, *Farm Policies*, 247.

36. Benedict, *Farm Policies*, 252.

37. Benedict, *Farm Policies*, 264.

38. Snyder, "Huey Long," 134–35 (crop was estimated at 15.584 million bales with a 6.4 million bale carryover and worldwide consumption around 12 million bales).

39. Snyder, "Huey Long," 136; Benedict, *Farm Policies*, 252.

40. Benedict, *Farm Policies*, 266–69; Perkins, *Crisis*, 27–35; Finegold, "From Agrarianism," 11–12.

41. Slichter, "Franklin D. Roosevelt," 239.

42. Kile, *Farm Bureau*, 179; Fite, *Cotton Fields*, 123–24; Snyder, "Huey Long," 139–43.

43. Snyder, "Huey Long," 147–50.

44. Snyder, "Huey Long," 144–45, 154–59.

45. Snyder, "Huey Long," 160.

46. Snyder, "Huey Long," 146.

47. Winders, *Politics*, 48; Hansen, *Gaining Access*, 70–71; Benedict, *Farm Policies*, 267–68; Campbell, *Farm Bureau*, 42; Kile, *Farm Bureau*, 170.

48. Benedict, *Farm Policies*, 266–68; Kile, *Farm Bureau*, 179.

49. Fite, "Farmer Opinion"; Finegold, "From Agrarianism," 20–22.

50. Slichter, "Franklin D. Roosevelt," 249–55; Finegold, "From Agrarianism," 18–21.

51. Finegold, "From Agrarianism," 17–20, 22–25; Slichter, "Franklin D. Roosevelt," 257–58.

52. Winders, *Politics*, 54 (reporting that Roosevelt won 89 percent of the electoral vote and 57 percent of the popular vote, carrying every state in the Wheat and Corn Belts; Democratic majorities in the House went to 313–117 and in the Senate to 59–36).

53. Cong. Rec., December 22, 1932, 907–8.

54. Cong. Rec., January 16, 1933, 1885; Cong. Rec., January 24, 1933, 2378–79 (conference report submitted in, and agreed to by, the Senate), 2418–19 (submitted to and agreed to by the House).

55. Cong. Rec., January 11, 1933; Cong Rec., March 3, 1933; Cong. Rec., April 13, 1933.

56. Campbell, *Farm Bureau*, 44–46, 52–53; May, "Marvin Jones," 423–24.

57. Conrad, *Forgotten Farmers*, 25.

58. These were W. J. Spillman (USDA), Beardsley Ruml (Rockefeller Foundation), John D. Black (Harvard), M. L. Wilson (Montana State College), Mordecai Ezekiel (Federal Farm Board), Rexford Tugwell (Columbia), and Henry A. Wallace. See Fite, *George N. Peek*, 229–42; Conrad, *Forgotten Farmers*, 21; Lord, *Wallaces*, 305–12, 330–31; Kile, *Farm Bureau*, 173–76; Campbell, *Farm Bureau*, 49–58. Legislative drafting was reportedly handled by Frederic P. Lee, who had been head of the Senate's legislative drafting service and worked on the McNary-Haugen bills. See Conrad, *Forgotten Farmers*, 21–22; Kile, *Farm Bureau*, 193.

59. Rasmussen, "New Deal," 355 (quoting *Agricultural Adjustment Relief Plan: Hearings on H.R. 13991 Before the Senate Comm. on Agriculture and Forestry*, 72nd Cong. 15 [1933]).

60. May, "Marvin Jones," 422.

61. May, "Marvin Jones," 427.

62. May, "Marvin Jones," 436.

63. Hollis, "Cotton Ed," 236.

64. Hollis, "Cotton Ed," 245.

65. Hollis, "Cotton Ed," 249–51.

66. Hollis, "Cotton Ed," 255–56.

67. Campbell, *Farm Bureau*, 52–55; Kile, *Farm Bureau*, 197–99; Perkins, *Crisis*, 26–75.

68. Cong. Rec., March 22, 1933, 766 (the bill passed by a vote of 315 to 98 [1 voting present; 17 not voting]).

69. Marvin Jones, House Report to Accompany Agricultural Adjustment Act of 1933, H.R. Rep. No. 73-6, at 7 (1933) (quoting President Roosevelt's message to Congress).

70. Campbell, *Farm Bureau*, 555–57; Hansen, *Gaining Access*, 78–81; Winders, *Politics*, 56–58.

71. Cong. Rec., April 12, 1933, 1548–51 (Senator George W. Norris [R-NE]), 1548–54 (Senators John Bankhead II [D-AL], Arthur Vandenberg [R-MI], and Thomas Gore [D-OK]), 1558 (Majority Leader Joseph Robinson [D-AR]). See also Cong. Rec., April 13, 1933, 1626–35 (Senators Alban Barkely [D-KY], Charles McNary, and Josiah Bailey [D-NC]).

72. Perkins, *Crisis*, 68–69; Cong. Rec., April 12, 1933, 1548–51.

73. Cong. Rec., April 12, 1933, 1548–51; Perkins, *Crisis*, 63–65, 68–69.

74. Perkins, *Crisis*, 70–71.

75. Cong. Rec., April 12, 1933, 1551–53 (Senator Bankhead), 1558 (Senator Robinson), 1552 (Senator Alben Barkley [D-KY]).

76. Cong. Rec., April 12, 1933, 1558–59 (statement of Senator [and Majority Leader] Joseph Robinson [D-AR]).

77. Cong. Rec., April 12, 1933, 1548–51, 1554 (Bankhead and Senator Arthur Vandenberg [R-MI]). See also Cong. Rec., April 13, 1933, 1634–35.

78. Cong. Rec., April 13, 1933, 1631 (Senator Josiah Bailey [D-NC]).

79. H.R. Rep. No. 73-6, at 62 (comments by John Simpson, NFU President).

80. Perkins, *Crisis*, 62.

81. Perkins, *Crisis*, 75.

82. Cong. Rec., May 9, 1933, 3079–80 (defeated cost of production by a vote of 190 to 283 [40 not voting]); and Cong. Rec., May 10, 1933, 3121 (passing the conference report without the cost-of-production provision by a vote of 53 to 28 [14 not voting]).

83. Perkins, *Crisis*, 75; Agricultural Adjustment Act of 1933, P.L. No. 73-10, 48 Stat. 31 (1933).

84. Campbell, *Farm Bureau*, 45–46 (explaining the penny sale); Kirkendall, "New Deal," 86 (discussing the unrest).

85. Breimyer, "Agricultural Philosophies," 341; Perkins, *Crisis*, 75.

86. P.L. No. 73-10.

87. P.L. No. 73-10, sec. 2 (the basic commodities were wheat, cotton, corn, hogs, rice, tobacco, and milk).

88. Kirkendall, "New Deal," 88–89; Perkins, *Crisis*, 182–85 (same), 129–31, 175.

89. Conrad, *Forgotten Farmers*, 22.

90. Fite, *George N. Peek*, 255.

91. Perkins, *Crisis*, 145–46 (quoting Secretary Wallace).

92. Addison Theyer Cutler and W. Powell, "Tightening the Cotton Belt," *Harper's*, February 1934, 310, https://harpers.org/archive/1934/02/tightening -the-cotton-belt/.

93. Whatley, "Labor," 924–25; Fite, *Cotton Fields*, 184; Daniel, "Crossroads," 447.

94. Fite, *Cotton Fields*, 214–19.

95. Daniel, "Crossroads," 433, 436.

96. Whatley, "Labor," 908.

97. Fite, *Cotton Fields*, 150.

98. Vance, "Human Factors," 262–66; Woodman, "Southern Agriculture," 322–24; Kirby, "Southern Exodus," 591; Daniel, "Crossroads," 455.

99. Kirby, "Transformation," 265–66.

100. Conrad, *Forgotten Farmers*, 77. See also Kirby, "Transformation," 261–66; Whatley, "Labor," 909, 924–25; Daniel, "Crossroads," 437; Fite, *Cotton Fields*, 184.

101. Conrad, *Forgotten Farmers*, 77. See also Kirby, "Transformation," 261–66; Whatley, "Labor," 909, 924–25; Daniel, "Crossroads," 437. See, generally, Fite, *Cotton Fields*.

102. Woodman, "Southern Agriculture," 337.

103. Woodman, "Southern Agriculture," 324–26; Daniel, "Crossroads," 31; Vance, "Human Factors," 263.

104. Daniel, "Crossroads," 431.

105. Hoffsommer, "AAA," 494.

106. Whatley, "Labor," 905; Daniel, "Crossroads," 431; Hoffsommer, "AAA," 494.

107. Hoffsommer, "AAA," 494.

108. Kirby, "Transformation," 268.

109. Woodman, "Southern Agriculture," 325–29; Vance, "Human Factors," 267–68.

110. Mann, "Sharecropping," 418. See also Woodman, "Southern Agriculture," 332–36; Vance, "Human Factors," 263.

111. Conrad, *Forgotten Farmers*, 4–7; Fite, *Cotton Fields*, 136, 234–38; Woodman, "Southern Agriculture," 331–37; Vance, "Human Factors," 264–68.

112. See, e.g., P.L. No. 73-10, sec. 8 ("General Powers" provided "for the reduction in the acreage or reduction in the production for market, or both, of any basic agricultural commodity, through agreements with producers or by other voluntary methods"); Kirby, "Transformation," 63–64.

113. Conrad, *Forgotten Farmers*, 24.

114. Whatley, "Labor," 913–14.

115. *Agricultural Emergency Act to Increase Farm Purchasing Power: Hearing Before the Committee on Agriculture and Forestry, United States Senate, on H.R. 3835*, 73rd Cong. 38 (1933) (Secretary Wallace answering that tenant issues were "a matter of regulation" but that under the previous allotment plan "you had it set out as a joint undertaking." Senator Bankhead [D-AL] said USDA would deal with "whichever had control," but Senator McGill [D-KS] said his understanding was that the "tenant has the same advantages under this act as the landlord." Secretary Wallace responded, "Of course, that is a matter of regulation"); Conrad, *Forgotten Farmers*, 27.

116. Conrad, *Forgotten Farmers*, 50–51.

117. Conrad, *Forgotten Farmers*, 52.

118. Conrad, *Forgotten Farmers*, 51–52 ("allow landlords to lease their land to the government for a safe and sure rental and leave tens of thousands of tenants and their families to 'idleness and beggary'"; and one of his closest advisors confirmed, "Some Southern farmers had already told him that when they rented part of their land to the government they would readjust by having fewer tenants to handle the remaining acres").

119. Conrad, *Forgotten Farmers*, 45 (discussing Cully Cobb and those "familiar with the Southern scene and not anxious to reform it radically"), 52–53, 114; Lord, *Wallaces*, 324.

120. Hoffsommer, "AAA," 496 (landlords' power was "maintained only through the subordination of the tenant group," and "any system of aid which comes between" them, such as "by granting of government relief [directly to sharecroppers,] obviously diminishes this subordination"); Conrad, *Forgotten Farmers*, 52–53.

121. In 1939, after his time at USDA, Johnston founded the National Cotton Council of America, which remains the main lobbying arm of the cotton industry to this day. See Nelson, "Oscar Johnston," 400–401; Nelson, *King Cotton's Advocate*, chapter 3 (45), chapter 6 (77); Conrad, *Forgotten Farmers*, 116.

122. The British-owned Delta and Pine Land Company (D&PL) of Scott, Mississippi, farmed approximately twenty-five thousand acres of cotton using more than a thousand sharecroppers. See Conrad, *Forgotten Farmers*, 116; Nelson, "Oscar Johnston," 400–401 (it was majority owned by Fine Cotton Spinners' and Doublers' Association, Ltd., of Manchester, England,

"a cotton textile spinning corporation consisting of 50 British mills"); Cong. Rec., March 13, 1934, 4439.

123. It was reported D&PL received $124,000 from AAA in 1934 and $318,000 total from 1933 to 1935 while another plantation Johnston managed received $78,000. See Daniel, "Crossroads," 238; Nelson, "Oscar Johnston," 404–7.

124. Conrad, *Forgotten Farmers*, 71.

125. Cutler and Powell, "Tightening," 315. See also Hoffsommer, "AAA," 501.

126. Conrad, *Forgotten Farmers*, 16. See also Nelson, *King Cotton's Advocate*, chapter 6.

127. Conrad, *Forgotten Farmers*, 75–76.

128. Conrad, *Forgotten Farmers*, 116 ("AAA had neither the authorization nor the capability to reform Southern tenancy"). See also Nelson, *King Cotton's Advocate*, chapter 6.

129. Nelson, *King Cotton's Advocate*, 54 ("broke the spirit and letter of the 1933 contract").

130. Nelson, *King Cotton's Advocate*, 89 (dispute with local compliance committee resulted in threat to pull out of the program); Nelson, "Oscar Johnston," 413 (efforts to reform the cotton program were met by threats from Johnston "to take his company out of the government program").

131. See, e.g., Nelson, *King Cotton's Advocate*, 85–91.

132. Hoffsommer, "AAA," 497–98 (quoting Calvin B. Hoover, Human Problems of Acreage Reduction in the South, AAA, Washington DC); Conrad, *Forgotten Farmers*, 55–56 ("Much more thought and work concerning tenants went into" the drafting of the second contract); Nelson, *King Cotton's Advocate*, 83–84 (Johnston and Cobb led the effort, working with Alger Hiss of the USDA legal department).

133. Conrad, *Forgotten Farmers*, 56–58 (quoting contract, paragraph 7, landlord "shall, insofar as possible, maintain on his farm the normal number of tenants and other employees," and the landlord shall permit tenants to occupy the house "unless any such tenant shall so conduct himself as to become a nuisance or a menace to the welfare of the producer" [i.e., the landlord]).

134. Conrad, *Forgotten Farmers*, 55–59 (contract was with the "owner, landlord, cash-tenant or managing share-tenant who operates or controls a cotton farm"), 70 (managing share tenant as "one who furnished work stock, equipment and labor and who manages" the farm operation); Hoffsommer, "AAA," 498–99 (managing share tenant was "assumed to be a step higher on the agricultural ladder than the cropper," and many were subsequently downgraded to sharecropper to avoid dividing payments with them).

135. Conrad, *Forgotten Farmers*, 98–104 (when a member of the Cotton Section "investigated" a farm for violations of the contract provisions, the

conclusion of the investigation was that "nothing could be done because the tenants on the . . . plantation were not managing share-tenants").

136. Conrad, *Forgotten Farmers*, 55–59; Hoffsommer, "AAA," 498–99. See also Nelson, *King Cotton's Advocate*, chapter 6.

137. See, e.g., Conrad, *Forgotten Farmers*, 47 ("was particularly appealing because all payments would be made to them and they could collect on debts from their tenants before giving them their share of the benefits"); Mann, "Sharecropping," 418 ("perpetual state of economic bondage"); Nelson, *King Cotton's Advocate*, 88–89 ("maintain social control").

138. See, e.g., Fite, *George N. Peek*, 253 (Secretary Wallace considered the AAA a "major social experiment"), 256 (liberals wanted to use farm policy "as an instrument of long-range social and economic reform"); Conrad, *Forgotten Farmers*, 45, 103–19; Lord, *Wallaces*, 342.

139. Fite, *George N. Peek*, 256 (considering it "an advantage that the labor costs of contract signers would be reduced, and although it occurred to them that this might work a severe hardship on farm workers they did not allow it to affect their plans").

140. Fite, *Cotton Fields*, 141–42 (that "by reducing acreage they did not need so many sharecroppers"); Conrad, *Forgotten Farmers*, 43 ("anxious to get some kind of cotton program under way").

141. Perkins, *Crisis*, 123–24; Kirkendall, "New Deal," 88–89.

142. Finegold, "From Agrarianism," 7, 11.

143. Fite, *Cotton Fields*, 186; Whatley, "Labor," 907.

144. Fite, *Cotton Fields*, 188–89, 207–8 (black farmers "were affected more than whites because a larger percentage of black farmers had been on small acreages"); Daniel, "Crossroads," 440.

145. Fite, *Cotton Fields*, 207–8 ("blacks who operated small units, often on poor land, simply could not compete"); Whatley, "Labor," 926; Daniel, "Crossroads," 429, 441–42; Vance, "Human Factors," 265.

146. Whatley, "Labor," 924–25; Fite, *Cotton Fields*, 184; Daniel, "Crossroads," 447.

147. Whatley, "Labor," 909, 921–24.

148. Whatley, "Labor," 928.

2. Adjusting to the New Deal and War

1. See, e.g., Bureau of the Census, *Historical Statistics of the United States, 1789–1957*, Chapter E: Agriculture, Series E 1–269, 99 (1949), http://www2.census.gov/library/publications/1949/compendia/hist_stats_1789-1945/hist_stats_1789-1945-chE.pdf; Benedict, *Farm Policies*, 289–90, 298–301, 311–14.

2. Benedict, *Farm Policies*, 281–82.

3. Kirkendall, "New Deal," 88–89; Conrad, *Forgotten Farmers*, chapter 3; Perkins, *Crisis*, 182–85.

4. Kirkendall, "New Deal," 91–92 (internal quotations omitted).

5. Davis, "Development of Agriculture Policy," 326.

6. Kirkendall, "New Deal," 83–89.

7. See, e.g., Perkins, *Crisis*, 174–75.

8. Cong. Rec., April 13, 1934, 6550; Cong. Rec., April 17, 1934, 6775. See, e.g., Daniel, "Crossroads," 441.

9. Cong. Rec., March 10, 1934, 4191 (comments of Representative Wall Doxey [D-MS] and Representative William Bankhead [D-AL]); Cong. Rec., March 13, 1934, 4430–31 (comments of House Agriculture Committee Chairman Marvin Jones [D-TX]); Cong. Rec., March 15, 1934, 4633–35 (Rep. Bankhead); Cong. Rec., March 24, 1934, 5300–5310 (comments of Senator John Bankhead [D-AL]).

10. Cong. Rec., March 15, 1934, 4633–35 (comments of Rep. Bankhead, "slacker"); Cong. Rec., March 16, 1934, 4701 (Chairman Jones, "chiseler").

11. Nelson, *King Cotton's Advocate*, 66–67; Heacock, "Bankhead," 347. See also Angela Jill Cooley, "John Hollis Bankhead II," *Encyclopedia of Alabama*, published January 9, 2008, http://www.encyclopediaofalabama.org /article/h-1424.

12. Nelson, *King Cotton's Advocate*, 67.

13. Cong. Rec., March 15, 1934, 4633–35 (Representative Bankhead).

14. Cong. Rec., March 10, 1934, 4194 (Representative Doxey).

15. Cong. Rec., March 26, 1934, 5409–14 (comments of Chairman Smith).

16. Cong. Rec., March 26, 1934, 5409–14 (comments of Senate Agriculture Committee Chairman Ellison D. Smith [D-SC]).

17. Cong. Rec., March 10, 1934, 4197 (comments of Rep. Leroy T. Marshall [R-OH]); Cong. Rec., March 13, 1934, 4440–41 (Representative Lawrence Ellzey [D-MS]); Cong. Rec., March 15, 1934, 4633–35 (comments). See also Fite, *Cotton Fields*, 172.

18. Nelson, *King Cotton's Advocate*, 54; Cong. Rec., March 13, 1934, 4439 (Representative Fulmer [D-TN] submitted into the record issues with Mr. Johnston's management from an earlier hearing).

19. Nelson, *King Cotton's Advocate*, 66.

20. Nelson, *King Cotton's Advocate*, 73–74.

21. Nelson, *King Cotton's Advocate*, 70–71.

22. Nelson, *King Cotton's Advocate*, 70–71 (because certificates would be "issued to landowners, most of who did not reside on their farms. . . . Tenants would get the short end of the deal").

23. Nelson, *King Cotton's Advocate*, 76.

24. Nelson, *King Cotton's Advocate*, 67–68, 75.

25. Breimyer, "Agricultural Philosophies," 345. See also Cong. Rec., April 13, 1934, 6550; Cong. Rec., April 17, 1934, 6775.

26. Cong. Rec., March 13, 1934, 4435–36 (comments of Representative Clifford Hope [R-KS]); Cong. Rec., March 26, 1934, 5403–8 (statement of Senator William Borah [R-ID]), 5311 (Senator Simeon Fess [R-OH]).

27. Vance, "Human Factors," 259–60; Fite, *Cotton Fields*, 132.

28. Cong. Rec., March 19, 1934, 5710.

29. Fite, *Cotton Fields*, 175.

30. Cong. Rec., March 17, 1934, 4722.

31. Cong. Rec., March 13, 1934, 4435–36 (Representative Hope).

32. See, e.g., Cong. Rec., March 13, 1934, 4451–53 (comments of William Lemke [nonpartisan R-ND]).

33. Cong. Rec., March 26, 1934, 5409–14 (comments of Senate Agriculture Committee Chairman Ellison D. Smith [D-SC]); Vance, "Human Factors," 259.

34. Daniel, "Crossroads," 435.

35. Cong. Rec., March 15, 1934, 4633–35 (Representative Bankhead); Fite, *Cotton Fields*, 174–45. See also Daniel, "Crossroads," 433.

36. Fite, *Cotton Fields*, 178, 161–62, 176–77; Cong. Rec., March 10, 1934, 4209 (Representative George B. Terrell [D-TX]).

37. Fite, *Cotton Fields*, 150, 205.

38. Fite, *Cotton Fields*, 143, 148, 185, 205.

39. Fite, *Cotton Fields*, 161–62, 176–77, 178.

40. Conrad, *Forgotten Farmers*, 67–68.

41. Cong. Rec., March 10, 1934, 4209.

42. Cong. Rec., March 10, 1934, 4202–3.

43. P.L. No. 73-169 (1934) (referendum of "persons who have the legal or equitable right as owner, tenant, share-cropper, or otherwise to produce cotton"); Cong. Rec., April 17, 1934, 6770 (conference reported as submitted in the House indicated a change made in conference, substituting the provision for the House requirement of two-thirds of the "persons who own, rent, share-crop, or control cotton land").

44. See, e.g., Woodman, "Southern Agriculture," 325–26.

45. Cong. Rec., March 10, 1934, 4204 (comments of Representative Whittington about Johnston's assistance in fixing defects in the bill); Cong. Rec., March 15, 1934, 4652–53 (Chairman Jones offered the amendment and noted, "Those who would have charge of the administration of the act felt it would be much more difficult to ascertain when they had gotten the two thirds of the individuals, than it would be when they had gotten the representatives of two thirds of the acreage").

46. Cong. Rec., March 15, 1934, 4652.

47. Cong. Rec., March 13, 1934, 4442 (an instance of dairy standing with cotton).

48. Cong. Rec., March 15, 1934, 4633–35 (Representative Bankhead). See also Cong. Rec., March 10, 1934, 4196 (Representative Doxey), 4204 (Representative William Whittington [D-MS]); Cong. Rec., March 13, 1934, 4432 (Chairman Jones and Representative Paul Kvale [DFL-MN]).

49. P.L. No. 73-169 (the secretary "may make regulations protecting the interests of share-croppers and tenants in the making of allotments and the issuance of tax-exemption certificates under this Act").

50. Conrad, *Forgotten Farmers*, 67.

51. The certificates of exemption from the tax were to be issued to the producer, a classification that Congress again left undefined but that generally did not include the sharecropper. See Nelson, *King Cotton's Advocate*, 74 ("a matter of contractual relationship between land-owner and tenant").

52. P.L. No. 73-169.

53. Nelson, *King Cotton's Advocate*, 76.

54. See, generally, Conrad, *Forgotten Farmers*, chapter 5.

55. See, generally, Conrad, *Forgotten Farmers*, chapters 8 and 9.

56. See, generally, Conrad, *Forgotten Farmers*, chapter 8.

57. Nelson, *King Cotton's Advocate*, 74.

58. Fite, *Cotton Fields*, 140; Whatley, "Labor," 913–14, 924; Finegold, "From Agrarianism," 26.

59. Cong. Rec., April 17, 1934, 6771–72.

60. Egan, *Worst Hard Time*, 198.

61. Benedict, *Farm Policies*, 122–23; Williams, "Soil Conservation," 370.

62. Egan, *Worst Hard Time*, 227–28.

63. An Act to Provide for the Protection of Land Resources against Soil Erosion, and for Other Purposes, P.L. No. 74-46, 49 Stat. 163 (1935).

64. Morgan, *Governing*, 52–53.

65. U.S. v. Butler, 297 U.S. 1 (1936).

66. Benedict, *Farm Policies*, 349.

67. Soil Conservation and Domestic Allotment Act of 1936, P.L. No. 74-461 (1936).

68. Benedict, *Farm Policies*, 350; Morgan, *Governing*, 41.

69. Benedict, *Farm Policies*, 350–51; P.L. No. 74-461, sec. 13.

70. Cong. Rec., February 20, 1936, 2513.

71. Cong. Rec., February 19, 1936, 2394.

72. Cong. Rec., February 21, 1936, 2552.

73. Cong. Rec., February 21, 1936, 2571.

74. Cong. Rec., February 20, 1936, 2469.

75. Cong. Rec., February 15, 1936, 2165.

76. Cong. Rec., February 19, 1936, 2373.

77. Cong. Rec., February 20, 1936, 2525 (Maverick amendment requiring "equitable distribution"); Cong. Rec., February 21, 1936, 2551 (Tarver amendment requiring consideration of "the contribution in services"), 2554 (Whelchel amendment added "including tenants and sharecroppers" to the apportionment of payments).

78. Cong. Rec., February 21, 1936, 2551.

79. Cong. Rec., February 21, 1936, 2552 (Maverick supported the addition and withdrew his amendment).

80. P.L. No. 74-461, sec. 8(b) ("Secretary shall not have the power to enter into any contract binding upon any producer or to acquire any land or any right or interest therein"); Soil Conservation and Domestic Allotment Act, H.R. Rep. No. 74-2079, at 10 (1936) (Conf. Rep.); Cong. Rec., February 26, 1936, 2805 (report language that USDA was to "protect the interests of tenants and sharecroppers" but only "in so far as practicable").

81. Cong. Rec., February 27, 1936, 2923.

82. Cong. Rec., February 27, 1936, 2934.

83. Cong. Rec., February 27, 1936, 2936.

84. Benedict, *Farm Policies*, 346–51; Morgan, *Governing*, 83–90.

85. Cong. Rec., November 29, 1937, 268 (comments of Senate Ag Committee Chairman Ellison D. Smith), 431–36 (comments of Senator John Bankhead [D-AL]), 470 (comments of Representative August H. Andresen [R-MN]).

86. Cong. Rec., November 29, 1937, 461–62 (House Ag Chairman Marvin Jones [D-TX]).

87. Kile, *Farm Bureau*, 234–39.

88. Kile, *Farm Bureau*, 234–37.

89. Cong. Rec., November 23, 1937, 266–67 (comments of Chairman Smith); Cong. Rec., February 11, 1938, 1822–24 (Senator James Pope [D-ID]). See also, Kile, *Farm Bureau*, 239–40.

90. Kile, *Farm Bureau*, 239–40.

91. Cong. Rec., November 29, 1937, 444–45 (comments of Senator John Bankhead [D-AL] describing the president's response), 448–51 (comments of Senator George S. McGill's [D-KS]).

92. Kile, *Farm Bureau*, 240.

93. Cong. Rec., November 30, 1937, 507–8 (Senator Allen Ellender [D-LA]); Cong. Rec., February 8, 1938, 1654–55 (Chairman Jones). See also Cong. Rec., November 29, 1937, 466 (comments of Chairman Jones); Cong. Rec., February 14, 1938, 1877–81 (comments of Senators Pope and Barkley).

94. Cong. Rec., November 29, 1937, 451–52, 465.

95. Cong. Rec., November 29, 1937, 465.

96. Cong. Rec., November 29, 1937, 479–80, 453 (Senator McGill).

97. Cong. Rec., November 29, 1937, 441–44 (Chairman Smith stated that there).

98. Cong. Rec., November 29, 1937, 441–44.

99. Cong. Rec., November 29, 1937, 470 (Representative August Andresen [R-MN]), 483 (comments of Representative Harry Sauthoff [Progressive-WI]), 436–37 (colloquy between Senator Bankhead [D-AL] and Senator McNary [R-OR]).

100. Cong. Rec., November 29, 1937, 436–37 (colloquy between Senator Bankhead [D-AL] and Senator McNary [R-OR]).

101. Cong. Rec., November 29, 1937, 438.

102. Cong. Rec., November 29, 1937, 436–38.

103. Cong. Rec., November 29, 1937, 268–69 (Senator Warren Austin [R-VT] questioning Chairman Smith).

104. Cong. Rec., November 29, 1937, 277–29.

105. Cong. Rec., November 24, 1937, 347–48; Cong. Rec., November 29, 1937, 451–52.

106. Cong. Rec., November 29, 1937, 451–55 (comments of Senator Pope [D-ID] and colloquy with Senator McGill [D-KS]).

107. Cong. Rec., November 29, 1937, 427–29 (cotton provisions), 438.

108. Cong. Rec., November 29, 1937, 436–37.

109. Cong. Rec., November 29, 1937, 438 (statement of Chairman Smith). See also Cong. Rec., November 29, 1937, 436–37 (comments of Senator Bankhead and Senator Pope).

110. Cong. Rec., November 29, 1937, 442–43. See also Burford, "Federal Cotton," 226.

111. Cong. Rec., November 29, 1937, 482 (Representative Harold Knutson [R-MN]); Cong. Rec., November 30, 1937, 515 (Senator Josiah Bailey [D-NC]).

112. Cong. Rec., November 29, 1937, 442–43 (Senator Bankhead), 461–62 (House Ag Chairman Jones).

113. Cong. Rec., November 29, 1937, 472–73 (Representative August Andresen [R-MN]). See also Cong. Rec., November 29, 1937, 474–79 (House debate); Cong. Rec., November 30, 1937, 518–23 (Senate debate).

114. Fite, *Cotton Fields*, 155.

115. Cong. Rec., December 7, 1937, 1052; Cong. Rec., December 8, 1937, 1093.

116. Cong. Rec., December 7, 1937, 1052; Cong. Rec., December 8, 1937, 1093.

117. See, e.g., Cong. Rec., November 30, 1937, 516–23 (Senator James Byrnes [D-SC]).

118. Cong. Rec., November 23, 1937, 277–29; Cong. Rec., November 24, 1937, 335; Cong. Rec., November 29, 1937, 454 (colloquy between Senators Pope and McGill).

119. Cong. Rec., November 24, 350–51.

120. Cong. Rec., November 29, 1937, 430 (Senator McNary), 41 (Senator Bankhead).

121. Cong. Rec., December 7, 1937, 1050–52 (comments of Rep. Scott Lucas [D-IL]).

122. Cong. Rec., December 7, 1937, 1050–52.

123. Cong. Rec., December 7, 1937, 1053 (Representative Wall Doxey [D-MS]); Cong. Rec., December 8, 1937, 1093 (Chairman Jones), 1095.

124. Cong. Rec., December 8, 1937, 1095.

125. Cong. Rec., December 8, 1937, 1093, 1096 (comments of Chairman Jones). See also Cong. Rec., December 8, 1937, 1096–1108 (debate surrounding an amendment by Representative Patman [D-TX]).

126. Cong. Rec., December 8, 1937, 1112–14.

127. Cong. Rec., December 10, 1937, 1290 (passage by a vote of 267 to 130 [3 present; 30 not voting]); Cong. Rec., December 17, 1937, 1768 (passage by a vote of 59 to 29).

128. Cong. Rec., February 8, 1938, 1659–60 (Comments of Representative Doxey); Cong. Rec., February 10, 1938, 1768 (Senate Ag Chairman Smith).

129. Agricultural Adjustment Act of 1938, P.L. No. 75-430, 52 Stat. 31 (1938) (hereafter cited as 1938 AAA). See also Cong. Rec., February 10, 1938, 1748; Cong. Rec., February 14, 1938, 1871 (Senator McNary).

130. Cong. Rec., February 8, 1938, 1658–64 (House debate over cotton assistance); Cong. Rec., February 10, 1938, 1759–60; Cong. Rec., February 14, 1938, 1869 (Senate debate).

131. Cong. Rec., February 9, 1938, 1666 (Representative Boileau).

132. Cong. Rec., February 8, 1938, 1670–71 (remarks of Representative Harry Coffee [D-NE]).

133. See, e.g., Cong. Rec., February 8, 1938, 1658 (Representative Andresen), 1655, 1662–66; Cong. Rec., February 10, 1938, 1769 (Senate Ag Chairman Smith), 1772–75 (debate surrounding the point of order raised by Senator Lewis Schwellenbach [D-WA]); Cong. Rec., February 11, 1938, 1822–24 (remarks of Senator Pope).

134. See, e.g., Cong. Rec., February 8, 1938, 1660–61, 1668 (Representative John Nichols [D-OK]); Cong. Rec., February 9, 1938, 1725 (Representative Clifford Hope [R-KS]).

135. 1938 AAA; Cong. Rec., February 10, 1938, 1749–57; Cong. Rec., February 11, 1822 (Senator Pope).

136. It also included a definition of parity income, defined as the per capita net income from farming that bears the same relation to non-farming per capita net income in the 1909–14 base period. See 1938 AAA, Title III, subtitle A, sec. 301(a).

137. Cong. Rec., February 8, 1938, 1656 (House Ag Chairman Jones); Cong. Rec., February 9, 1938, 1725 (Representative Hope); Cong. Rec., February 10, 1938, 1752.

138. Cong. Rec., February 8, 1938, 1663 (Representative Doxey), 1669–70 (Representative Scott Lucas [R-IL]); Cong. Rec., February 9, 1938, 1718–21 (Representatives Fred Gilchrist [R-IA] and Francis Case [R-SD]).

139. Cong. Rec., February 10, 1938, 1760–61.

140. Cong. Rec., February 10, 1938, 1761, sec. 508; Cong. Rec., November 23, 1937, 270 (Senator Pope); Cong. Rec., February 8, 1938, 1657, 1664.

141. Cong. Rec., February 10, 1938, 1727 (House passed it by a vote of 264 to 135 [30 not voting]), 1747 (the debate over an anti-lynching bill created a potential for a filibuster in the Senate); Cong. Rec., February 14, 1938, 1881–82 (Senate passed it by a vote of 56 to 31 [9 not voting]). The bill was signed into law on February 16, 1938.

142. Cong. Rec., February 8, 1938, 1657 (Representative August Andresen [R-MN]), 1669–70 (comments of Rep. Harold Cooley [D-NC]); Cong. Rec., February 10, 1938, 1768 (Senator McNary), 1870 (Senator McNary).

143. Cong. Rec., February 8, 1938, 1658–59 (Representative Andresen); Cong. Rec., February 9, 1938, 1827–37 (Senator Arthur Vandenberg [R-MI]); Cong. Rec., February 14, 1938, 1871–77.

144. Cong. Rec., February 8, 1938, 1660 (Representative Wall Doxey [D-MS]). See also Cong. Rec., February 10, 1938, 1766–67 (Senate Ag Chairman Smith); Cong. Rec., February 9, 1938, 1726 (House Ag Committee Chairman Jones).

145. Fite, *Cotton Fields*, 185.

146. See, e.g., Daniel, "Crossroads," 446.

147. See, e.g., Whatley, "Labor," 924–25; Fite, *Cotton Fields*, 141–42, 184; Burford, "Federal Cotton," 227, 236.

148. Whatley, "Labor," 909, 928; Daniel, "Crossroads," 437.

149. 1938 AAA; Cong. Rec., February 7, 1938, 1585–86 (Conf. Rep.).

150. See, e.g., Cong. Rec., November 18, 1937, 118 (comments of Senator Walter F. George [D-GA]); Cong. Rec., November 29, 1937, 444–45 (comments of Senator Bankhead); Cong. Rec., February 10, 1938, 1747 (the debate over an anti-lynching bill created a potential for a filibuster in the Senate).

151. Cong. Rec., December 17, 1937, 1741–42 (Senator Lynn Frazier [Nonpartisan League, R-ND]).

152. Cong. Rec., December 15, 1937, 1546.

153. Cong. Rec., December 17, 1937, 1730.

154. Cong. Rec., December 7, 1937, 1033.

155. Cong. Rec., November 29, 1937, 465.

156. Cong. Rec., December 2, 1937, 793–94 (Representative Harold Cooley [D-NC]).

157. Cong. Rec., December 2, 1937, 761.

158. Cong. Rec., December 1, 1937, 649.

159. Cong. Rec., December 6, 1937, 974.

160. Cong. Rec., December 2, 1937, 782–83; Cong. Rec., December 3, 1937, 847.

161. Cong. Rec., December 3, 1937, 847–48.

162. Cong. Rec., December 3, 1937, 845–46.

163. Cong. Rec., December 3, 1937, 845–46.

164. Cong. Rec., December 3, 1937, 837–38 (Representative Andresen [R-MN]). See also Cong. Rec., December 3, 1937, 851 (Representative Lucas).

165. Cong. Rec., December 6, 1937, 974 (Tarver debating Representative Cooley).

166. Cong. Rec., December 3, 1937, 850; Cong. Rec., December 6, 1937, 974–75 ("the allotment is made to the owner as the producer and not to the tenant who pays a part of the proceeds of the crop as rent").

167. Cong. Rec., December 3, 1937, 848.

168. Cong. Rec., December 3, 1937, 972 (small white landowner farmers diversified but were hurt by the Bankhead Act).

169. Cong. Rec., December 6, 1937, 975.

170. 1938 AAA, sec. 103 (amending sec. 8 of the 1936 act); Cong. Rec., February 7, 1938, 1602.

171. 1938 AAA ("provisions do not apply if on investigation the local committee finds that the change is justified and approves such change in the relationship"); Cong. Rec., February 7, 1938, 1602.

172. Cong. Rec., February 7, 1938, 1602 (federal payments to tenants or sharecroppers "may be assigned by him, in writing, to his landlord as security for cash or advances").

173. Smith, FDR, 373–74 (reporting on Roosevelt's landslide victory with 60.79 percent of the vote and a margin of more than eleven million votes. Roosevelt won 98.6 percent of the vote in South Carolina, 98 percent in Mississippi, and 87.1 percent in Georgia. Democrats had a 331 to 89 majority in the House and a 76 to 16 majority in the Senate).

174. McCoy, "McGill," 2.

175. Hansen, Gaining Access, 90.

176. Smith, FDR, 479 (reporting that FDR won 449 Electoral College votes to Wendell Wilkie's 82, with a nearly 5 million vote victory. Democrats also picked up 6 House seats but lost 3 in the Senate).

177. Benedict, Farm Policies, 389–90.

178. Kile, Farm Bureau, 274.

179. Kile, Farm Bureau, 244, 275–79; Benedict, Farm Policies, 389–90; Bowers, Rasmussen, and Baker, "History of," 14–15.

180. The move toward a high, fixed loan rate began when Senator Bankhead [D-AL] introduced a bill to amend the 1938 act to make loans on the major or basic commodities mandatory at 100 percent of parity. See *Commodity Loans and Marketing Quotas: Hearings Before the Comm. on Agriculture and Forestry, U.S. Senate*, 75th Cong. (1941); Cong. Rec., May 14, 1941, 4020.

181. Cong. Rec., April 29, 1941, 3413 (remarks of Representative Stephen Pace [D-GA]), 3420 (remarks of Representative Walter Pierce [D-OR]), 3417 (remarks of Representative Harry Coffee [D-NE]).

182. Cong. Rec., March 27, 1941, 2633 (Senate Ag Committee Chairman Ellison D. Smith [D-SC]); *Commodity Loans and Marketing Quotas*, 169.

183. Cong. Rec., April 29, 1941, 3409 (Representative Wall Doxey [D-MS]), 3417 (comments of Representative Coffee).

184. Cong. Rec., April 29, 1941, 3418 (Representative Reid Murray [R-WI]), 3428–30 (comments of Representative Jeannette Rankin [R-MT], Representative Hampton Fulmer [D-SC], and Representative James O'Connor [D-MT]), 3432 (the vote was 123 to 15); Cong. Rec., May 14, 1941, 4021–22 (Senators Bankhead and Russell).

185. Cong. Rec., May 6, 1941, 3607 (Senator Bankhead), 3613 (Senator Arthur Capper [R-KS]); Cong. Rec., May 7, 1941, 3716–18; Cong. Rec., May 12, 1941, 3940; Cong. Rec., May 13, 1941, 3981–85 (Conf. Rep., agreed to by the House by a vote of 277 to 63 [91 not voting]); Cong. Rec., May 14, 1941, 4020–23 (agreed to by the Senate by a vote of 75 to 2 [18 not voting]). See also P.L. No. 77-74 (May 26, 1941).

186. Cong. Rec., May 13, 1941, 3983–84 (Representative Wall Doxey [D-MS]). See also "Commodities: Farm Staples Spurt on Congress Approval of 85% Parity Loans," *Wall Street Journal*, May 15, 1941, 1.

187. Kile, *Farm Bureau*, 280–81.

188. Benedict, *Farm Policies*, 402–10.

189. Benedict, *Farm Policies*, 410–25, 446–51.

190. Wickard v. Filburn, 317 U.S. 111, 113 (1942).

191. *Wickard*, 317 U.S. at 118–19.

192. *Wickard*, 317 U.S. at 124.

193. *Wickard*, 317 U.S. at 130–33.

194. Chen, "Wickard," 22.

195. See, generally, Rasmussen, "New Deal Agricultural Policies"; Saloutos, "New Deal."

196. Bureau of the Census, *Historical Statistics*, 99.

197. Christenson, *Brannan Plan*, 9.

198. Smith, *FDR*, 636.

199. Matusow, *Farm Policies*, 1–14.

200. Matusow, *Farm Policies*, 63.

201. Bureau of the Census, *Historical Statistics*, 99.

202. See, e.g., Saloutos, "New Deal," 403; Breimyer, "Agricultural Philosophies," 351–52.

203. For example, 21.5 percent of the U.S. workforce was employed in agriculture in 1930, but that percentage had fallen to 16 percent by 1945. See Dimitri, Effland, and Conklin, "20th Century," 2 (sidebar), 5 (figure 3).

204. Dimitri, Effland and Conlin, "20th Century," 12; Saloutos, "New Deal," 403.

205. Fite, *Cotton Fields*, 168–69, 94.

206. Whatley, "Labor," 908, 928.

207. Fite, *Cotton Fields*, 186–87.

208 Fite, *Cotton Fields*, 169.

209. Fite, *Cotton Fields*, 219; Daniel, "Crossroads," 437.

210. Daniel "Crossroads," 448; Fite, *Cotton Fields*, 171.

211. See, generally, Fite, *Cotton Fields*; Winders, *Politics*, chapter 5.

212. Olson, "Federal Farm," 5.

213. Matusow, *Farm Policies*, 115–18; Christenson, *Brannan Plan*, 13; Benedict, *Farm Policies*, 465–72.

214. Benedict, *Farm Policies*, 471–72; Cong. Rec., May 17, 1948, 5896.

215. Matusow, *Farm Policies*, 118; Christenson, *Brannan Plan*, 14.

3. Transition and Turbulence after War

1. Heinz, "Political Impasse," 968.

2. Cochrane and Ryan, *American Farm Policy*, 26–27; Benedict, *Farm Policies*, 469–72; Christenson, *Brannan Plan*, 9–11.

3. Benedict, *Farm Policies*, 460–62; Winders, *Politics*, 147; Cong. Rec., July 21, 1949, 9941 (Representative Helen G. Douglas [D-CA] discussing the Marshall Plan).

4. Matusow, *Farm Policies*, 125–31; Benedict, *Farm Policies*, 461–65.

5. Cochrane and Ryan, *American Farm Policy*, 26–27; Benedict, *Farm Policies*, 469–71.

6. Matusow, *Farm Policies*, 110–15.

7. Matusow, *Farm Policies*, 120–22.

8. Matusow, *Farm Policies*, 135–38.

9. Matusow, *Farm Policies*, 136–37.

10. Matusow, *Farm Policies*, 138; Hansen, *Gaining Access*, 112–16; Kile, *Farm Bureau*, 332–33.

11. For example, in the twenty-year period from 1950 to 1970, U.S. agricultural output shot up 38 percent compared to USDA expert predictions for a much smaller increase in output. See Cochrane and Ryan, *American Farm Policy*, 28.

12. Christenson, *Brannan Plan*, 15; Matusow, *Farm Policies*, 61–62. See also Cong. Rec., June 15, 1948, 8301 (opening statement of Senator George Aiken [R-VT]).

13. Kansas Historical Society, "Arthur Capper," *kansapedia*, last modified July 2016, https://www.kshs.org/kansapedia/arthur-capper/12001.

14. Cong. Rec., June 15, 1948, 8301 (Chairman Aiken).

15. Forsythe, "Clifford Hope," 407.

16. Connell B. Gallagher, "About George Aiken," Detailed Bio, March 2000, College of Engineering and Mathematical Sciences, University of Vermont, https://learn.uvm.edu/aiken/about-george-aiken/; Alan Krebs, "George Aiken, Longtime Senator and G.O.P. Maverick, Dies at 92," *New York Times*, November 20, 1984, http://www.nytimes.com/1984/11/20/obituaries/george-aiken-longtime-senator-and-gop-maverick-dies-at-92.html.

17. Matusow, *Farm Policies*, 140–43; Hansen, *Gaining Access*, 114–16.

18. Cong. Rec., June 11, 1948, 7901 (Representative Virgil Chapman [D-KY]).

19. Cong. Rec., June 11, 1948, 7900 (Chairman Hope), 7891–92 (Representative Ross Rizley [R-OK]).

20. Cong. Rec., June 11, 1948, 7901 (Chairman Hope); Cong. Rec., June 12, 1948, 8014 (comments of Representative Sam Rayburn [D-TX]).

21. Cong. Rec., June 11, 1948, 7896 (statement of Representative Henry Talle [R-IA]).

22. Cong. Rec., June 11, 1948, 7894 (statement of Representative George Sadowski [D-MI]). See also Cong. Rec., June 11, 1948, 7894–95 (Representatives Adolph Sabath [D-IL] and Helen Douglas [D-CA]).

23. Matusow, *Farm Policies*, 140–43.

24. Cong. Rec., June 12, 1948, 8014 (the House passed its extension without a recorded vote).

25. Cong. Rec., June 15, 1948, 8302–5; Cong. Rec., June 16, 1948, 8447–48.

26. Cong. Rec., June 15, 1948, 8202–3.

27. Cong. Rec., June 16, 1948, 8439–42 (Senator Ellender), 8444–45 (Senator Milton Young [R-ND] and Ellender), 8593–95 (Senator Scott Lucas [D-IL]).

28. Cong. Rec., June 17, 1948, 8569–72, 8584 (Senator Absalom Robertson [D-VA]); Caro, *Master of the Senate*, 164–202.

29. Cong. Rec., June 17, 1948, 8598 (the vote was 27 to 55 [with 14 not voting]).

30. He wanted to include the cost of labor in the parity formula, which, given cotton's slow pace of mechanization, would have likely benefitted cotton farmers. See Cong. Rec., June 17, 1948, 8577, 8599, 8444, 8600–8603 (the Senate rejected the amendment by a vote of 23 to 59 [14 not voting] and again defeated 28 to 51 [17 not voting]).

31. Cong. Rec., June 15, 1948, 8308 (Senator Burnet Maybank [D-SC]), 8589 (Senator Maybank).

32. See, e.g., Cong. Rec., June 17, 1948, 8590 (comments of Senator James Eastland [D-MS]).

33. Cong. Rec., June 17, 1948, 8612 (it passed by a vote of 79 to 3 [14 senators not voting]).

34. Matusow, *Farm Policies*, 143.

35. Matusow, *Farm Policies*, 143; Cong. Rec., June 19, 1948, 9338–47 (Chairman Hope).

36. Cong. Rec., June 19, 1948, 9344–45 (Representative John Flannagan [D-VA]), 9345–46 (Representatives Harold Cooley [D-NC] and Stephen Pace [D-GA]).

37. Cong. Rec., June 19, 1948, 9157 (the Senate passed it by voice vote), 9347 (the conference report was agreed to by the House on a vote of 147 to 70).

38. Agricultural Act of 1948, P.L. No. 80-897 (1948); Christenson, *Brannan Plan*, 15–16; Matusow, *Farm Policies*, 143.

39. Matusow, *Farm Policies*, 170–88; Christenson, *Brannan Plan*, 17; Benedict, *Farm Policies*, 478 (noting that Harold D. Cooley [D-NC] and Senator Elmer Thomas [D-OK] replaced Hope and Capper, respectively, as chairmen).

40. Hansen, *Gaining Access*, 116; Matusow, *Farm Policies*, 185–90.

41. Winders, *Politics*, 234n75.

42. Matusow, *Farm Policies*, 173–79; Christenson, *Brannan Plan*, 16–17; Benedict, *Farm Policies*, 332; Cochrane and Ryan, *American Farm Policy*, 137. See also "Commodity Credit Corporation," U.S. Department of Agriculture, Farm Service Agency, http://www.fsa.usda.gov/about-fsa/structure-and-organization/commodity-credit-corporation/index.

43. The president's remarks were made at the National Plowing Contest in Dexter, Iowa, before eighty thousand farmers on September 18, 1948. See Cochrane and Ryan, *American Farm Policy*, 28; Hansen, *Gaining Access*, 116–19; Matusow, *Farm Policies*, 178–81.

44. Matusow, *Farm Policies*, 170–90 (reporting that Truman won Minnesota, Missouri, Illinois, Iowa, Wisconsin, and Ohio, giving him 101 electoral votes to seal his victory; Truman gave credit to labor for his come-from-behind, miraculous victory but others gave credit to the farmers); Hansen, *Gaining Access*, 116.

45. Hansen, *Gaining Access*, 116–17.

46. Dean, "Charles F. Brannan," 30–32.

47. Christenson, *Brannan Plan*, 17; Matusow, *Farm Policies*, 125.

48. Christenson, *Brannan Plan*, 18.

49. By 1946, with less labor than before the war, farmers were producing 33 percent more than they had in the late 1930s; productivity increased even more into the 1950s and 1960s. See, e.g., Matusow, *Farm Policies*, 110–30.

50. Matusow, *Farm Policies*, 110–30; Dean, "Farm Policy Debate," 33–34 (quoting Wayne Rasmussen, "The Impact of Technological Change on American Agriculture, 1862–1962," *Journal of Economic History* 22 [December 1962]).

51. Dean, "Farm Policy Debate," 33–34.

52. Benedict, *Farm Policies*, 478.

53. Cong. Rec., January 23, 1974, 487–89 (Representative Lawrence H. Fountain [D-NC] remarks for the record in honor of Rep. Cooley upon his passing); L. Walter Seegers, "Cooley, Harold Dunbar," in *Dictionary of North Carolina Biography*, ed. William S. Powell (Chapel Hill: University of North Carolina Press, 1979), reprinted online at http://www.ncpedia.org /biography/cooley-harold-dunbar.

54. Eric Manheimer, "The Public Career of Elmer Thomas" (PhD thesis, University of Oklahoma, 1952).

55. He had been appointed to the controversial Resettlement Administration in 1935 and moved to the equally controversial Farm Security Administration as assistant administrator in 1944. He moved up to assistant secretary of agriculture under Secretary Claude Wickard. Brannan had been on the campaign trail for a dozen weeks and gave eighty speeches in support of President Truman. For farm policy his views were squarely in the postwar abundance camp with an emphasis on conservation—views with clear partisan political angles to lock in a labor-farmer coalition for the Democratic Party. See Christenson, *Brannan Plan*, 19–22; Matusow, *Farm Policies*, 134, 196–201.

56. Christenson, *Brannan Plan*, 28, 34–36, 92–97; Matusow, *Farm Policies*, 204–5.

57. Cong. Rec., July 20, 1949, 9845–46 (comments of Representative Clifford Hope [R-KS]).

58. Matusow, *Farm Policies*, 200–205; Christenson, *Brannan Plan*, 53–54.

59. See, e.g., Cong. Rec., July 20, 1949, 9845 (Representative Edward Cox [D-GA]), 9855. See also Christenson, *Brannan Plan*, 54–55, 88, 143–58; Matusow, *Farm Policies*, 204–5.

60. See, e.g., Cong. Rec., July 20, 1949, 9836–37 (Representatives Adolph Sabath [D-IL] and James Sutton [D-TN]), 9843 (House Ag Chairman Harold Cooley [D-NC]), 9850 (Representative Sutton); Cong. Rec., July 21, 1949, 9925 (Representative W. R. Poage [D-TX]).

61. Cong. Rec., July 20, 1949, 9838 (Representative August Andresen [R-MN]), 9842 (House Ag Committee Chairman Cooley), 9852 (Representative Charles B. Hoeven [R-IA]), 9853 (Representative Usher L. Burdick [nonpartisan, R-ND]).

62. Cong. Rec., July 20, 1949, 9842 (Chairman Cooley); Cong. Rec., July 21, 1949, 9926 (Representative Poage).

63. Agricultural Act of 1949, H.R. Rep. No. 81-998, at 47 (1949); Cong. Rec., July 20, 1949, 9844 (Chairman Cooley), 9850 (Representative Sutton), 9856–57 (Representative Pace).

64. Cong. Rec., July 20, 1949, 9854–56 (Representative Pace), 9842–43 (Chairman Cooley); Cong. Rec., July 21, 1949, 9941 (Representative Helen G. Douglas [D-CA]). See also Cong. Rec., July 20, 1949, 9854 (Representative Pace), 9859 (Representative Norris Cotton [R-NH]).

65. Cong. Rec., July 20, 1949, 9854–55 (Representative Pace); Cong. Rec., July 21, 1949, 9931.

66. Cong. Rec., July 20, 1949, 9855–59 (Representative Pace).

67. Cong. Rec., July 20, 1949, 9839–40 (Representatives Noah Mason [R-IL] [brain child], Sid Simpson [R-IL], W. Sterling Cole [R-NY], and James Wadsworth [R-NY]), 9852 (Rep. Charles Hoeven [R-IA], camel's nose), 9859–60 (Representative Norris Cotton, camel's head).

68. Cong. Rec., July 20, 1949, 9838 (Representative August Andresen [R-MN] on cheap food), 9845–46 (Representative Hope made the labor connection).

69. Cong. Rec., July 20, 1949, 9844, 9849.

70. Cong. Rec., July 21, 1949, 9926 [Rules Committee Chairman Adolph Sabath [D-IL]). See also Cong. Rec., July 21, 1949, 9844–45 (statement of Representative Hope).

71. Cong. Rec., July 20, 1949, 9857–59.

72. Cong. Rec., July 20, 1949, 9859.

73. Cong. Rec., July 21, 1949, 9925.

74. Cong. Rec., July 21, 1949, 9936–37 (the amendments: [1] to provide 60 percent to 90 percent support on cottonseed; [2] provisions for Maryland tobacco producers; and [3] a full repeal of the Aiken provisions).

75. Cong. Rec., July 21, 1949, 9962–63 (the Pace substitute to the Gore substitute was defeated by a vote of 152 to 222; the Gore substitute amendment [as amended by the Sutton amendment to repeal the Aiken provisions] was agreed to by a vote of 239 to 170 [23 not voting]; final passage of the bill was by a vote of 384 to 25 [23 not voting]).

76. Stabilization of Agricultural Prices, S. Rep. No. 81-1129 (1949); Cong. Rec., September 30, 1949, 13580–81, 13609 (letter from F. K. Woolley, Deputy Administrator, USDA Production and Marketing Administration, dated September 30, 1949, submitted to the record by Senator Anderson).

77. Cong. Rec., October 12, 1949, 14300–14309 (Chairman Thomas). See also Cong. Rec., October 4, 1949, 13742 (Senator Milton Young [R-ND]).

78. Cong. Rec., October 3, 1949, 13614–15 (Senator Anderson; Senator Scott Lucas [D-IL] opposed), 13615–16 (comments of Senator A. Willis Robertson [D-VA]).

79. Cong. Rec., October 3, 1949, 13627.

80. Cong. Rec., October 3, 1949, 13650–51.

81. Cong. Rec., October 3, 1949 13647–48 (Senator Aiken).

82. Cong. Rec., October 4, 1949, 13754–55 (Senator Anderson).

83. Cong. Rec., October 4, 1949, 13754 (Senator Anderson). See also Cong. Rec., October 13, 1949, 13647–48 (Senator George Aiken [R-VT]).

84. Cong. Rec., October 3, 1949, 13615; Cong. Rec., October 4, 1949, 13742, 13745 (Senator Aiken); Cong. Rec., October 7, 1949, 14096–98 (debating this point).

85. Cong. Rec., October 4, 1949, 13744.

86. Cong. Rec., October 4, 1949, 13754–56.

87. Cong. Rec., October 4, 1949, 13757 (Senator Edward Thye [R-MN] and Senator Russell colloquy).

88. Cong. Rec., October 7, 1949, 14106–8 (Senator Russell).

89. Cong. Rec., October 4, 1949, 13756–57; Cong. Rec., October 7, 14099–14100 (Senator Young).

90. See, e.g., Cong. Rec., October 11, 1949, 14187–96 (amendment by Senator Lester Hunt [D-WY]); Cong. Rec., October 4, 1949, 13742, 13786 (Senator Thye).

91. Cong. Rec., October 3, 1949, 13615–16 (Senators Anderson and Aiken); Cong. Rec., October 4, 1949, 13623–24 (Senators Lucas and Anderson), 13790 (Senator Young), 14102–6 (Senators Lucas and Russell).

92. Cong. Rec., October 7, 1949, 14097 (Senator Young and Senator Aiken), 14105–7 (Senators Young and Russell).

93. Cong. Rec., October 7, 1949, 14099–14100 (Senator Young), 14108–9 (newly elected Senator Hubert Humphrey [D-MN], who had defeated a sitting Republican).

94. Cong. Rec., October 4, 1949, 13773–89, 13792.

95. Cong. Rec., October 6, 1949, 13984; Cong. Rec., October 7, 1949, 14096–97, 14115 (the Young-Russell amendment was defeated by a vote of 26 to 45 [25 senators not voting]).

96. Cong. Rec., October 12, 1949, 14324 (by a voice vote).

97. Cong. Rec., October 19, 1949, 15062 (House Ag Chairman Cooley), 14999–15001 (Senator Aiken and Majority Leader Lucas).

98. Cong. Rec., October 19, 15063 (Representative Pace).

99. Cong. Rec., October 19, 1949, 14994–99 (Senators Anderson and Young).

100. Cong. Rec., October 19, 1949, 14997 (Senator Anderson).

101. Cong. Rec., October 19, 1949, 15065; P.L. 81-439, 63 Stat. 1051, sec. 416.

102. See, e.g., Cong. Rec., October 19, 1949, 14997 (colloquy between Senators Fulbright and Anderson).

103. Cong. Rec., October 19, 1949, 15064 (Representative Pace), 15077 (Representative Poage), 15002 (Senator Aiken).

104. Cong. Rec., October 19, 1949, 14993, 15008 (by a vote of 46 to 7 [43 senators not voting]), 15077 (by a vote of 175 to 34 [233 members present]).

105. Agricultural Act of 1949, P.L. No. 81-439, 63 Stat. 1051 (1949).

106. See, e.g., Agricultural Act of 2014.

107. Cong. Rec., October 19, 1949, 15004 (Senator Young), 15074 (Representative August Andresen [R-MN]), 15075 (Representative Reid Murray [R-WI]), 15077 (Representative W. R. Poage [D-TX]).

108. Cong. Rec., October 19, 1949, 15074 (Representative August Andresen [R-MN]); quote is from Hansen, *Gaining Access*, 126.

109. Cong. Rec., October 19, 1949, 15062 (Chairman Cooley), 15067 (Representative Francis Case [R-SD]).

110. Cong. Rec., October 19, 1949, 14999 (Senator Anderson).

111. Dean, "Farm Policy Debate," 43.

112. Christenson, *Brannan Plan*, 167–69; Winders, *Politics*, 85–86.

113. See, e.g., Cong. Rec., March 15, 1934, 4633–35 (Representative Bankhead); Cong. Rec., November 29, 1937, 443 (Senator Bankhead); Cong. Rec., August 17, 1954, 14838 (Representative Cooley).

114. Winders, *Politics*, 96–97.

115. Winders, *Politics*, 85–92.

116. Planted acres fell 60 percent from 1926 to 1946, a difference of over 27 million acres; farmers planted 45.839 million acres of cotton in 1926 but only 18.638 in 1946, according to USDA NASS statistics. See NASS, Quick Stats.

117. Fite, *Cotton Fields*, 178 (quoting from a 1944 article by Frank H. Jeter in *Farm Journal* that a National Cotton Council member argued concerns about sharecroppers and farm labor "should not be allowed to stand in the way of progress," and quoting E. D. White, assistant to the Secretary of Agriculture, at the Beltwide Cotton Mechanization Conference that farming was not responsible for the unemployed left in the wake of mechanization; the unemployed should be dealt with by "means other than the slowing down of mechanization"); Fite, *Cotton Fields*, 183–87, 220; Whatley, "Labor," 907.

118. Fite, *Cotton Fields*, 221–22 ("the civil rights movement gained momentum, the attitudes of white landowners towards blacks hardened," and some "intensified their move to mechanize as a means of eliminating the need for any kind of hired labor").

119. See, generally, Fite, *Cotton Fields*; Winders, *Politics*, chapter 5. See also Burford, "Federal Cotton," 227

120. Christenson, *Brannan Plan*, 169–72.

121. Christenson, *Brannan Plan*, 172.

4. A Surplus of Problems and Disagreement

1. Hadwiger and Talbot, *Pressures and Protests*, 7.

2. See, e.g., Cong. Rec., June 30, 1954, 9385 (Representative Charles Hoeven [R-IA]).

3. U.S. Department of Agriculture, "The Balance Sheet of Agriculture, 1956," *Federal Reserve Bulletin*, August 1956, 826, Table 3, https://fraser.stlouisfed .org/docs/publications/frb/pages/1955-1959/15304_1955-1959.pdf. But see Agricultural Act of 1956, S. Rep. 84-1484, at 63 (1956), 63, graph 4.

4. Cong. Rec., August 5, 1954, 13421 (Senator Young and Chairman Aiken), 13441 (Senator Anderson).

5. Cochrane and Ryan, *American Farm Policy*, 30, 179, 203, 225.

6. McCullough, *Truman*, 775.

7. Matusow, *Farm Policies*, 220; Christenson, *Brannan Plan*, 164. See also Dean, "Why Not," 268.

8. Agricultural Act of 1949.

9. "Farm Price Supports," in CQ *Almanac 1952*, 8th ed., 81–82 (Washington DC: Congressional Quarterly, 1953), http://library.cqpress.com/cqalmanac /cqal52-1380585.

10. See, e.g., Hansen, *Gaining Access*, 126; Cong. Rec., June 30, 1954, 9366 (Representatives Abernethy [D-MS] and Whitten [D-MS]).

11. Cong. Rec., June 30, 1954, 9366 (Representative Harold Cooley [D-NC]); Cong. Rec., August 4, 1954, 13421 (Senator Milton Young [R-ND]), 13422 (Senate Ag Chairman George Aiken [R-VT]); Cong. Rec., August 5, 1954, 13438 (Senator Young); Cong. Rec., April 11, 1956, 6111 (Representative Cooley), 6081 (Cooley).

12. Cong. Rec., June 30, 1954, 9366 (Representatives Abernethy and Whitten).

13. See, e.g., Cong. Rec., August 5, 1954, 13407 (Senator Hubert Humphrey [D-MN]), 13417 (minority reviews in the committee report), 13455; Cong. Rec., August 6, 1954, 13521 (Senator Spessard Holland [D-FL]), 13535 (Holland), 13557 (Senator Andrew Schoeppel [R-KS]), 13715 (Senator Thye).

14. Cong. Rec., August 5, 1954, 13535 (Senator Holland); Cong. Rec., August 6, 1954, 13519, 13547 (Senator James Eastland [D-MS]).

15. Matusow, *Farm Policies*, 221.

16. Democrats held 263 seats to 171 Republicans in the Eighty-First Congress (1949–51) compared to 235 Democrats and 199 Republicans in the Eighty-Second Congress (1951–53). Congress Profiles, History, Art and Archives, U.S. House of Representatives, http://history.house.gov/Congressional-Overview/ Profiles/81st/. In the Senate, Democrats held 54 seats to Republicans' 42 in the Eighty-First Congress, which fell to 49 Democrats and 47 Republicans in the Eighty-Second Congress. Party Division, Art and History, U.S. Sen-

ate, https://www.senate.gov/history/partydiv.htm. See also Hansen, *Gaining Access*, 125.

17. Winders, *Politics*, 81–82; Hansen, *Gaining Access*, 126; Cochrane and Ryan, *American Farm Policy*, 30.

18. Hansen, *Gaining Access*, 169; Winders, *Politics*, 119–24.

19. Cochrane and Ryan, *American Farm Policy*, 30–31.

20. Eisenhower was the first Republican president since Roosevelt had defeated Hoover in 1932. He brought with him majorities in both chambers of Congress. See Winders, *Politics*, 78; Hansen, *Gaining Access*, 125–26.

21. Hansen, *Gaining Access*, 127–28;

22. Cochrane and Ryan, *American Farm Policy*, 31; Schapsmeier and Schapsmeier, "Eisenhower and Agricultural Reform," 51.

23. Cong. Rec., July 28, 1953, 10075 (S.2475), 10078 (Senator Andrew Schoeppel [R-KS]).

24. Cong. Rec., July 28, 1953, 10078–88.

25. Cong. Rec., July 28, 1953, 10079 (Senator Milton Young [R-ND]), 10082 (Senator Hubert Humphrey [D-MN]), 10084 (Senator Young), 10086 (Senator Spessard Holland [D-FL]); Cong. Rec., June 16, 1954, 8370–80. See also Cong. Rec., June 15, 1954, 8286; Cong. Rec., June 16, 1954, 8361 (House Ag Committee Chairman Clifford Hope [R-KS]); Cong. Rec., June 30, 1954, 9335–38, 9528–29 (Senator Holland).

26. See, e.g., Winders, *Politics*, 148–51.

27. Winders, *Politics*, 147–49.

28. Cong. Rec., June 30, 1954, 9528–38.

29. "Farm Price Supports," *CQ Almanac 1955*, 11th ed., 169–72 (Washington DC: Congressional Quarterly, 1956), http://library.cqpress.com/cqalmanac/cqal55-1352801; Schapsmeier and Schapsmeier, "Farm Policy from FDR," 367–68.

30. Schapsmeier and Schapsmeier, *Ezra Taft Benson*, 39–40; "Farm Price Supports" (1955). See also Watson, "Federal Farm Subsidies," 287.

31. See, e.g., "Farm Price Supports" (1955); Cong. Rec., June 30, 1954, 9385 (Representative Charles Hoeven [R-IA]).

32. Cong. Rec., June 30, 1954, 9363; Comm. on Agriculture, Agricultural Act of 1954, H.R. Rep. No. 83-1927 (1954). See also Winders, *Politics*, 82.

33. Cong. Rec., June 30, 1954, 9372–73.

34. Cong. Rec., June 30, 1954, 9373–74.

35. Cong. Rec., June 30, 1954, 9372–73. See also Cong. Rec., June 30, 1954, 9366 (Cooley blamed USDA).

36. Cong. Rec., August 4, 1954, 13236; Comm. on Agriculture and Forestry, Agricultural Act of 1954, S. Rep. No. 83-1810 (1954).

37. Cong. Rec., August 4, 1954, 13221–23.

38. Cong. Rec., July 1, 1954, 9532, 9562–63 (Harrison amendment agreed to by the House by 179 to 164 on a division vote and 179 to 154 on tellers); Hansen, *Gaining Access*, 131.

39. Cong. Rec., August 4, 1954, 1323–30; Cong. Rec., August 9, 1954, 13703, 13712 (passed by a vote of 49 to 44 [3 not voting]).

40. Cong. Rec., August 10, 1954, 13931–38 (Senator Holland offered the amendment; final passage by a vote of 62 to 28 [6 not voting]).

41. Cong. Rec., August 17, 1954, 14784 (Senate), 14827–28 (House), 14835–36 (Chairman Hope), 14784 (Chairman Aiken).

42. Cong. Rec., August 17, 1954, 14789 (Senator Spessard Holland [D-FL]).

43. Hansen, *Gaining Access*, 131–32.

44. Cong. Rec., July 2, 1954, 9635–39 (Chairman Hope).

45. See, e.g., Cong. Rec., July 2, 1954, 9678 (House vote on the Harrison amendment that passed 228 to 170 [35 not voting; 1 voting present]). See Cong. Rec., August 5, 1954, 13421–41 (Chairman Aiken).

46. It was dropped in conference. Cong. Rec., August 6, 1954, 13522 (Senator Young); Cong. Rec., August 17, 1954, 14784 (Senate), 14827–28 (House).

47. Cong. Rec., August 17, 1954, 14851 (passed the House 208 to 100 [124 not voting]), 14800 (passed the Senate around midnight by a vote of 44 to 28 [24 not voting]).

48. See, e.g., Cong. Rec., June 30, 1954, 9406 (Representative John Phillips [R-CA]); generally, Cong. Rec., July 1, 1954.

49. Cong. Rec., August 5, 1954, 13425–28 (Senator Anderson), 13524 (Senator Holland); Cong. Rec., August 6, 1954, 13522–23 (Senators Holland and Young), 13532, 13549–50 (Senator Eastland), 13535 (Senator Holland); Cong. Rec., August 9, 1954, 13692 (Senator Howard Smith [R-NJ]).

50. Cong. Rec., August 5, 1954, 13421–41 (Senator Aiken); Cong. Rec., August 6, 1954, 13519, 13547 (Senator Eastland).

51. Cong. Rec., August 17, 1954, 148785–87 (Senators Young, Mundt, and Thye).

52. Cong. Rec., August 5, 1954, 13421 (Chairman Aiken).

53. Cong. Rec., August 5, 1954, 13441 (Senator Anderson).

54. See, e.g., Cong. Rec., Jun. 30, 1954, 9386–90 (Representative Hoeven [R-IA]), 9537, 9636 (Chairman Hope).

55. Cong. Rec., August 17, 1954, 14836 (Rep. Cooley), 14785 (Senator Young).

56. Hansen, *Gaining Access*, 137; Pennock, "Party and Constituency," 202; Hardin, "Farm Price Policy," 622.

57. Hansen, *Gaining Access*, 137–38.

58. "Farm Price Supports" (1955).

59. Senator Ellender first served as chair of the Senate Ag Committee from 1951–53, handing the gavel back to Senator Aiken after Republicans

won a Senate majority in 1952. See, generally, Becnel, *Senator Ellender*; Jeremy Alford, "Ellender Maintains Stature in U.S. Senate History," houmatoday.com, November 1, 2009, http://www.houmatoday.com/news/20091101/ellender-maintains-stature-in-us-senate-history.

60. Cong. Rec., May 3, 1955, 5455–56 (Cooley), 5467 (Representative H. Carl Andersen [R-MN]).

61. Comm. on Agriculture, Price Support Programs for Basic Commodities, Wheat, and Dairy Products, H.R. Rep. No. 84-203, at 5 (1955).

62. Cong. Rec., May 3, 1955, 5476–77 (Representative Clifford Hope [R-KS]), 5464–66 (Representative Harrison), 5471–72 (Representatives Ralph Harvey [R-IN] and Charles Hoeven [R-IA]).

63. Cong. Rec., May 3, 1955, 5457 (Representative James Van Zandt [R-PA]), 5464–68 (Representatives Robert Harrison [R-NE] and W. R. Poage [D-TX]), 5476 (Representative Clifford Hope [R-KS]); Cong. Rec., May 4, 1955, 5659 (Representative William Green Jr. [D-PA]), 5662 (Representative John McMillan [D-SC]), 5668 (Representative Omar Burleson [D-TX], 5671 (Representative Jamie Whitten [D-MS]), 5673 (peanut amendment agreed to by the House on a vote of 186 to 150), 5677 (Representative Leslie Arends [R-IL]), 5802 (Chairman Cooley); Cong. Rec., May 5, 1955, 5760 (Poage blamed House Minority Leader Halleck [R-IN]), 5802–3 (Chairman Cooley and Speaker Sam Rayburn [D-TX]).

64. Cong. Rec., May 5, 1955, 5803 (separate vote on the peanut amendment defeated by a vote of 193 to 215 [24 not voting; 2 voting present]), 5804 (motion to recommit the bill defeated by a vote of 199 to 212 [20 not voting; 3 voting present]), 5806 (final passage with price supports at 90 percent of parity by a vote of 206 to 201 [22 not voting; 5 voting present]).

65. "Farm Price Supports" (1955) (quoting the 1955 State of the Union address).

66. "Farm Price Supports" (1955).

67. "President Outlines 9-Point Farm Program," CQ *Almanac 1956*, 12th ed., 10-52-10-58 (Washington DC: Congressional Quarterly, 1957) http://library.cqpress.com/cqalmanac/cqal56-1348268.

68. Cong. Rec., February 22, 1956, 3125–26 (Senator Edward Thye [R-MN]); Cong. Rec., February 23, 1956, 3152–53 (Chairman Ellender).

69. Cong. Rec., February 24, 1956, 3319 (Senator Aiken); "President Outlines."

70. Hansen, *Gaining Access*, 133–35.

71. Cong. Rec., February 22, 1956, 3117–21.

72. Cong. Rec., February 22, 1956, 3120 (Chairman Ellender). See also Hansen, *Gaining Access*, 136–47.

73. Cong. Rec., February 24, 1956, 3317–18.

74. Cong. Rec., February 22, 1956, 3120 (Chairman Ellender).

75. Cong. Rec., February 22, 1956, 3117–18 (Chairman Ellender); "Soil Bank Enacted after Farm Bill VETO," CQ *Almanac 1956*, 12th ed., 02-375-02-392 (Washington DC: Congressional Quarterly, 1957), http://library.cqpress.com /cqalmanac/cqal56-1348982.

76. Cong. Rec., March 8, 1956, 4305 (Senator Bourke Hickenlooper [R-IA]), 4316 (Senator Karl Mundt [R-SD]).

77. Cong. Rec., February 22, 1956, 3117–18 (Senator Hickenlooper); Cong. Rec., March 8, 1956, 4287 (Senator Clinton Anderson [D-NM]), 4305 (Senator Hickenlooper).

78. Cong. Rec., March 8, 1956, 4287–88.

79. Cong. Rec., March 8, 1956, 4289 (Senator Spessard Holland [D-FL]).

80. Cong. Rec., February 24, 1956, 3320–23. See also Cong. Rec., March 8, 1956, 4287–88 (Senator Anderson), 4290–91 (Senator Holland).

81. Cong. Rec., February 23, 1956, 3162; Cong. Rec., February 22, 1956, 3121–22 (Chairman Ellender), 3181–82 (Senator Milton Young [R-ND]).

82. Cong. Rec., February 23, 1956, 3128, 3162–64 (Chairman Ellender); Cong. Rec., March 8, 1956, 4288–89 (Chairman Ellender), 4295 (Senator Hubert Humphrey [D-MN]), 4297 (Senator Alben Barkley [D-KY] and Senator Young), 4304 (by a vote of 54 to 41).

83. Cong. Rec., March 9, 4413–27 (Vice President Nixon broke the 45 to 45 tie); Cong. Rec., March 13, 1956, 4507–9, 4576–78 (Senator Frank Carlson [R-KS] amendment), 4579 (Senator Young), 4585 (Senator Humphrey), 4579–80 (Senator Dirksen), 4594 (Carlson amendment agreed to by a vote of 54 to 39 [2 not voting]).

84. Cong. Rec., March 8, 1956, 4307–8; Cong. Rec., March 9, 1956, 4379 (Ellender), 4384–85.

85. Cong. Rec., March 9, 1956, 4384–85.

86. Cong. Rec., March 8, 1956, 4305–6.

87. Cong. Rec., March 19, 1956, 5065.

88. Cong. Rec., March 8, 1956, 4308.

89. Cong. Rec., February 24, 1956, 3320–23 (Senator Aiken); Cong. Rec., February 22, 1956, 3121–22 (Senator Aiken); Cochrane and Ryan, *American Farm Policy*, 179, 180, 183, 203, 225.

90. Cong. Rec., February 22, 1956, 3121–22 (Senator Aiken stated), 3129 (Chairman Ellender); Cong. Rec., February 24, 1956, 3325 (Aiken), 4289–90 (Senator Holland); Cong. Rec., March 8, 1956, 4305 (Senator Hickenlooper).

91. Cong. Rec., February 24, 1956, 3320–21.

92. Cong. Rec., February 22, 1956, 3123 (Senator Edward Thye [R-MN] and Senator Aiken).

93. Cong. Rec., March 8, 1956, 4308 (Senator Hubert Humphrey [D-MN]), 4297 (Senator Hickenlooper amendment), 4305–6, 4313 (Senator Everett

Dirksen [R-IL]; Cong. Rec., March 9, 1956, 4386–89 (Chairman Ellender), 4308 (Senator Milton Young [R-ND]), 4380–81 (colloquy among Senators Young [R-ND], Holland [D-FL], Anderson [D-NM], and Humphrey [D-MN]), 4384 (Senator Young).

94. Cong. Rec., March 8, 1956 4316–27 (Humphrey amendment defeated by a vote of 44 to 46 [5 not voting]); Cong. Rec., March 9, 1956, 4374–76 (Senator Daniel [D-TX] amendment), 4390 (Hickenlooper amendment modified and agreed to by the Senate on a voice vote).

95. Cong. Rec., March 19, 1956, 5042–45 (Senator Holland amendment), 5047–49 (Senator Hickenlooper), 5044–47 (Chairman Ellender and Senator Young), 5050 (agreed to by the Senate on a vote of 48 to 46 [2 not voting]).

96. Cong. Rec., March 19, 1956, 5078 (by a vote of 93 to 2 [1 not voting]).

97. Cong. Rec., March 12, 1956, 4476–78 (Senator John J. Williams [D-DE] amendment); Cong. Rec., February 23, 1956, 3152 (Chairman Ellender).

98. Cong. Rec., March 12, 1956, 4481–82 (Chairman Ellender), 4485 (Senator Anderson), 4496–97 (negotiations on the Williams amendment; passed 84 to 9 [20 not voting]), 4507 (amended limits on loans at $100,000 was agreed to by the Senate on a vote of 78 to 11 [6 not voting]).

99. Hansen, *Gaining Access*, 133–35 (quoting Representative Poage at a 1956 House hearing).

100. Cong. Rec., February 22, 1956, 3123; "Soil Bank Enacted" (reporting it was agreed to by an 8 to 2 vote on April 6, 1956); Cong. Rec., April 11, 1956, 6111–25 (Conf. Rep.), 6081–82 (Chairman Ellender), 6084 (Senator Young).

101. Cong. Rec., April 11, 1956, 6078 (Chairman Ellender), 6079 (soil bank was voluntary, except for corn and feed grains), 6086 (Senator Aiken), 6126–27 (Cooley explained), 6157 (Representative W. R. Poage), 6126 (Representative Leslie Arends [R-IL]), 6132 (Representative Halleck), 6136–37 (Representative Arends).

102. Cong. Rec., April 11, 1956, 6132–39 (Representatives Arends, Hope, August H. Andresen [R-MN], and Halleck). See also Cong. Rec., April 11, 1956, 6102 (Senator Aiken).

103. Cong. Rec., April 11, 1956, 6126–28 (Chairman Cooley, Representative Charles Halleck [R-IN], and House Minority Leader Joseph Martin [R-MA]), 6157–58 (in the House the motion to recommit was defeated by a vote of 181 to 238 [14 not voting], and the conference report was passed 237 to 181 [15 not voting]), 6108 (the Senate passed the conference report by a relatively close vote of 50 to 35 [11 not voting]).

104. Cong. Rec., April 11, 1956, 6128 (Chairman Cooley), 6133–34 (Representative Hope [R-KS]).

105. Cong. Rec., April 11, 1956, 6132 (Representative Halleck), 6101 (Senator Holland), 6128 (Representative Karl King [R-PA]).

106. Schapsmeier and Schapsmeier, *Ezra Taft Benson*, 161; "Soil Bank Enacted"; Cong. Rec., April 18, 1956, 6541 (veto override failed by a vote of 202 to 211 [20 not voting]).

107. Cochrane and Ryan, *American Farm Policy*, 185–86.

108. Soth, "Farm Policy," 132.

109. Cong. Rec., May 23, 1956, 8823–24.

110. Cong. Rec., May 2, 1956, 7345 (House); "Soil Bank Enacted."

111. Cong. Rec., May 2, 1956, 7346.

112. Cong. Rec., May 3, 1956, 7417 (Representative Poage); Cong. Rec., May 2, 1956, 7347 (Representative Poage); Cong. Rec., May 3, 1956, 7449–50 (House passed by a vote of 314 to 78 [41 not voting]); Cong. Rec., May 18, 1956, 8472–74 (Senator Holland amendments), 8476–77 (Chairman Ellender; Holland amendments were agreed to by a vote of 73 to 14 [8 not voting]), 8514 (Senate passed bill by voice vote); Cong. Rec., May 22, 1956, 8684 (Senators agreed to the conference report by voice vote); Cong. Rec., May 23, 1956, 8821–23 (Representative Poage), 8825–27 (Representatives Carl Anderson [R-MN] and Poage), 8830–32 (passage by a vote of 305 to 59 [69 not voting]).

113. Agricultural Act of 1956, P.L. No. 84-540, 70 Stat. 188 (1956).

114. Schapsmeier and Schapsmeier, *Ezra Taft Benson*, 163; Hansen, *Gaining Access*, 136.

115. Schapsmeier and Schapsmeier, *Ezra Taft Benson*, 177–79 (he won the Electoral College 457 to 74 over Adlai Stevenson; he did lose votes in five farm states [Iowa, Kansas, Minnesota, Oklahoma, and South Dakota]).

116. Hansen, *Gaining Access*, 140.

117. Hansen, *Gaining Access*, 141; Schapsmeier and Schapsmeier, *Ezra Taft Benson*, 190–91.

118. Hanson, *Gaining Access*, 141.

119. "Soil Bank Enacted" (quoting President Eisenhower).

120. Schapsmeier and Schapsmeier, *Ezra Taft Benson*, 170, 182–83.

121. Schapsmeier and Schapsmeier, *Ezra Taft Benson*, 184–90 (quoting from Benson's letter to Chairman Ellender).

122. *Hearings on the Agricultural Situation, Operation of the Soil Bank, Corn Program, and Disposal of Surplus Commodities*, 85th Cong. 4, 10–11, 61 (1957) (hereafter cited as "1957 Senate Hearings"); *Farm Program: Hearings Before a Subcommittee of the Comm. on Agriculture and Forestry*, 85th Cong. 485 (1958).

123. 1957 Senate Hearings, 11 (Secretary Benson), 93; *Present Conditions in Agriculture: Hearings Before the Subcommittee on Department of Agriculture and Related Agencies Appropriations*, 85th Cong. 2–3 (1957); *Corn Program: Hearing Before a Subcommittee of the Comm. on Agriculture and Forestry*, 85th Cong. 4–5 (1957) (Senator Everett Dirksen [R-IL]).

124. Schapsmeier and Schapsmeier, *Ezra Taft Benson*, 184–90, 190–91, 199; Cong. Rec., March 13, 1958, 4214 (Chairman Ellender); Cong. Rec., March 20, 1958, 4906 (Chairman Cooley).

125. *Wheat: Hearings Before the Subcommittee on Wheat*, 85th Cong. 59–60 (1958). See also Schapsmeier and Schapsmeier, *Ezra Taft Benson*, 184–90.

126. See, e.g., *Corn Program*.

127. 1957 Senate Hearings, 93; *Present Conditions in Agriculture*, 2–3; *Cotton Programs: Hearings on Various Bills Relating to Price Support and Acreage Allotments for Upland Cotton*, 85th Cong. (1957); *Farm Program: Hearings on Cotton Price Supports and Acreage Allotments*, 85th Cong. 341–43 (1958).

128. *Farm Program*, 522, 534.

129. Cong. Rec., March 20, 1958, 4906 (Chairman Cooley). See also Cong. Rec., March 13, 1958, 4259 (the joint resolution to prevent loan rate reductions passed the Senate by a vote of 50 to 43 [3 not voting]); Cong. Rec., March 20, 1958, 4939–40 (motion to recommit defeated 173 to 211 [46 not voting]), 4940 (by a vote of 211 to 172 [47 not voting]).

130. Cong. Rec., March 20, 1958, 4923.

131. Cong. Rec., March 21, 1958, 5757–58 (veto message).

132. Cong. Rec., July 23, 1958, 14772–76 (Chairman Ellender), 14784 (Senator William Proxmire [D-WI]), 15049; Cong. Rec., July 24, 1958, 14909 (Senator Hubert Humphrey [D-MN]); Cong. Rec., July 25, 1958, 15047 (Senator William Symington [D-MO]), 15049 (Senator George Aiken [R-VT]), 15130 (Senator Herman Talmadge [D-GA]).

133. Cong. Rec., July 23, 1958, 14775 (Senator Young), 14783–84 (Senators Proxmire and Humphrey); Cong. Rec., July 24, 1958, 14893 (Senator Humphrey), 14909 (Senator Wayne Morse [D-OR] accused the President), 14910 (Senator Humphrey); Cong. Rec., July 25, 1958, 15157 (Senator Paul Douglas [D-IL]), 15162 (passage by a vote of 62 to 11 [23 not voting]).

134. Cong. Rec., August 4, 1958, 16048 (Representative Sidney Simpson [R-IL] reported the vote as 28 to 0 in committee).

135. Cong. Rec., August 4, 1958, 16040–43 (Chairman Cooley), 16045 (Representative Joseph Martin [R-MA]), 16045–46 (Representative H. Carl Anderson [R-MN]).

136. Hansen, *Gaining Access*, 169.

137. Cong. Rec., August 4, 1958, 16046 (Representative H. Carl Anderson); Cong. Rec., August 14, 1958, 17626–31.

138. Cong. Rec., August 6, 1958, 16418 (the vote was 201 to 186 [34 not voting]); Cong. Rec., August 14, 1958, 17625 (Chairman Cooley noted that the bill), 17635 (passed by voice vote after request for yeas and nays was refused).

139. Cong. Rec., August 18, 1958, 18200–18201 (Chairman Ellender; Senator Karl E. Mundt [R-SD]), 18204 (Senator Humphrey), 18207 (the bill as passed by the House passed Senate on a voice vote).

140. Cong. Rec., August 18, 1958, 18199–18200 (Chairman Ellender); Cong. Rec., January 20, 1959, 958–960 (address of Senator Spessard Holland [D-FL] before the American Farm Bureau Federation's 40th Annual meeting, December 10, 1958, Boston MA, submitted by Senator Eastland on January 20, 1959); Harkin and Harkin, "Roosevelt to Reagan," 505; Nordin and Scott, *From Prairie Farmer*, 168.

141. "Agriculture," in *CQ Almanac 1960*, 16th ed., 2–8 (Washington DC: Congressional Quarterly, 1960), http://library.cqpress.com/cqalmanac/cqal60 -1329875.

142. Hansen, *Gaining Access*, 147–48.

143. Hansen, *Gaining Access*, 169; Winders, *Politics*, 119–24.

144. Hadwiger and Talbot, *Pressures and Protests*, 246; Hansen, *Gaining Access*, 167.

145. Hansen, *Gaining Access*, 173 (referring to Baker v. Carr [1962] and Wesberry v. Sanders [1964]).

146. Jensen, "Food Insecurity," 1215–28; Landers, "Food Stamp Program," 1945–51.

147. Ferejohn, "Logrolling," 228; Landers, "Food Stamp Program," 1945–51.

148. Cong. Rec., April 11, 1956, 6149.

149. Cong. Rec., March 7, 1961, 3396 (House Agriculture Committee Chairman Harold Cooley [D-NC]); Cochrane and Ryan, *American Farm Policy*, 79.

150. Hansen, *Gaining Access*, 148; Hadwiger and Talbot, *Pressures and Protests*, 27; Cochrane and Ryan, *American Farm Policy*, 79; Caro, *Passage of Power*, 48, 150–51.

151. Hansen, *Gaining Access*, 148.

152. Hansen, *Gaining Access*, 148–49.

153. Cong. Rec., March 8, 1961, 3506 (Representative Poage). See also *Emergency Feed Grain Program: Hearings on H.R. 4510*, 87th Cong. 1–2 (1961); Cong. Rec., March 8, 1961, 3467–81 (Senate Ag Committee Chairman Allen Ellender [D-LA]); Cong. Rec., March 10, 1961, 3748 (agreed to by a vote of 52 to 26).

154. Cong. Rec., March 7, 1961, 3400–3401 (Representative Charles Hoeven [R-IA]).

155. Cong. Rec., March 7, 1961, 3400–3401 (Rep. Hoeven); Cong. Rec., March 8, 1961, 3499 (Representative Quie [D-MN]), 3506 (Representative Poage), 3667–68 (Representative Hoeven), 3503 (Representative Leslie Arends [R-IL]), 3660–62 (Representative Paul Findley [R-IL]).

156. Cong. Rec., March 7, 1961, 3395–96 (Chairman Cooley); Cong. Rec., March 8, 1961, 3668 (amendment to strike the sale provision was rejected by a vote of 132 to 163 [on tellers]), 3670 (motion to recommit instructed that the bill be reported back without the provision on selling surplus commodities; it was defeated 196 to 214 [22 not voting]), 3670–71 (the bill was passed on a vote of 209 to 202 [21 not voting]).

157. Cong. Rec., March 21, 1961, 4405–7 (Representative Poage), 4408–9 (Representative Hoeven and Representative Findley), 4410 (Representative Quie), 4507 (Senator Roman L. Hruska [R-NE]), 4412 (the House agreed to the conference report on March 21, 1961, by a vote of 231 to 185 [15 not voting]); Cong. Rec., March 22, 1961, 4498, 4501 (Chairman Ellender), 4509 (the Senate passed the conference report on March 22, 1961, by a vote of 58 to 21 [11 not voting]).

158. Cong. Rec., May 21, 1962, 8783–86 (Chairman Ellender), 8797 (Senator George Aiken [R-VT]); Cong. Rec., May 23, 1962, 8997 (Senators Herman Talmadge [D-GA] and Richard Russell [D-GA]), 9011–15 (Senator Eugene McCarthy [D-MN]), 9021–27 (Senator Hubert Humphrey [D-MN]).

159. Cong. Rec., May 21, 1962, 8786 (Chairman Ellender); Cong. Rec., May 22, 1962, 8949–50 (Ellender and Ellender colloquy with Senator Mundt); Cong. Rec., May 23, 1962, 8998 (Senator Proxmire and Chairman Ellender), 9007 (Senator Carlson).

160. Cong. Rec., May 21, 1962, 8785 (Chairman Ellender); Cong. Rec., May 23, 1962, 8989 (Senator Eastland), 8992–93 (Senators Eastland and Ellender debate), 8997–98 (Senators Proxmire and Ellender); Cong. Rec., May 24, 1962, 9182 (Senator Russell).

161. Cong. Rec., May 21, 1962, 8784–85 (Chairman Ellender); Cong. Rec., May 23, 1962, 8995 (Chairman Ellender colloquy with Senator John O. Pastore [D-RI]).

162. See, e.g., Cong. Rec., May 23, 1962, 8990 (Senator James Eastland [D-MS]).

163. Cong. Rec., May 22, 1962, 8948 (Senator Mundt); Dallek, *An Unfinished Life*, 500–501; Hadwiger and Talbot, *Pressures and Protests*, 101, 196; "Billie Sol Estes Investigation Reports Released," in CQ *Almanac 1964*, 20th ed., 998–99 (Washington DC: Congressional Quarterly, 1965), http://library .cqpress.com/cqalmanac/cqal64-1302889.

164. Cong. Rec., May 21, 1962, 8784 (Chairman Ellender), 8795 (Chairman Ellender); Cong. Rec., May 23, 1962, 8995–97 (Chairman Ellender; Senator Herman Talmadge [D-GA]); Cong. Rec., May 24, 1962, 9196–97 (Senator Everett Dirksen [R-IL]).

165. Cong. Rec., May 22, 1962, 8950 (Senator Karl Mundt [R-SD]).

166. Cong. Rec., May 22, 1962, 8953 (Senator Proxmire); Cong. Rec., May 23, 1962, 8990 (Senator George Aiken [R-VT]), 8993 (Senator James Eastland [D-MS]), 8998 (Senator Milton Young [R-ND]), 9004 (Senators Proxmire and Spessard Holland [D-FL]); Cong. Rec., May 24, 1962, 9196–97 (Senator Everett Dirksen [R-IL]), 9217–18 (Senator Proxmire).

167. Cong. Rec., May 21, 1962, 8785 (Senator Holland; Chairman Ellender acknowledged it), 8808 (Senator Proxmire), 8992–93 (Senators Eastland and Ellender debate), 8997–98 (Senators Proxmire and Russell); Cong. Rec., May 23, 1962, 8989–91 (Senator Eastland); Cong. Rec., May 24, 1962, 9182 (Senator Russell).

168. Hadwiger and Talbot, *Pressures and Protests*, 138–44; Cong. Rec., May 21, 8807–8, 8785–95 (Ellender amendment); Cong. Rec., May 22, 1962, 8950; Cong. Rec., May 23, 1962, 8994 (Senator Mansfield [D-MT] and Ellender colloquy), 9004; Cong. Rec., May 24, 1962, 9192 (Senator Young), 9198 (Senator Mundt).

169. Cong. Rec., May 23, 1962, 8997–98, 9004–5 (Senator Proxmire), 9006–7 (Senator Carlson [R-KS]); Cong. Rec., May 21, 8808; Cong. Rec., May 22, 1962, 8952–53, 8949–50 (Senator Mundt); Cong. Rec., May 23, 1962, 9003; Cong. Rec., May 24, 1962, 9200 (the Senate agreed to Ellender's wheat amendment by a vote of 53 to 36 [11 not voting]), 9236–37 (the Senate agreed to Ellender's amendment for feed grains by a vote of 46 to 37 [17 not voting]).

170. Cong. Rec., June 19, 1962, 10957–60, 10966 (Representative W. R. Poage).

171. Hadwiger and Talbot, *Pressures and Protests*, 121–22, 144–69.

172. Cong. Rec., June 19, 1962, 10955–57.

173. Cong. Rec., June 20, 1962, 11172–75 (Representatives Albert Quie [R-MN] [quoted] and Leslie Arends [R-IL]), 11197 (Representative Bob Dole [R-KS]), 11191–92 (Representative Neal Smith [D-IA]), 11200 (Representative John Kyl [R-IA]).

174. Cong. Rec., June 19, 1962, 10963–64 (Ranking Member Hoeven); Cong. Rec., June 20, 1962, 11189 (Representative Paul Findley [R-IL]), 11193–98 (Representative Ralph F. Beerman [R-NE]).

175. Cong. Rec., June 20, 1962, 11200 (Representative Harlan), 11189 (Representative Findley).

176. Cong. Rec., June 19, 1962, 10963–64 (Ranking Member Hoeven).

177. Cong. Rec., June 21, 1962, 11315 (Representative Harlan Hagen [D-CA]). See also Cong. Rec., June 21, 1962, 11319 (Representative Albert Quie [R-MN]).

178. Cong. Rec., June 20, 1962, 11197.

179. Cong. Rec., June 20, 1962, 11177 (Representative Thomas Abernethy [D-MS]).

180. Hadwiger and Talbot, *Pressures and Protests*, 48–67; Cochrane and Ryan, *American Farm Policy*, 152–53.

181. Hadwiger and Talbot, *Pressures and Protests*, 48–67, 99–105 (quoting Kennedy's State of the Union address); Cochrane and Ryan, *American Farm Policy*, 152–53.

182. Cong. Rec., May 24, 1962, 9181–82 (Senator Keating [R-NY]).

183. Cong. Rec., May 24, 1962, 9302–3.

184. Cong. Rec., May 24, 1962, 9302–3. See also Cong. Rec., May 21, 1962, 8797.

185. Cong. Rec., May 22, 1962, 8920–23 (Senators Eastland and Talmadge).

186. See, e.g., Winders, *Politics*, chapter 5; Fite, *Cotton Fields*, 205, 218, 221.

187. Cong. Rec., May 22, 1962, 8920–23 (Senator Eastland); Cong. Rec., May 24, 1962, 9303–5 (Chairman Ellender, Senators Williams, Dirksen, and Holland), 9198 (Senator Karl Mundt [R-SD] attacked and threatened southerners), 9318 (Majority Leader Mansfield offered a motion to table the Keating amendment, which the Senate agreed to by a narrow vote of 43 to 40 [17 not voting]), 9320–35 (the farm bill, as amended on the floor, passed by a vote of 42 to 38 [20 not voting]).

188. Cong. Rec., June 19, 1962, 10961–63 (Representative Charles B. Hoeven [R-IA]), 10968–69 (Representatives Page Belcher [R-OK], Representatives Odin Langen [R-MN] and Quie); Cong. Rec., June 20, 1962, 11171 (Minority Leader Charles Halleck [R-IN]), 11207 (Representative Findley amendment to ensure federal funds were used for "nonsegregated" recreational facilities); Cong. Rec., June 21, 1962, 11737 (Hoeven feed grain amendment defeated [on a division] 122 to 224).

189. Hadwiger and Talbot, *Pressures and Protests*, 194–97.

190. Hadwiger and Talbot, *Pressures and Protests*, 203.

191. Hadwiger and Talbot, *Pressures and Protests*, 197.

192. Hadwiger and Talbot, *Pressures and Protests*, 198–202; Dallek, *Unfinished Life*, 500.

193. Cong. Rec., June 20, 1962, 11207–8 (Representative Poage and Chairman Cooley; amendment was rejected first on division 101 to 145 and then on tellers 106 to 142), 11210–11 (a similar amendment was also defeated).

194. Cong. Rec., June 21, 1962, 11383–84 (roll call vote on the motion to recommit by Representative Findley was agreed to by a vote of 215 to 205 [17 not voting]); Cong. Rec., June 23, 1962, 11450 (Senator Williams reported on the defeat in the House); Hadwiger and Talbot, *Pressures and Protests*, 206–9.

195. Hadwiger and Talbot, *Pressures and Protests*, 206–9.

196. Hadwiger and Talbot, *Pressures and Protests*, 213–15; Cong. Rec., July 12, 1962, 13462 (revised farm bill); Cong. Rec., July 19, 1962, 14158 (Representative Poage). See also Cong. Rec., August 20, 1962, 17116, 17121 (Chairman Ellender attempted to reinsert mandatory controls but was again rejected by his committee); Cong. Rec., August 21, 1962, 17204–20 (chairman's amend-

ment; Senators John Cooper [R-KY], Mundt, and others debated), 17221 (Chairman Ellender moved to table Cooper's motion to reconsider the vote on Ellender's amendment, which was agreed to by a vote of 58 to 23 [19 not voting]).

197. Cong. Rec., July 19, 1962, 14149. The House also voted to kill the Senate bill, which required the Senate to pass the farm bill again in order to go to conference. See Cong. Rec., July 19, 1962, 14152 (Representative Hoeven), 14162–63 (Representative Dole), 14149 (Chairman Cooley); Cong. Rec., August 20, 1962, 17116 (Chairman Ellender); Cong. Rec., August 22, 1962, 17275–80 (floor statements of Senator Bourke Hickenlooper [R-IA] and Senator Mundt), 17286 (Senator Cooper), 17289 (the bill passed the Senate by a vote of 47 to 37).

198. Cong. Rec., July 19, 1962, 14150–52 (Ranking Member Hoeven), 14198 (by a vote of 229 to 163 [43 not voting]); Cong. Rec., August 21, 1962, 17240–41 (Senator Cooper offered the amendment, and it was agreed to on voice vote after Chairman Ellender stated he was fine with it); Hadwiger and Talbot, *Pressures and Protests*, 227–28.

199. Cong. Rec., September 17, 1962, 19677–78; Cong. Rec., September 20, 1962, 20094–20105.

200. Cong. Rec., September 20, 1962, 20129–30 (it nearly failed 199 to 200 before votes were switched to pass it by 202 to 197 [36 not voting]). See also Cong. Rec., September 25, 1962, 20610 (Senator Hickenlooper), 20619 (Senator Clinton Anderson [D-NM]), 20632 (the vote was 52 to 41 [7 not voting]).

201. Hadwiger and Talbot, *Pressures and Protests*, 213, 234; Food and Agriculture Act of 1962, P.L. 87-703 (1962).

202. Hadwiger and Talbot, *Pressures and Protests*, 248–53, 257–64.

203. Hadwiger and Talbot, *Pressures and Protests*, 297–99, 304, 309, 312–13.

204. Hadwiger and Talbot, *Pressures and Protests*, 316–19.

205. Hansen, *Gaining Access*, 152. See also Hadwiger and Talbot, *Pressures and Protests*, 318–20.

206. Hadwiger and Talbot, *Pressures and Protests*, 322–23. See also Dallek, *Unfinished Life*, 694; Caro, *Passage of Power*, 312–18.

207. Cong. Rec., March 5, 1964, 4441 (Senator George McGovern [D-SD]); Cong. Rec., December 3, 1963, 23036, 23039 (Chairman Harold Cooley [D-NC]).

208. Cong. Rec., March 3, 1964, 4143–47 (Chairman Ellender), 4153–54 (Senator Aiken disputed the harm to the domestic textile industry). Congress had also put a floor under cotton acres, but yield increases resulted in more surplus. See Cong. Rec., March 2, 1964, 4091 (Senator John Stennis [D-MS]).

209. Cong. Rec., March 3, 1964, 4145–46.

210. Cong. Rec., March 2, 1964, 4091–92 (Senator Stennis), 4093 (Chairman Ellender), 4113 (Senators George McGovern [D-SD] and Jack Miller

[R-IA]), 4115–16 (Senator Miller); Cong. Rec., March 3, 1964, 4151 (Senator John Tower [R-TX]); Cong. Rec., March 4, 1964, 4342 (Senator Quentin Burdick [D-ND]), 4345 (Senator Humphrey); Cong. Rec., March 5, 1964, 4443 (Senator Keating); Cong. Rec., March 6, 1964, 4596 (Senator Hubert Humphrey [D-MN]).

211. Cong. Rec., December 3, 1963, 23039, 23042 (Chairman Cooley), 23044 (Representative Ralph Beermann [R-NE]); Cong. Rec., December 4, 1963, 23325 (passed House by a vote of 216 to 182 [7 present; 29 not voting]); Cong. Rec., February 24, 1964, 3380 (Senator Jack Miller [R-IA]); Cong. Rec., February 27, 1964, 3851–54 (Senator Kenneth Keating [R-NY]; senators voted to consider the cotton-wheat bill by a vote of 57 to 19 [24 not voting]); Cong. Rec., March 5, 1964, 4441 (Senator McGovern), 4448–49 (Senator Jacob K. Javits [R-NY]); Cong. Rec., March 3, 1964, 4149–50 (comments of Senator John O. Pastore [D-RI]); Cong. Rec., March 6, 1964, 4589–90 (Senator Pastore), 4655 (passed Senate by a vote of 53 to 35 [12 not voting]).

212. Cong. Rec., April 8, 1964, 7303 (Representative Clarence Brown [R-OH]), 7311 (Minority Leader Charles Halleck [R-IN]).

213. Cong. Rec., April 8, 1964, 7303, 7307 (consideration of the food stamp bill; the motion to recommit that bill was defeated 195 to 223 [15 not voting]), 7310 (Representative Brown [R-OH]), 7312 (Representative Emanuel Celler [D-NY]). See also Cong. Rec., March 2, 1964, 4111.

214. Cong. Rec., April 8, 1964, 7308 (the food stamp bill passed by the House by a vote of 229 to 189 [15 not voting]), 7329 (House agreed to Senate bill by a vote of 211 to 203 [4 voting present; 15 not voting]). President Johnson signed the bill into law on April 11, 1964. P.L. 88-297 (1964)).

215. Ferejohn, "Logrolling," 226–28.

216. Ferejohn, "Logrolling," 228–30.

217. Ferejohn, "Logrolling," 229–30.

218. Ferejohn, "Logrolling," 230.

219. Cong. Rec., September 9, 1965, 23315 (Senator Herman Talmadge [D-GA]); Cong. Rec., August 17, 1965, 20699–20706 (House Ag Chairman Harold Cooley [D-NC]).

220. Cong. Rec., August 17, 1965, 20701 (Representative Sisk), 20723–24 (Representative Thomas Abernethy [D-MS]); Cong. Rec., August 18, 1965, 20908 (Representative Paul Jones [D-MO]).

221. Cong. Rec., August 17, 1965, 20712 (Representative Harlan Hagen [D-CA]). See also Cong. Rec., August 17, 1965, 20701 (Representative Bernice Sisk [D-CA]), 20702 (Representative Charles Teague [R-CA]); Cong. Rec., August 18, 1965, 20913 (Representative Whitten), 20940–41 (Representative Gerald R. Ford [R-MI]).

222. Cong. Rec., August 17, 1965, 20699–20704, 20710 (House Ag Committee Ranking Member Paul Dague [R-PA]); Cong. Rec., September 9, 1965, 23295–97 (Chairman Allen Ellender [D-LA]).

223. Cong. Rec., August 17, 1965, 20718.

224. Cong. Rec., August 18, 1965, 20912–13.

225. Cong. Rec., August 17, 1965, 20726 (Chairman Cooley), 20743–44 (Representative Findley); Cong. Rec., August 18, 1965, 20909 (Representative Paul Jones); Cong. Rec., September 9, 1965, 23306–8 (Chairman Ellender).

226. Cong. Rec., August 18, 1965, 20941 (Representative Poage acknowledged it), 20908 (Representative Jones [D-MO]).

227. Cong. Rec., August 19, 1965, 21013–14 (Representative Maston O'Neal [D-GA] attacked California), 21017–18 (Representative Whitten and Chairman Cooley; Representative Hagen), 21028 (Representative William Dickinson [R-AL]).

228. Cong. Rec., August 17, 1965, 20703; Cong. Rec., August 18, 1965, 20943–44 (Representative Poage's amendment), 20729 (Representative Ezekiel Gathings [D-AR]).

229. Cong. Rec., August 18, 1965, 20945 (Representative Poage), 20956 (the Poage amendment was agreed to by the House on a voice vote).

230. Cong. Rec., August 17, 1965, 20732–33 (Representative Graham Purcell Jr. [D-TX], chair of the wheat subcommittee; Representative Catherine May [R-WA]; Representative Clair Callan [D-NE]), 20705, 20218 (Representative Findley), 20700 (Representative Bernice Sisk [D-CA]), 20701–2 (Chairman Cooley and Representative Thomas O'Neill [D-MA]); Cong. Rec., August 19, 1965, 21031–33 (Cooley amendment), 21042 (Representative Findley).

231. Cong. Rec., September 9, 1965, 23311; Cong. Rec., September 10, 1965, 23440.

232. Cong. Rec., October 8, 1965, 26391 (House took up the conference report), 26400–26401, 26404; Cong. Rec., October 12, 1965, 26678 (Chairman Ellender), 26674 (Chairman Ellender).

233. Cong. Rec., October 8, 1965, 26401–2 (Chairman Cooley), 26415 (House passed it by a vote of 219 to 150 [63 not voting]); Cong. Rec., October 12, 1965, 26672 (Senate debate began), 26689 (the bill passed on voice vote); P.L. 89-321 (1965).

234. Hansen, *Gaining Access*, 154, 158.

235. Cong. Rec., August 17, 1965, 20700–20701, 20742 (Chairman Cooley acknowledged the shift in a colloquy with Representative Michel [R-IL]), 20727–28 (Representative Poage explained), 20729 (Representative Gathings); Cong. Rec., September 9, 1965, 23307–9 (Chairman Ellender opposed the bill), 23313–16, 23317–22 (he received support from Senator John Pastore [D-RI], Senator Leverett Saltonstall [R-MA], Senator Thomas Kuchel

[R-CA], Senator Edmund Muskie [D-ME], Senator Thomas Dodd [D-CT], and Senator Thomas McIntyre [D-NH]), 23324–27 (Senators John Sparkman [D-AL], Walter Mondale [D-MN], and Edward Kennedy [D-MA]); Cong. Rec., September 10, 1965, 23412 (Senator John Tower [R-TX] supported it). See also Cong. Rec., September 9, 1965, 23351 (Senator James Eastland [D-MS]; Cong. Rec., September 10, 1965, 23410 (Senator John Stennis [D-MS], 23413–17 (Chairman Ellender).

236. Cong. Rec., September 9, 1965, 23351–52; Cong. Rec., September 10, 1965 23417, 23354 (Senator Stennis).

237. Cong. Rec., August 18, 1965, 20913–14 (Representative Whitten), 20916–17 (Representative Abernethy), 20923; Cong. Rec., September 13, 1965, 23550 (Senator Daniel Brewster [D-MD]); Cong. Rec., September 14, 1965, 23712, 23721, 23725–26 (Chairman Ellender), 23728 (Brewster amendment defeated 33 to 56 [11 not voting]), 23730 (Senator Williams limit of $50,000 lost 42 to 49 [9 not voting]).

238. Cong. Rec., September 10, 1965, 23425–26 (the Senate agreed to the Talmadge amendment by a vote of 62 to 24 [14 not voting]); Cong. Rec., September 14, 1965, 23768 (it passed by a vote of 72 to 22 [6 not voting]). The House had passed it on August 19, 1965. See Cong. Rec., August 19, 1965, 21070–71 (motion to recommit the bill was defeated 169 to 224 [2 voting present; 39 not voting], and the bill passed 221 to 172 [3 voting present; 37 not voting]).

239. Hansen, *Gaining Access*, 204; Ferejohn, "Logrolling," 231.

240. Wolfgang Saxon, "Ex Congressman W.R. Poage, 87; Texan Headed Agricultural Panel," *New York Times*, January 4, 1987, http://www.nytimes .com/1987/01/04/obituaries/ex-congressman-wr-poage-87-texan-headed -agriculture-panel.html.

241. See, generally, Lerner, "Robert Harold Duke"; Briscoe, "LBJ"; Caro, *Passage of Power*, 533.

242. Lerner, "Robert Harold Duke"; Briscoe, "LBJ."

243. Ferejohn, "Logrolling," 232–33.

244. Ferejohn, "Logrolling," 233–34 (quoting testimony).

245. Ferejohn, "Logrolling," 235–36.

246. Ferejohn, "Logrolling," 235–36; Hansen, *Gaining Access*, 206–7.

247. Hansen, *Gaining Access*, 204; Ferejohn, "Logrolling," 232–33.

248. Hansen, *Gaining Access*, 205–6.

249. Cochrane and Ryan, *American Farm Policy*, 167.

250. See, e.g., Fite, *Cotton Fields*, 194–95, 200, 226; Winders, *Politics*, 115.

251. See, e.g., Hadwiger and Talbot, *Pressures and Protests*, 246; Hansen, *Gaining Access*, 167, 173.

252. Fite, *Cotton Fields*, 205.

253. Fite, *Cotton Fields*, 205, 222; Winders, *Politics*, chapter 5.

254. See, e.g., Hansen, *Gaining Access*, 169.

255. See, e.g., Dallek, *Unfinished Life*, 481–96, 510–18.

256. Nelson, *King Cotton's Advocate*, 74.

257. Hadwiger and Talbot, *Pressures and Protests*, 100–103.

258. Dallek, *Unfinished Life*, 491.

259. Hadwiger and Talbot, *Pressures and Protests*, 100–103.

260. Cong. Rec., April 11, 1956, 6149.

261. See, e.g., Hansen, *Gaining Access*, 171.

5. Commodity "Roller Coaster" and the Crash

1. Cochrane and Ryan, *American Farm Policy*, 61–62.

2. Rapp, *How the U.S.*, 51.

3. Cong. Rec., August 4, 1970, 27131; Agricultural Act of 1970, H.R. Rep. No. 91-1329, at 9 (1970), 9; Cong. Rec., September 15, 1970, 31775–77 (Senator Jacob Javitts amendment for food stamps).

4. Cong. Rec., August 4, 1970, 27146 (Representative Gerald R. Ford [R-MI]), 27150 (Representative Keith G. Sebelius [R-KS]), 27315 (Representative Robert Price [R-TX]); Cong. Rec., September 15, 1970, 31767–68 (Senator George McGovern [D-SD]).

5. Cochrane and Ryan, *American Farm Policy*, 56–62; Cong. Rec., August 4, 1970, 27127 (Representative Delbert Latta [R-OH], ranking member on the House Rules Committee).

6. Percival, "Presidential Management," 985–86.

7. Agriculture and Consumer Protection Act of 1973, H.R. Rep. No. 93-337, at 7–8 (1973) (explaining the 1970 farm bill).

8. Cong. Rec., August 4, 1970, 27131.

9. Cong. Rec., August 4, 1970, 27189 (Representative Paul Findley [R-IL]), 27152 (Representative Sebelius).

10. Cong. Rec., August 4, 1970, 27136 (Representative George Goodling [R-PA]), 27146 (Representative Gerald Ford), 27152 (Representative Sebelius), 27139 (Representative Thomas Kleppe [R-ND]); Cong. Rec., September 14, 1970, 31624 (Senate Ag Chairman Allen Ellender [D-LA]); Cong. Rec., September 15, 1970, 31772 (Senator Harold Hughes [D-IA]).

11. Cong. Rec., August 4, 1970, 27134.

12. Cong. Rec., August 4, 1970, 27188, 27191 (Representative Graham Purcell [D-TX]); Cong. Rec., September 14, 1970, 31624 (corn), 31624–25 (cotton); Cong. Rec., September 15, 1970, 31765–66 (amendment by Senator Joseph Montoya [D-NM] and Senator Walter Mondale [D-MN]), 31767 (Senator William Symington [D-MO] and Senator McGovern), 31774–75 (the Montoya amendment failed 31 to 38 [28 not voting], followed by Senator Sym-

ington amendment agreed to by voice vote); Cong. Rec., October 13, 1970, 36584 (by a vote of 191 to 145 [92 not voting; 1 voting present]); Cong. Rec., November 19, 1970, 38119 (Senator Holland).

13. Cong. Rec., August 4, 1970, 27136 (Representative Goodling), 27189 (Representative Findley), 27149 (Representative Sebelius), 27146 (Representative Gerald Ford), 27140–41 (Representative John Melcher [D-MT]); Cong. Rec., September 14, 1970, 31626 (Senator Young [R-ND]); Cong. Rec., September 15, 1970, 31768 (Senator McGovern), 31769–70 (Chairman Ellender), 31771 (Senator Young), 31772 (Senator Harold Hughes [D-IA]), 31803–5 (Senators Ralph Smith [R-IL] and Charles Percy [R-IL]); Cong. Rec., November 19, 1970, 38100–38104 (Senators Ellender, Talmadge, and Young refused to sign the conference report), 38106 (Senator McGovern opposed), 38107 (Senator Talmadge), 38118 (Senator Young), 38119 (Senator Holland).

14. Cong. Rec., November 19, 1970, 38129 (by a vote of 48 to 35 [14 not voting]). See also Cong. Rec., October 12, 1970, 36134–36; Cong. Rec., November 19, 1970, 38100–38102.

15. See, e.g., Cong. Rec., August 4, 1970, 27136 (Representative Goodling); Cong. Rec., August 5, 1970, 27459 (Representative Silvio Conte [R-MA]), 27460 (Representative Florence Dwyer [R-NJ]).

16. Cong. Rec., August 5, 1970, 27459–60 (Representative Findley).

17. Cong. Rec., August 4, 1970, 27188–89 (Representatives Findley and Abernethy).

18. Cong. Rec., August 4, 1970, 27132 (Chairman Poage); Cong. Rec., September 14, 1970, 31783 (Chairman Ellender), 31788–90 (Senator Herman Talmadge [D-GA]).

19. Cong. Rec., August 4, 1970, 27137 (Representative May [R-WA]), 27141 (Representative Conte); Cong. Rec., August 5, 1970, 27453 (Representatives Conte and Allard Lowenstein [D-NY]), 27460–61 (Representative Abernethy); Cong. Rec., September 14, 1970, 31624, 31783 (Chairman Ellender); Cong. Rec., September 15, 1970, 31778 (Senator Ralph Smith [R-IL] amendment), 31785–88 (Senator John Williams [R-DE]).

20. Cong. Rec., August 5, 1970, 27470–71, 27492 (Representative Conte), 27508 (motion to recommit defeated 167 to 218 [45 not voting] and final passage by a vote of 212 to 171 [47 not voting]), 27499 (Findley); Cong. Rec., September 15, 1970, 31793–94 (Senate rejected tighter limits by a vote of 21 to 44 [27 not voting]), 31815 (Senate passed the bill by a vote of 65 to 7 [27 not voting]).

21. Schuh, "New Macroeconomics," 803–4; Orden, "Exchange Rate Effects," 303.

22. Sandra Kollen Ghizoni, "Nixon Ends Convertibility of U.S. Dollars to Gold and Announces Wage/Price Controls," federalreservehistory.org, November 22, 2013, http://www.federalreservehistory.org/Events/DetailView/33.

23. Schuh, "New Macroeconomics," 805.

24. Rapp, *How the U.S.*, 52–53; Collin and Collin, *Mr. Wheat*, 140.

25. See, e.g., Cong. Rec., July 10, 1973, 23080.

26. Rapp, *How the U.S.*, 52–53.

27. See, e.g., Winders, *Politics*, 131, 156–57.

28. Hansen, *Gaining Access*, 189–91.

29. Bonnen, "Implications for," 391; Winders, *Politics*, 131.

30. Agriculture and Consumer Protection Act of 1973, S. Rep. No. 93-173, at 11 (1973) (reprinting the secretary's speech before the National Agricultural Advertising and Marketing Association, Memphis, TN, May 1, 1973).

31. "Omnibus Farm Bill: New Target Price System Voted," in CQ *Almanac 1973*, 29th ed., 287–307 (Washington DC: Congressional Quarterly, 1974), http//:library.cqpress.com/cqalmanac/cqal73-1228233; Bonnen, "Implications for," 391–92 (quoting from President Nixon's message on "Natural Resources and Environment," delivered on February 15, 1973).

32. Hackbart-Dean, "Greatest Civics Lesson," 316; Minchin, "Historic Upset," 156–97.

33. "Omnibus Farm Bill" (quoting Chairman Talmadge).

34. Barton, "Coalition Building," 144; Hansen, *Gaining Access*, 209.

35. Cong. Rec., June 5, 1973, 18084 (Chairman Talmadge).

36. "Omnibus Farm Bill"; Collin and Collin, *Mr. Wheat*, 138; S. Rep. No. 93-173, at 23.

37. Milton Young, "On Capitol Hill with Senator Young: A Personal Report to the People of North Dakota," May 14, 1973 (on file with author); Collin and Collin, *Mr. Wheat*, 141–42.

38. Collin and Collin, *Mr. Wheat*, 142.

39. Cong. Rec., July 19, 1990, 18165 (Senator Larry Pressler [R-SD]).

40. See, e.g., Cong. Rec., June 5, 1973, 18088–90 (Senator Young), 18092 (Senator Henry Bellmon [R-OK]).

41. S. Rep. No. 93-173, at 1.

42. Cong. Rec., June 5, 1973, 18089 (Senate Ag Committee Ranking Member Carl Curtis [R-NE]).

43. Cong. Rec., June 5, 1973, 18088 (Senator Young), 18085 (Chairman Talmadge), 18091 (Senator Young).

44. S. Rep. No. 93-173, 8, 23–24.

45. Cong. Rec., June 5, 1973, 18085 (Chairman Talmadge), 18088–91 (Senator Young and Senator Curtis), 18092–93 (Senator Henry Bellmon [R-OK]), 18103–4 (Senator Hubert Humphrey [D-MN]), 18106 (Senator Robert Dole [R-KS]); Cong. Rec., June 6, 1973, 18308–9 (Senator George McGovern [D-SD]). See also the Agriculture and Consumer Protection Act of 1973, P.L. No. 93-86, 87 Stat. 221 (1973), 3–4, 224, 233.

46. Cong. Rec., June 6, 1973, 18308 (Senator McGovern).

47. Cong. Rec., June 8, 1973, 18865–66.

48. Cong. Rec., June 8, 1973, 18866–69 (Senator Lloyd Bentsen [D-TX]; Bayh amendment agreed to by a vote of 45 to 37 [13 not voting]).

49. Ferejohn, "Logrolling," 242–43.

50. Cong. Rec., June 7, 1973, 18641 (passed by a vote of 56 to 38 [6 not voting], 18665–66 (Senator Edward Kennedy [D-MA] food stamp amendment); Cong. Rec., June 8, 1973, 18839, 18841–43, 18846.

51. Cong. Rec., June 8, 1973, 18846–48 (Dole conservation reserve program amendment passed by voice vote), 18892 (bill passed by a vote of 78 to 9 [12 not voting]).

52. Agriculture and Consumer Protection Act of 1973, H.R. Rep. No. 93-337, at 7–9 (1973).

53. *General Farm Bill Hearing Before Subcommittee on Livestock and Grains of the Comm. on Agriculture*, 93rd Cong. 409–13, 418, 487–89 (1973) (statements of Dr. Carroll G. Brunthaver, Assistant Secretary for International Affairs and Commodity Programs, and Clifford McIntyre, AFBF, Legislative Director).

54. *General Farm Bill Hearing*, 192–93; Cong. Rec., July 10, 1973, 23034 (Representative Charles Teague [R-CA]); Cong. Rec., July 11, 1973, 23180 (Representative Robert Michel [R-IL]; his amendment to strike it was defeated by a vote of 186 to 220 [27 not voting]).

55. Cong. Rec., July 10, 1973, 23033 (Poage), 23036 (Representative Wiley Mayne [R-IA]), 23041 (Representative Tom Foley [D-WA]), 23043 (Representative Keith Sebelius [R-KS]); H.R. Rep. No. 93-337, 10–13, 20–29.

56. "Omnibus Farm Bill"; Barton, "Coalition Building," 153; Cong. Rec., July 10, 1973, 23034 (Chairman Poage), 23077 (Findley payment limits amendment passed 246 to 163); Cong. Rec., July 11, 1973, 23157–62 (Cotton Inc. amendment).

57. Cong. Rec., July 16, 1973, 23970 (Bergland amendment to strike cotton), 23971–72 (amendment passed by a vote of 207 to 190 [37 not voting]).

58. Cong. Rec., July 16, 1973, 23971 (Representatives Mayne and Conte).

59. "Omnibus Farm Bill."

60. Cong. Rec., July 19, 1973, 24922–28 (Foley amendment to expand food stamps passed 238 to 173 [22 not voting]), 24928–24940 [Dickinson amendment to prohibit food stamps to striking workers passed 213 to 203 [17 not voting]), 24957 (Conte amendment to eliminate the Dickinson amendment and food stamps), 24960 (Conte amendment passed 250 to 165 [18 not voting], but Dickinson tried and won again). See also Cong. Rec., July 16, 1973, 23993–94.

61. Vote records available at https://www.govtrack.us/congress/votes/93 -1973/h252.

62. Barton, "Coalition Building," 151–57; Cong. Rec., July 19, 1973, 24969 (by a vote of 226 to 182 [25 not voting] after much confusion).

63. "Omnibus Farm Bill"; Cong. Rec., July 31, 1973, 26897. See also Marshall A. Martin, Bob F. Jones, William J. Uhrig, Paul R. Robbins, and D. H. Doster. "The Food and Agriculture Act of 1977," Historical Documents of the Purdue Cooperative Extension Service, paper 680, January 1, 1978, https://docs.lib.purdue.edu/cgi/viewcontent.cgi?article=1679&context=agext.

64. Cong. Rec., July 31, 1973, 26897, 26906–7 (Chairman Talmadge explained), 26908 (Senator Helms amendment), 26910–11 (Helms defeated by Senator Humphrey's motion to table the amendment on a vote of 58 to 34 [8 not voting]; final passage was by 85 to 7 [8 not voting]).

65. Cong. Rec., August 3, 1973, 28122 (Representative Teague explained and opposed the tactic), 28123 (House Minority Leader, Representative Gerald R. Ford [R-MI], Representatives Findley and Conte). See also Cong. Rec., July 31, 1973, 27001 (the entire report), 28121 (Poage amendment); Cong. Rec., August 3, 1973, 28124 (Poage's amendment passed by a vote of 349 to 54 [30 not voting], 28126 (the Senate-passed substitute as passed by the House by a vote of 252 to 151 [30 not voting]), 28033 (the Senate concurred in the amendment); Agricultural and Consumer Protection Act of 1973, P.L. No. 93-86 (1973).

66. Porter, "Congress," 15–22; Walter, "Impacts of," 37.

67. Porter, "Congress," 15–22; Walter, "Impacts of," 37–38; CBO website: http://www.cbo.gov/about/overview.

68. See, e.g., Schudson, "Notes on Scandal," 1232.

69. Fortier and Ornstein, "Presidential Succession," 1002.

70. Uslaner and Conway, "Responsible Congressional," 788.

71. Lyons and Taylor, "Farm Politics," 131; Cong. Rec., September 16, 1977, 29570 (Poage).

72. He ascended to Speaker after Jim Wright was forced to resign over ethics scandals. Foley was the first Speaker to be unseated since the Lincoln presidency, a victim of the Gingrich-led Republican revolution in 1994. See Emily Langer, "Thomas S. Foley," Washington Post, October 18, 2013.

73. Lyons and Taylor, "Farm Politics," 131.

74. Cochrane and Ryan, American Farm Policy, 84.

75. See, generally, Blanpied, Politics of Soil.

76. Rapp, How the U.S., 54, 153.

77. Hansen, Gaining Access, 198; Lyons and Taylor, "Farm Politics," 139–45; "1977 Farm Bill Raises Crop Price Supports," in CQ Almanac 1977, 33rd ed., 417–34 (Washington DC: Congressional Quarterly, 1978), http://library.cqpress.com/cqalmanac/cqal77-1203052.

78. See, e.g., Cong. Rec., May 23, 1977, 16043 (Ranking Member, Senator Robert J. Dole [R-KS]); "1977 Farm Bill"; Bowers, Rasmussen, and Baker, "History of," 31; Peters, "1977 Farm Bill," 23–35.

79. Cong. Rec., July 21, 1977, 24400 (Representative Glenn English [D-OK]).

80. Cong. Rec., September 9, 1977, 28523–25 (Senator Hubert Humphrey [D-MN]).

81. "1977 Farm Bill"; Claffey and Stucker, "Food Stamp Program," 44–45.

82. Cong. Rec., September 9, 1977, 28528–30 (Senator Dole [ranking member]).

83. Cong. Rec., September 9, 1977, 28351.

84. Cong. Rec., September 9, 1977, 28536, 28351 (Senator Dole); Cong. Rec., September 16, 1977, 29570 (Representative Wampler quoting Sebelius), 29564 (Chairman Foley).

85. Cong. Rec., May 23, 1977, 16039 (Chairman Talmadge), 16044 (Senator Dole), 16048 (Senator Henry Bellmon [R-OK]), 16050 (Senator Carl Curtis [R-NE] opposed); Cong. Rec., May 24, 1977, 16271 (Senator Strom Thurmond [R-SC] striking workers amendment), 16277 (Senator Curtis amendment on food stamps to help prevent fraud and abuse), 16280–81 (Curtis amendment defeated 37 to 57 [6 not voting]), 16289 (motion to table Thurmond amendment agreed to by a vote of 56 to 38 [6 not voting]), 16303 (Curtis purchase requirement amendment), 16304 (Senator Patrick Leahy [D-VT] opposed), 16307 (Curtis amendment rejected by a vote of 31 to 64 [5 not voting]), 16322 (Curtis amendment on eligibility defeated by a vote of 27 to 64 [9 not voting]), 16323–25 (Senator Samuel Hayakawa [R-CA] college students amendment rejected by a vote of 26 to 63 [11 not voting]).

86. Cong. Rec., May 23, 1977, 16037–38 (Chairman Talmadge), 16043–44 (Ranking Member Robert Dole [R-KS]); "1977 Farm Bill."

87. Cong. Rec., May 23, 1977, 16037.

88. Cong. Rec., May 23, 1977, 16046–47 (Senators Milton Young [R-ND] and Bellmon); Cong. Rec., May 24, 1977, 16273–24 (Senator Richard Clark [D-IA] amendment; Chairman Talmadge opposed; Senator Richard Lugar [R-IN] supported), 16289 (Clark amendment defeated 25 to 69 [6 not voting]), 16293–95 (Senator Edmund Muskie [D-ME], Chairman of the Senate Budget Committee, tried to lower the wheat target price), 16301 (Muskie amendment defeated by a vote of 46 to 50 [4 not voting]), 16351 (passage agreed to by a vote of 69 to 18 [12 not voting]).

89. Cong. Rec., July 20, 1977, 24054–55 (Chairman Foley).

90. Cong. Rec., July 20, 1977, 24063, 24087 (Representative Max Baucus [D-MT]); Cong. Rec., July 21, 1977, 24399 (Foley amendment), 24407 (Foley amendment accepted by the House without a recorded vote), 24415 (Foley amendment), 24417 (agreed to without a recorded vote). See also

Cong. Rec., July 20, 1977, 24057 (Ranking Member William Wampler [R-VA]; Representative (and former chair) W. R. Poage [D-TX]); Cong. Rec., July 22, 1977, 24554, 24573–74.

91. Cong. Rec., July 20, 1977, 24054 (Chairman Tom Foley [D-WA] explained), 24097, 24100 (defeated by a vote of 183 to 230 [1 present; 19 not voting]; Representative Nolan), 24103 (Representative Charles Grassley [R-IA]), 24107 (Nolan amendment defeated 199 to 207 [1 present; 26 not voting]); Cong. Rec., July 21, 1977, 24384–85, 24387 (Grassley amendment rejected without a recorded vote), 24391 (Harking amendment rejected without a recorded vote).

92. Cong. Rec., July 19, 1977, 23700–23701 (Representatives Delbert Latta [R-OH], Steven Symms [R-ID], and Robert Bauman [R-MD]), 23699 (Foley); Cong. Rec., July 20, 1977, 24060 (Ranking Member Wampler), 24074 (Representative Frederick Richmond [D-NY]), 24053–57 (Foley explained); Cong. Rec., July 27, 1977, 25208–9 (amendment by Representative Robert Walker [R-PA] on nutritional foods), 25214 (Walker amendment rejected; Symms offered a similar amendment), 25217 (Symms amendment defeated by a vote of 185 to 227 [21 not voting]), 25218–19 (Mathis amendment to cap spending on food stamps; Representative Richmond [D-NY] opposed), 25221 (Mathis amendment agreed to by a vote of 242 to 173 [18 not voting]).

93. Cong. Rec., July 27, 1977, 25239 (Representative Steven Symms [R-ID] offered the amendment), 25241–42 (Representative Poage supported the amendment and opposed elimination of the purchase requirement), 25248 (Symms amendment defeated by a vote of 102 to 317 [14 not voting]).

94. Cong. Rec., July 27, 1977, 25224 (Kelly amendment), 25234 (Kelly amendment defeated by a vote of 170 to 249 [14 not voting]); Cong. Rec., July 28, 1977, 25480 (Foley substitute on food stamps passed by a vote of 320 to 91 [22 not voting]).

95. Cong. Rec., July 28, 1977, 25530–31 (Findley motion to recommit narrowly defeated 201 to 210 [1 present; 21 not voting; the House passed the 1977 farm bill by a vote of 294 to 114 [25 not voting]). See also Cong. Rec., September 9, 1977, 28541 (conference report passed by a vote of 63 to 8 [29 not voting]); Cong. Rec., September 16, 1977, 29572 (Representative Robert Giaimo [D-CT] on budget concerns), 29579–80 (Representative Findley on further harm to farmers), 29588–89 (conference report passed House by a vote of 283 to 107 [44 not voting].

96. "Pared-Down Emergency Farm Bill Enacted," in CQ Almanac 1978, 34th ed., 436–46 (Washington DC: Congressional Quarterly, 1979), http://library.cqpress.com/cqalmanac/cqal78-1239064; "Grain Target Prices," in CQ Almanac 1980, 36th ed., 98 (Washington, DC: Congressional Quarterly, 1981), http://library.cqpress.com/cqalmanac/cqal80-1175422.

97. Hurt, Problems of Plenty, 138.

98. Hurt, *Problems of Plenty*, 139.

99. Infanger, Bailey, and Dyer, "Agricultural Policy in Austerity," 3; Rapp, *How the U.S.*, 37–38; Harkin and Harkin, "Roosevelt to Reagan," 507.

100. Harkin and Harkin, "Roosevelt to Reagan," 507; Infanger, Bailey, and Dyer, "Agricultural Policy in Austerity," 2.

101. Sinclair, "Agenda Control," 291.

102. Born in 1921 in Monroe, North Carolina, he was the son of the chief of police in the segregated South. He dropped out of college to become a reporter for the *News and Observer* in Raleigh and served in the Navy during World War II. See Steven A. Holmes, "Jesse Helms Dies at 86; Conservative Force in the Senate," *New York Times*, July 5, 2008, http://www.nytimes.com /2008/07/05/us/politics/00helms.html; David Broder, "Jesse Helms, White Racist," *Washington Post*, July 7, 2008, http://www.washingtonpost.com/wp -dyn/content/article/2008/07/06/ar2008070602321.html.

103. Infanger, Bailey, and Dyer, "Agricultural Policy in Austerity," 2.

104. Harkin and Harkin, "Roosevelt to Reagan," 507; Infanger, Bailey, and Dyer, "Agricultural Policy in Austerity," 2.

105. Sam Roberts, "Kika de la Garza, Texas Congressman and Farmers' Ally, Dies at 89," *New York Times*, March 15, 2017, https://www.nytimes.com /2017/03/15/us/politics/kika-de-la-garza-dead.html?_r=0.

106. "Reagan Seeks Sharp Cut in Federal Role," in CQ *Almanac 1981*, 37th ed., 278–80 (Washington DC: Congressional Quarterly, 1982), http:// library.cqpress.com/cqalmanac/cqal81-1172455; Miller and Range, "Reconciling," 12–13.

107. "Reagan Seeks Sharp Cut"; "Fiscal 1982 Reconciliation: $35.2 Billion," in CQ *Almanac 1981*, 27th ed., 256–66 (Washington DC: Congressional Quarterly, 1982), http://library.cqpress.com/cqalmanac/cqal81-1172403; Peters, "1981 Farm Bill," 161–62; "New Farm Bill Clears by Two-Vote Margin," in CQ *Almanac 1981*, 37th ed., 535–48 (Washington DC: Congressional Quarterly, 1982), http://library.cqpress.com/cqalmanac/cqal81-1171658; Harkin and Harkin, "Roosevelt to Reagan," 508.

108. Infanger, Bailey, and Dyer, "Agricultural Policy in Austerity," 4; William Greider, "The Education of David Stockman," *The Atlantic*, December 1981, http://www.theatlantic.com/magazine/archive/1981/12/the-education -of-david-stockman/305760/.

109. Lynch, "Budget Reconciliation."

110. Miller and Range, "Reconciling," 7–15.

111. Cong. Rec., April 2, 1981, 6302–3; Steven V. Roberts, "Rules Chairman Says Reagan Uses Budget as a Method to 'Tyrannize,'" *New York Times*, June 17, 1981, http://www.nytimes.com/1981/06/17/us/rules-chairman-says -reagan-uses-budget-as-a-method-to-tyrannize.html.

112. Rapp, *How the U.S.*, 37.

113. Bodnick, "'Going Public' Reconsidered," 13–28.

114. Margot Hornblower and T. R. Reid, "After Two Decades, the 'Boll Weevils' Are Back, and Whistling Dixie," *Washington Post*, April 26, 1981, https://www.washingtonpost.com/archive/politics/1981/04/26/after-two-decades-the-boll-weevils-are-back-and-whistling-dixie/c256b8cd-840f-4ce9-bec3-790e832e7e84/?utm_term=.dfe4096eb278; Steven V. Roberts, "The Importance of Being a Boll," *New York Times*, June 14, 1981, http://www.nytimes.com/1981/06/14/weekinreview/the-importance-of-being-a-boll-w-stenholm-and-dan-daniel.html; David Frum, "Righter than Newt," *The Atlantic*, March 1995, https://www.theatlantic.com/past/docs/politics/policamp/gramm.htm; Julia Malone, "Reagan's Boll Weevil Friends Get Warning," *Christian Science Monitor*, January 5, 1983, https://www.csmonitor.com/1983/0105/010542.html; Bodnick, "'Going Public' Reconsidered," 16.

115. Key players were Phil Gramm (D-TX), John Breaux (D-LA), and Charles Stenholm (D-TX). Republicans had a seven-vote majority in the Senate but needed at least twenty-seven Democrats in the House. See Weatherford and McDonnell, "Ronald Reagan," 1–29.

116. Miller and Range, "Reconciling," 18, 20; Infanger, Bailey, and Dyer, "Agricultural Policy in Austerity," 4–5; Cong. Rec., May 7, 1981, 8966, 9016 (the vote was 253 to 176 [1 not voting]); Bodnick, "'Going Public' Reconsidered," 18; Omnibus Budget Reconciliation Act of 1981, P.L. No. 97-35, 95 Stat. 357 (1981). See also Harkin and Harkin, "Roosevelt to Reagan," 509.

117. Cong. Rec., June 25, 1981, 14065–85 (debate and votes); Bodnick, "'Going Public' Reconsidered," 19; Cong. Rec., June 26, 1981, 14565, 14570 (Chairman Kika de la Garza [D-TX]), 14568 (Representative Fred Richmond [D-NY]), 14569 (Representative Berkley Bedell [D-IA]), 14681–82 (by a vote of 217 to 211), 14564–65 (Representative Tom Coleman [R-MO]). See also Schick, "How the Budget," 16, 20–22; Miller and Range, "Reconciling," 20–21; "Fiscal 1982 Reconciliation."

118. Cong. Rec., June 26, 1981, 14681 (comments of Jim Wright [D-TX], majority leader of the House); Greider, "Education of David Stockman"; Bodnick, "'Going Public' Reconsidered," 19–20.

119. Thomas B. Edsall, "Democrats Watch GOP Checks Bounce," *Washington Post*, October 25, 1981, https://www.washingtonpost.com/archive/business/1981/10/25/democrats-watch-gop-checks-bounce/b43c7d5f-5563-4a87-b0eb-d6a4f024036b/?utm_term=.4934645ca267.

120. See, e.g., Peters, "1981 Farm Bill," 170; Hornblower and Reid, "After Two Decades."

121. Cong. Rec., June 10, 1981, 11962; Infanger, Bailey, and Dyer, "Agricultural Policy in Austerity," 6.

122. See, e.g., "Reagan Seeks Sharp Cut."

123. Cong. Rec., June 24, 1981, 13615–16; Ferejohn, "Logrolling," 244-46; Infanger, Bailey, and Dyer, "Agricultural Policy in Austerity," 6.

124. Food and Agriculture Act of 1981, H.R. Rep. No. 97-106 (1981); "New Farm Bill"; Infanger, Bailey, and Dyer, "Agricultural Policy in Austerity," 6.

125. Peters, "1981 Farm Bill," 164–66 (non-metro regions and no city larger than 25,000).

126. Peters, "1981 Farm Bill," 166–68.

127. See, e.g., Cong. Rec., September 14, 1981, 20462 (Senator Howell Heflin [D-AL]), 20466 (Senator Dole).

128. Cong. Rec., September 14, 1981, 20460, 20468–69; Cong. Rec., October 2, 1981, 22940–43 (Chairman de la Garza; Representative Berkley Bedell [D-IA] amendment), 22971 (Representative Bedell); Cong. Rec., October 7, 1981, 23496–97 (Representative Tom Daschle [D-SD]).

129. See, e.g., Cong. Rec., September 14, 1981, 20454 (Senator Huddleston [D-KY)] ranking member on Senate Ag), 20466 (Senator Dole), 20489 (Senator Edward Zorinsky [D-NE]); Cong. Rec., October 2, 1981, 22953 (Representative Byron Dorgan [D-ND]), 22958–59 (Representative Pat Roberts [R-KS]), 22946 (Representative William Wampler [R-VA]), 22957 (Representative Foley [D-WA]); Cong. Rec., October 7, 1981, 23498 (Representative Dorgan).

130. Cong. Rec., September 17, 1981, 21059 (Senator Zorinsky), 21062 (Senator James Exon [D-NE]). Zorinsky's amendment was defeated on Senator Dole's motion to table. See Cong. Rec., September 17, 1981, 21064 (by a vote of 60 to 35 [5 not voting]).

131. Cong. Rec., September 14, 1981, 20422 (Chairman Helms), 20448–53.

132. Cong. Rec., September 14, 1981, 20460–61 (Lugar amendment), 20465 (Senator Zorinsky), 20468–69.

133. Cong. Rec., September 18, 1981, 21190–91.

134. Cong. Rec., September 14, 1981, 20484–86 (Chairman Helms).

135. Cong. Rec., September 14, 1981, 20461.

136. Cong. Rec., September 14, 1981, 20460; Cong. Rec., September 17, 1981, 21062.

137. Cong. Rec., September 14, 1981, 20469; Cong. Rec., September 15, 1981, 20560–62 (Lugar peanut amendment); Cong. Rec., September 16, 1981, 20795 (Senator Melcher), 20796 (Senator Larry Pressler [R-SD] agreed); Cong. Rec., September 18, 1981, 21190–91, 21206–7, 21213 (Senator Huddleston).

138. Cong. Rec., September 18, 1981, 21214 (Senators voted to table the Lugar amendment by 45 to 43 [12 not voting]), 21224 (Dole amendment to reduce target prices), 21227–28 (Senator Max Baucus [D-MT] amendment), 21195 (Senator John Melcher [D-MT]), 21228–29 (Dole moved to table the

Baucus amendment and the Senate agreed by 59 to 28 [13 not voting], and then the Senate agreed to the Dole amendment by 46 to 39 [15 not voting]).

139. See, e.g., Cong. Rec., September 15, 1981, 20567 (Senator Exon); Cong. Rec., September 16, 1981, 20768 (Chairman Helms), 20772 (Senator John East [R-NC]), 20773 (Senator Paul Tsongas [D-MA]), 20777 (Senator Roger Jepsen [R-IA] supported Lugar).

140. Cong. Rec., September 16, 1981, 20785 (Majority Leader Howard Baker [R-TN] moved to table the Lugar amendment, but the motion to table failed on a vote of 42 to 56 [2 not voting]), 20796 (Senator John Warner [R-VA]), 20799 (Senator Lugar), 20801 (Senator Lugar and Senator Mack Mattingly [R-GA] debated the compromise), 20802–3 (Senator Heflin, Chairman Helms, and Senator Dole), 20805 (the modified Mattingly amendment narrowly passed the Senate by a vote of 51 to 47 [2 not voting], and the Lugar amendment as amended by Mattingly was agreed to without a recorded vote).

141. Cong. Rec., September 17, 1981, 20940 (Senator Dan Quayle [R-IN]), 20944 (Senator Gordon Humphrey [R-NH]), 20960 (Senator Exon; Chairman Helms moved to table the Quayle amendment, and the Senate agreed to the motion, tabling and thus killing the amendment by a vote of 61 to 33 [1 present; 5 not voting]), 21002–3, 21007 (Senator Mark Hatfield [R-OR] amendment), 21033 (Senate agreed to table the Hatfield amendment by a vote of 53 to 42 [5 not voting]), 21017; Cong. Rec., September 18, 1981, 21247 (passed by a vote of 49 to 32 [19 not voting]).

142. Steven V. Roberts, "Farm Bloc Facing Unusual Coalition," *New York Times*, October 14, 1981, http://www.nytimes.com/1981/10/14/us/farm-bloc -facing-unusual-coalition.html.

143. Edsall, "Democrats Watch" (quoting Representative Evans).

144. Cong. Rec., October 15, 1981, 24149–51 (Representative Glenn English [D-OK]), 24158–60 (Representative Charlie Stenholm [D-TX] feed grain amendment agreed to by voice vote), 24162–64 (Stenholm amendment rejected without a vote), 24163 (Representative David Bowen [D-MS]). See also Cong. Rec., October 15, 1981, 24160 (Representative Findley), 24168 (Representatives Stanley Lundine [D-NY] and Paul Findley [R-IL] amendment to eliminate the peanut program), 24170 (Representative Charles Rose [D-NC] and Findley debated), 24180 (House agreed to the Lundine amendment by a vote of 250 to 159 [24 not voting]), 24181–24209 (Representative Peter Peyser [D-NY] amendment to eliminate the sugar program agreed to by a vote of 213 to 190 [1 voting present; 29 not voting]).

145. Cong. Rec., October 21, 1981, 24724–25 (Representative Robert Shamansky [D-OH] amendment to eliminate the tobacco price support and allotment program), 24755–56 (Shamansky amendment defeated by a vote of 184 to 231 [2 voting present; 16 not voting]), 24757–58 (Representative

Findley amendment to make the tobacco program "self-financing"), 24760 (Findley amendment as amended by the Foley substitute to make sure the tobacco program ran at no cost to the taxpayers agreed to overwhelmingly by the House on a vote of 412 to 0 [21 not voting]); Edsall, "Democrats Watch"; David S. Broder, "Hill Democrats Grant Amnesty to Boll Weevils," *Washington Post*, September 17, 1981, https://www.washingtonpost.com/archive /politics/1981/09/17/hill-democrats-grant-amnesty-to-boll-weevils/94db7f22 -49b4-42d4-ae36-558cd8c549e4/?utm_term=.af7c447da57b.

146. Cong. Rec., October 7, 1981, 23501 (Bedell amendment passed by a vote of 400 to 14 [19 not voting]); Cong. Rec., October 14, 1981, 23881–23899.

147. See, e.g., Cong. Rec., October 2, 1981, 22955 (Representative Lynn Martin [R-IL]); Cong. Rec., October 21, 1981, 24778–90; Cong. Rec., October 22, 1981, 24934 (motion to recommit defeated, and the bill passed without a recorded vote).

148. Cong. Rec., September 17, 1981, 20977 (Senator William Armstrong [R-CO]); Mayer, "Farm Exports," 99–111.

149. See, e.g., Mayer, "Farm Exports," 100.

150. See, e.g., Cong. Rec., September 17, 1981, 20999 (Senator Patrick Leahy [D-VT] amendment), 21042 (Senator Charles Grassley amendment), 21046 (agreed to), 21070–71 (another Grassley conservation amendment agreed to without a recorded vote).

151. Cong. Rec., October 2, 1981, 22944–46.

152. Cong. Rec., December 10, 1981, 30381 (Chairman Helms explained).

153. Cong. Rec., December 16, 1981, 31822 (Representative Tom Daschle), 31826 (Representative Bedell), 31828 (Representative Dorgan), 31842 (Representative Jerome Traxler [D-MI]).

154. Cong. Rec., December 16, 1981, 31812–13.

155. Cong. Rec., December 16, 1981, 31817.

156. Cong. Rec., December 10, 1981, 30379, 30389 (Senator Thad Cochran [R-MS]), 30382–84 (Ranking Member Walter "Dee" Huddleston [D-KY]), 30397 (Senator Grassley).

157. Cong. Rec., December 10, 1981, 30394, 30397 (Senator Grassley).

158. Cong. Rec., December 10, 1981, 30387 (Senator David Boren [D-OK] and Senator Zorinsky [D-NE]), 30403 (Senator David Pryor [D-AR]), 30390–91 (Senator Jim Exon [D-NE]).

159. Cong. Rec., December 10, 1981, 30394–95 (Senator Dole), 30395–97 (Senator Leahy).

160. Cong. Rec., December 10, 1981, 30404 (by a vote of 67 to 32 [1 not voting]).

161. Cong. Rec., December 16, 1981, 31822 (Representative Dan Glickman [D-KS]), 31830 (Representative Tom Harkin [D-IA]), 31823 (Representative

Lundine), 31825–26 (Representative Findley [R-IL]), 31847 (passed by a vote of 205 to 203 [1 voted present and 24 not voting]).

162. Cong. Rec., December 16, 1981, 31833–37 (Representative Richmond), 31838. President Reagan signed the Agricultural and Food Act of 1981 on December 22, 1981. See Agricultural and Food Act of 1981, P.L. No. 97-98 (1981).

163. Orden, Paarlberg, and Roe, *Policy Reform*, 72–73; Rapp, *How the U.S.*, 38; Bowers, Rasmussen, and Baker, "History of," 38.

164. Spitze, "Agriculture and Food Act of 1981," 66; Orden, Paarlberg, and Roe, *Policy Reform*, 72–73.

165. Orden, Paarlberg, and Roe, *Policy Reform*, 73; Bowers, Rasmussen and Baker, "History of," 40.

166. Orden, Paarlberg, and Roe, *Policy Reform*, 73; GAO, "1983 Payment-in-Kind," i, 4.

167. GAO, "1983 Payment-in-Kind," 4; Bowers, Rasmussen, and Baker, "History of," 41.

168. U.S. House Joint Committee on Taxation, "Description of the Administration's," 3, 9.

169. Rapp, *How the U.S.*, 39, 147.

170. Orden, Paarlberg, and Roe, *Policy Reform*, 73; Bowers, Rasmussen, and Baker, "History of," 41; GAO, "1983 Payment-in-Kind," iii.

171. GAO, "1983 Payment-in-Kind," ii; Bowers, Rasmussen, and Baker, "History of," 41; Orden, Paarlberg, and Roe, *Policy Reform*, 74.

172. Rapp, *How the U.S.*, 132–33, 140–41; Bowers, Rasmussen, and Baker, "History of," 41–42.

173. Cong. Rec., October 25, 1985, 29130 (Senator Cochran); Cong. Rec., October 30, 1985, 29711–12 (Ranking Member Zorinsky); Cong. Rec., November 19, 1985, 32315 (Senator Boschwitz).

174. Rapp, *How the U.S.*, 21; "1985 Farm Bill Changes," in CQ *Almanac 1986*, 42nd ed., 302–4 (Washington DC: Congressional Quarterly, 1987), http://library.cqpress.com/cqalmanac/cqal86-1150429.

175. Cong. Rec., October 28, 1985, 29267 (Senator Jim Exon [D-NE]).

176. "1985 Farm Bill Changes."

177. Cong. Rec., October 1, 1985, 25450–53 (Representative Harold Volkmer [D-MO]); Rapp, *How the U.S.*, 20–21.

178. Rapp, *How the U.S.*, 15.

179. Bowers, Rasmussen, and Baker, "History of," 42–43; Orden, Paarlberg, and Roe, *Policy Reform*, 74.

180. Guither, "Tough Choices," 14; Rapp, *How the U.S.*, 19; "1985 Farm Bill Changes"; Orden, Paarlberg, and Roe, *Policy Reform*, 74; Spitze, "Evolution and Implications," 179–82.

181. Guither, "Tough Choices," 14; Rapp, *How the U.S.*, 19; "1985 Farm Bill Changes"; Orden, Paarlberg, and Roe, *Policy Reform*, 74; Spitze, "Evolution and Implications," 179–82.

182. Cong. Rec., September 20, 1985, 24543 (Representative Combest), 24553 (Representative Tom Daschle [D-SD]), 24561 (Representative Roberts), 24554 (Representative Arlan Strangeland [R-MN]), 24552(Representative Charlie Stenholm [D-TX]).

183. Guither, "Tough Choices," 1–3; Rapp, *How the U.S.*, 10, 15–17; Orden, Paarlberg, and Roe, *Policy Reform*, 74.

184. See, e.g., Rapp, *How the U.S.*, 22–24, 138; Guither, "Tough Choices," 16; "1985 Farm Bill Changes."

185. Rapp, *How the U.S.*, 19, 74–76; "1985 Farm Bill Changes"; Orden, Paarlberg, and Roe, *Policy Reform*, 74–75; Guither, "Tough Choices," 16; Spitze, "Evolution and Implications," 182.

186. Cong. Rec., October 1, 1985, 25456–57 (Representative Bill Alexander [D-AR]).

187. Guither, "Tough Choices," 13 (quoting Bergland's testimony before the House Ag Committee in February 1984).

188. Cong. Rec., September 20, 1985, 24533 (Chairman de la Garza), 24537 (Representative Pat Roberts [R-KS]), 24543 (Representative Larry Combest [R-TX]), 24560–61 (Representative Roberts).

189. Cong. Rec., December 17, 1985, 37070; Cong. Rec., December 18, 1985, 37570–71 (Ranking Member, Senator Ed Zorinsky [D-NE]), 37573–76 (Senators Dole and Cochran), 37774–79 (Chairman de la Garza [D-TX]; Representative Robert F. Smith [R-OR]).

190. Cong. Rec., September 26, 1985, 24975 (Rep. Downey sugar amendment), 24990 (Chairman de la Garza), 24991 (amendment rejected on a vote of 142 to 263 [2 voting present; 27 not voting], 25038 (Representative Tony Coelho [D-CA]); Cong. Rec., October 3, 1985, 25951 (Lundine peanut amendment), 25957 (Representatives Lundine, William Dickinson [R-AL], Charles Hatcher [D-GA], and Stenholm), 25966 (the amendment defeated by a vote of 195 to 228 [11 not voting]).

191. Food Security Act of 1985, H.R. Rep. No. 99-271 (1985).

192. Cong. Rec., October 1, 1985, 25426–27, 25438 (Representative Roberts).

193. Cong. Rec., October 1, 1985, 25418–23 (amendments from Representatives Glickman [D-KS] and Dorgan [D-ND]), 25428–29 (Representatives Daschle and Ed Madigan [R-IL]), 25437 (Representative English), 25432 (Chairman de la Garza opposed), 25424–25 (Representative Foley), 25439 (Strangeland amendment rejected by the House on a vote of 200 to 228 [6 not voting]), 25444 (Dorgan amendment rejected without a recorded vote).

194. Cong. Rec., October 3, 1985, 25949–50 (Madigan amendment to strike Bedell provisions); Cong. Rec., October 1, 1985, 25454 (Representative Harold Volkmer [D-TN]), 25464–65 (Madigan amendment), 25467 (Daschle and English; Emerson and Stenholm), 25459 (Representative Richard Durbin [D-IL]), 25463 (Representative Lane Evans [D-IL] amendment accepted by voice vote).

195. Cong. Rec., September 20, 1985, 24561 (Representative Roberts); Cong. Rec., October 1, 1985, 25436 (Representative Stenholm), 25463 (Volkmer amendment rejected without a recorded vote); Cong. Rec., October 2, 1985, 25736; Cong. Rec., October 3, 1985, 25892–93, 25895 (Representative Roberts), 25905–6 (Title X of the bill), 25950–51 (Madigan amendment agreed to by a vote of 251 to 174 [9 not voting]); Cong. Rec., October 8, 1985, 26561 (Representative Bill Alexander [D-AR]), 26575–76, 26577–78 (Chairman de la Garza and Ranking Member Madigan), 26656 (Representative Strangeland apologized), 26660–61 (Alexander amendment defeated 59 to 368 [1 present; 6 not voting]), 26665 (the House passed the farm bill without a recorded vote).

196. Orden, Paarlberg, and Roe, *Policy Reform*, 80–81; Guither, "Tough Choices," 51–55.

197. Rapp, *How the U.S.*, 70–73; Agriculture, Food, Trade, and Conservation Act of 1985, S. Rep. 99-145 (1985).

198. Cong. Rec., October 25, 1985, 29103–5 (Chairman Helms).

199. Cong. Rec., October 25, 1985, 29103–5 (Chairman Helms), 29109–10 (Chairman Helms), 29124 (Ranking Member Ed Zorinsky [D-NE]), 29130 (Senator Thad Cochran [R-MS]).

200. Cong. Rec., October 30, 1985, 29707 (Senator Lugar), 29721 (Senator Boren); Cong. Rec., Oct 25, 1985, 29102, 29107–8, 29123–24 (Ranking Member Ed Zorinsky [D-NE]).

201. Cong. Rec., October 28, 1985, 29267–68 (Senator Exon); Cong. Rec., October 30, 1985, 29706 (Lugar amendment).

202. Cong. Rec., October 30, 1985, 29708–9 (Senator Lugar).

203. Cong. Rec., October 30, 1985, 29708–9 (Chairman Helms).

204. Cong. Rec., October 30, 1985, 29712–13 (Senator David Pryor [D-AR]), 29716–17 (Senator Dale Bumpers [D-AR] and Senator Exon [D-NE]), 29714–15 (Senator Grassley [R-IA]), 29717 (Senator Andrews [R-ND]).

205. See, e.g., Guither, "Tough Choices," 79; "1985 Farm Bill Changes."

206. Cong. Rec., October 28, 1985, 29261–64 (Majority Leader Dole's placeholder amendment in the amendment tree), 29275 (Dole); Cong. Rec., October 30, 1985, 29724 (by a vote of 48 to 51 [1 not voting]), 29726 (Majority Leader Dole); Cong. Rec., October 31, 1985, 29973 ("smoke and mirrors"), 30025, 30048 (Dole in response to Senator Melcher), 30057–64, 30088–89 (summary of the Lugar/Dole compromise; 9:00 at night); Cong. Rec.,

November 1, 1985, 30262–65 (Dole motion to table was defeated by a vote of 45 to 49 [6 not voting]).

207. Cong. Rec., November 18, 1985, 32240 (Helms), 32242–46 (Ranking Member Zorinsky and Senator Melcher); Cong. Rec., November 19, 1985, 32348–50 (Dole), 32351 (Senator Harkin moved to table the Dole amendment, and the motion was agreed to by a vote of 88 to 8 [4 not voting]; Senator Pryor), 32352 (Senator Dole).

208. Cong. Rec., November 19, 1985, 32353 (Senators Melcher and Dole); Cong. Rec., November 20, 1985, 32665–69 (Senator Dole and Chairman Helms), 32733 (Dole modified amendment agreed to by a vote of 56 to 41 [3 not voting]; Cong. Rec., November 22, 1985, 33302–6 (Senator Boschwitz [R-MN] amendment), 33307–8 (Senators Melcher and Zorinsky), 33310–11 (Boschwitz amendment defeated by a vote of 42 to 48 [10 not voting]); Cong. Rec., November 23, 1985, 33476–78 (late night negotiations; Senator Byrd; Senator Dole), 33479 (the amendment was accepted by the Senate without a recorded vote [unanimous consent]), 33482 (passage by a vote of 61 to 28 [11 not voting]).

209. Cong. Rec., December 18, 1985, 37565 (Senate Ag Chairman Helms).

210. Guither, "Tough Choices," 110–11 (quoting Chairman Helms); Cong. Rec., December 17, 1985, 37071–72; Cong. Rec., December 18, 1985, 37618, 37775 (Chairman de la Garza).

211. Guither, "Tough Choices," 106; Cong. Rec., December 18, 1985, 37573 (Dole), 37566–67 (Helms).

212. Cong. Rec., October 3, 1985, 26033 (Representative Bill Emerson [R-MO]); Cong. Rec., October 7, 1985, 26439–41 (Representative Newt Gingrich [R-GA]), 26426 (Representative Leon Panetta [D-CA]), 26435 (Emerson amendment defeated 171 to 238 [25 not voting]), 26441–42 (Gingrich amendment was defeated by a vote of 183 to 227 [24 not voting]), 26436–37 (Representative Jim Jeffords [R-VT] amendment); Cong. Rec., October 25, 1985, 29106 (Chairman Helms on food stamp block granting), 29127 (Zorinsky); Cong. Rec., November 21, 1985, 33006 (Chairman Helms amendment); Cong. Rec., October 25, 1985, 33007–9 (Senator Cochran), 33017 (Senate agreed to table it by a vote of 68 to 30 [2 not voting]); Cong. Rec., November 21, 1985, 33038 (Senator Robert Stafford [R-VT] amendment); Cong. Rec., October 25, 1985, 33044 (motion to table his amendment was defeated by a vote of 36 to 63 [1 not voting], and the amendment was agreed to on voice vote); Cong. Rec., December 18, 1985, 37794 (Representative Ludine [D-NY]), 37566–67 (Chairman Helms).

213. Cong. Rec., December 18, 1985, 37570 (Senator Zorinsky [D-NE]).

214. Cong. Rec., December 18, 1985, 37586 (Senator Boschwitz), 37780 (Representative Pat Roberts), 37792–93 (Representative Larry Combest [R-TX]).

215. Cong. Rec., December 18, 1985, 37802 (Ranking Member Ed Madigan [R-IL]).

216. Cong. Rec., December 18, 1985, 37774–75 (Chairman de la Garza [D-TX]), 37573 (Senator Dole), 37576 (Senator Cochran), 37779 (Representative Dan Glickman [D-KS], "the long shadow of the President's veto pen"), 37780 (Representative Bedell [D-IA]), 37792 (Representative Byron Dorgan [D-ND]), 37783 (Representative Tom Daschle [D-SD]), 37797 (Representative Conyers [D-MI]), 37590 (Senator Dixon), 37592 (Senator Harkin), 37598 (Senator Grassley), 37579–80 (Senator James Exon [D-NE]).

217. Cong. Rec., December 18, 1985, 37590 (Senator Dixon), 37592 (Senator Harkin), 37598 (Senator Grassley), 37579–80 (Senator James Exon [D-NE]).

218. Cong. Rec., December 18, 1985, 37773–74 (Chairman de la Garza), 37575–37769 (Senator Melcher), 37803 (by a vote of 325 to 96 [13 not voting]), 37616–17 (by a vote of 55 to 38 [7 not voting].

219. See, e.g., Malone, "Historical Essay," 577; Angelo, "Corn, Carbon and Conservation," 624; McGranahan et al., "Historical Primer," 67A; Zachary Cain and Steven Lovejoy, "History and Outlook for Farm Bill Conservation Programs," *Choices*, Quarter 4 (2004), http://www.choicesmagazine .org/2004-4/policy/2004-4-09.htm.

220. Malone, "Historical Essay," 581–82.

221. Cong. Rec., December 18, 1985, 37565 (Chairman Helms), 37571 (Senator Zorinksy), 37573 (Senator Dole), 37774 (Chairman de la Garza).

222. Food Security Act of 1985, P.L. No. 99-198, 99 Stat. 1504 (1985); Orden, Paarlberg, and Roe, *Policy Reform*, 77; Spitze, "Evolution and Implications," 186; Malone, "Historical Essay."

223. See, e.g., Spitze, "Evolution and Implications," 186.

224. Cong. Rec., October 25, 1985, 29106 (Chairman Helms), 29128 (Senator Zorinsky).

225. Cong. Rec., October 3, 1985, 26011 (Representative Glickman), 26014 (Representative Ronald Marlenee [R-MT]; Representative Coleman), 26017 (Glickman amendment [as amended by the Coleman amendment] was agreed to by the House on a vote of 313 to 90 [31 not voting]). See, e.g., Spitze, "Evolution and Implications," 186.

226. See, e.g., Orden, Paarlberg, and Roe, *Policy Reform*, 77.

227. Lin, Riley, and Evans, *Feed Grains*; Hoffman, Chomo, and Schwartz, *Wheat*; Glade, Meyer, and MacDonald, *Cotton*; Schnepf and Just, *Rice*; Ash et al., *Oilseeds*; Cochrane, *Development of American Agriculture*, 332–33; Winders, *Politics*, 160.

228. USDA, ERS, "Farm Commodity Programs."

229. Note that the continued authorization for the PIK program also added to the overall costs and reach of the farm bill. See, e.g., Robinson, *Farm and Food Policies*, 24; Rapp, *How the U.S.*, 149–50.

230. See, e.g., Robinson, *Farm and Food Policies*, 24.

231. Rapp, *How the U.S.*, 31–32.

232. Norris, "1986 US Elections," 197.

233. E. J. Dionne Jr., "Elections; Democrats Gain Control of Senate, Drawing Votes of Reagan's Backers; Cuomo and D'Amato Are Easy Victors; What Awaits Congress; Broad G.O.P. Losses," *New York Times*, November 5, 1986; Norris, "1986 US Elections," 199.

234. "Plan to Cut Farm Programs Stalls," in CQ *Almanac 1995*, 51st ed., 3-47-3-56 (Washington DC: Congressional Quarterly, 1996), http://library .cqpress.com/cqalmanac/cqal95-1100229.

235. Rapp, *How the U.S.*, 24, 105.

236. Rapp, *How the U.S.*, 116–19, 122–23.

237. Rapp, *How the U.S.*, 25–26, 149–50.

238. Rapp, *How the U.S.*, 158–60.

239. Cochrane, *Development of American Agriculture*, 165–67.

240. Spitze, "Continuing Evolution," 127.

241. Milton Young, "On Capitol Hill with Senator Young: A Personal Report to the People of North Dakota," May 14, 1973 (on file with author); Collin and Collin, *Mr. Wheat*, 141–42; Cong. Rec., June 5, 1973, 18090 (Senator Curtis); Cong. Rec., June 7, 1973, 18620 (Senator McGovern).

242. See, e.g., Winders, *Politics*, 155–57.

243. See, e.g., Barton, "Coalition Building," 152–53.

244. Daniel, "Crossroads," 455; Fite, *Cotton Fields*, 168, 207.

245. See, e.g., Fite, *Cotton Fields*, 205, 218–22; Daniel, "Crossroads," 455.

246. See, e.g., Peters, "1977 Farm Bill," 163–64; Infanger, Bailey, and Dyer, "Agricultural Policy in Austerity," 4.

247. See, e.g., Cong. Rec., September 14, 1981, 20462.

248. See, e.g., Rapp, *How the U.S.*, 37; Peters, "1977 Farm Bill," 169; Steven V. Roberts "Congress: The Eclipse of the Boll Weevils," *New York Times*, March 26, 1983, http://www.nytimes.com/1983/03/26/us/congress-the-eclipse -of-the-boll-weevils.html.

249. See, e.g., Robinson, *Farm and Food Policies*, 24; Rapp, *How the U.S.*, 149–50.

250. See, e.g., Orden, Paarlberg, and Roe, *Policy Reform*, 79; Spitze, "Evolution and Implications," 178; Robinson, *Farm and Food Policies*, 22–23; Cong. Rec., December 18, 1985, 37565 (Chairman Helms).

251. See, e.g., Spitze, "Future of Agricultural," 11–12.

6. Revolution and Reform

1. See, e.g., Cong. Rec., July 19, 1990, 18149–50 (Senator Lugar), 18157 (Senator Thad Cochran [R-MS]).

2. Cong. Rec., July 19, 1990, 18163 (Senator David Boren [D-OK]).

3. Winders, *Politics*, 182–83.

4. Cong. Rec., July 19, 1990, 18149–50 (Senator Richard Lugar [R-IN]).

5. Stephen Knott, "George H. W. Bush: Campaigns and Elections," Miller Center of Public Affairs, University of Virginia, n.d., https://millercenter.org/president/bush/campaigns-and-elections.

6. "Congress Enacts Lean Farm Package," in CQ *Almanac 1990*, 46th ed., 323–51 (Washington DC: Congressional Quarterly 1991), http://library.cqpress.com/cqalmanac/cqal90-1112689.

7. "Biography: About Senator Leahy," n.d., https://www.leahy.senate.gov/about/.

8. Shaw, *Richard G. Lugar*, 22–25; Liz Halloran, "A Senate Legend, Undone by His Greatest Strength," NPR.org, May 8, 2012, http://www.npr.org/2012/05/08/152301533/a-senate-legend-undone-by-his-greatest-strength. See also Cong. Rec., March 18, 2009, S3350–51 (Senators McConnell and Reid congratulating Senator Lugar on his 12,000th vote); "Senator Richard Lugar," Lugar Center, n.d., http://www.thelugarcenter.org/about-lugar.html.

9. Shaw, *Richard G. Lugar*, 23.

10. Hershey, "Richard Lugar," 274–75.

11. Jerry Thornton, "Former Rep. Edward Madigan, 58," *Chicago Tribune*, December 8, 1994, http://articles.chicagotribune.com/1994-12-08/news/9412080108_1_edward-madigan-farm-bill-production-limits.

12. "Madigan, Edward Rell," Biographical Directory of the United States Congress, n.d., http://bioguide.congress.gov/scripts/biodisplay.pl?index=m000041.

13. Harris, "Legislative Parties," 189.

14. "Lean Farm Package."

15. Omnibus Budget Reconciliation Act of 1987, P.L. No. 100-203 (1987); Omnibus Budget Reconciliation Act of 1987, H.R. Rep. No. 100-391, at 14–15 (1987).

16. GAO, "Basic Changes Needed"; H.R. Rep. No. 100-391, at 14.

17. Cong. Rec., August 1, 1990, 21426–27 (Representative Silvio Conte [R-MA] amendment), 21428–29 (Conte amendment), 21432–33 (Representative Armey), 21435–37 (Representative Conte), 21441 (Representative Schumer).

18. Cong. Rec., October 22, 1990, 31976, 32193; Cong. Rec., October 23, 1990, 32685 (Chairman de la Garza); Cong. Rec., October 24, 1990, 33887 (Chairman Leahy); Cong. Rec., October 23, 1990, 32706–7 (conference report passes the House on a vote of 318 to 102 [13 not voting]; Cong. Rec., October 24, 1990, 33933 (Senate passed the conference report by a vote of 60 to 36 [1 live pair; 3 not voting]); Cong. Rec., October 26, 1990, 34544 (triple base language), 35253 (final passage of reconciliation by a vote of 228 to 200 [5 not voting]); Cong. Rec., October 27, 1990, 36278 (reconciliation passed the

Senate by a vote of 54 to 45 [1 not voting]); Food, Agriculture, Conservation, and Trade Act of 1990, P.L. No. 101-624, 104 Stat. 3359 (1990).

19. Cong. Rec., October 23, 1990, 32688 (Ranking Member Madigan and Representative Glickman), 32689 (Representative Huckaby); Cong. Rec., October 24, 1990, 33884 (Senate debate), 33885 (Senator Baucus opposed), 33902-3 (Senator Harkin opposed).

20. The CRP acreage cap was increased to between forty and forty-five million acres, which included one million acres specifically for wetlands. The conservation provisions in the bill provided Chairman de la Garza with the opportunity to argue that it was "probably the most environmentally progressive legislation to ever come out of the House and Senate Agriculture Committees"—appropriate for the twentieth anniversary of Earth Day. See Cong. Rec., October 23, 1990, 32685-86.

21. Cong. Rec., July 24, 1990, 18968 (Chairman Leahy).

22. Cong. Rec., July 19, 1990, 18147 (Chairman Leahy informed it was reported by a 15 to 4 vote, with "bipartisan broad support").

23. Cong. Rec., July 19, 1990, 18147-49 (Chairman Leahy).

24. Cong. Rec., July 19, 1990, 18157 (Senator Cochran), 18197 (Senator Grassley [R-IA]), 18161 (Senator Boschwitz); Cong. Rec., July 20, 1990, 18439 (Senator Boschwitz amendment), 18439-41 (Senator Boschwitz).

25. Ritchie, "'De-Coupled' Approach"; Ritchie, "Global Agricultural Trade"; Winders, *Politics*, 155-57.

26. Cong. Rec., July 20, 1990, 18440-41 (Senator Lugar), 18442 (Senator Bob Kerrey [D-NE]), 18444 (Senator Grassley), 18445 (Boschwitz amendment fails on a vote of 37 to 50 [13 not voting]); Cong. Rec., July 24, 1990, 18963-64 (Boschwitz), 18966 (Majority Leader George Mitchell [D-ME]; amendment agreed to without a recorded vote), 19011 (Senator Simon), 19013-14 (Senator Grassley; Senator Steven Symms [R-ID]), 19018 (Senator Boschwitz), 19019 (Senator Dole), 19017 (Leahy), 19020 (Senator Cochran; Senator Kerrey), 19021 (Senator Howell Heflin [D-AL]), 19023 (the Grassley amendment defeated on Leahy's motion to table by a vote of 66 to 30 [4 not voting]).

27. Cong. Rec., July 23, 1990, 18624 (amendment by Senator Kent Conrad [D-ND], supported by Senator Max Baucus [D-MT]); Cong. Rec., July 24, 1990, 18997-99 (Daschle amendment); Cong. Rec., July 25, 1990, 19283-86 (Baucus amendment).

28. Cong. Rec., July 19, 1990, 18155-56, 18165 (Senator Larry Pressler [R-SD]), 18172-73 (Senator Max Baucus [D-MT]); Cong. Rec., July 25, 1990, 19290-91 (Baucus).

29. Cong. Rec., July 19, 1990, 18150 (Senator Rudy Boschwitz [R-MN]), 18625 (Senator Lugar); Cong. Rec., July 25, 1990, 19287-89 (Senator Lugar), 19290 (Senator Baucus), 19288 (Lugar), 19291 (Senator Simon).

30. Cong. Rec., July 19, 1990, 18158 (Senator Cochran); Cong. Rec., July 24, 1990, 18999 (Senator Lugar), 19005–7 (Chairman Leahy), 19008 (Senator Daschle), 19010 (Senator Conrad); Cong. Rec., July 25, 1990, 19289–90 (Chairman Leahy), 19011 (Daschle amendment rejected 24 to 72 [4 not voting]), 19322 (Baucus amendment defeated by a vote of 26 to 72 [2 not voting]).

31. Cong. Rec., July 20, 1990, 18450 (Boschwitz amendment).

32. Cong. Rec., July 19, 1990, 18159–61.

33. Cong. Rec., July 24, 1990, 18966–67 (Senator Boschwitz).

34. Cong. Rec., July 24, 1990, 18968 (Chairman Leahy and Senator Lugar), 18970 (by a vote of 32 to 65 [3 not voting]).

35. Cong. Rec., July 19, 1990, 18152; Cong. Rec., July 26, 1990, 19514–15 (Senator Kit Bond [R-MO] and Senator Conrad), 19524 (Senator Grassley).

36. See, e.g., Cong. Rec., July 26, 1990, 19514–15 (Senator Bond), 19522–23 (Senator Bond).

37. Cong. Rec., July 26, 1990, 19515 (Senator Conrad), 19518 (Senator Daschle), 19532 (Senator Bond), 19535 (Bond moved to table Daschle's second-degree amendment, but the motion failed by a vote of 52 to 45 [3 not voting]; Bond amendment as amended by Daschle was agreed to without a recorded vote).

38. Cong. Rec., July 27, 1990, 19889 (passage by a vote of 70 to 21 [9 not voting]).

39. Cong. Rec., July 19, 1990, 18148–55 (the Leahy-Lugar amendment to reduce the bill's cost estimate was agreed to by the Senate without a recorded vote).

40. "Lean Farm Package."

41. See, e.g., Cong. Rec., July 23, 1990, 18741–72 (Representative Pat Roberts [R-KS]), 18729–31 (House Ag Committee Chairman Kika de la Garza [D-TX]), 18734 (Ranking Member Ed Madigan [R-IL]).

42. Cong. Rec., July 25, 1990, 19166 (Madigan amendment), 19173–74 (Representative Roberts; Madigan amendment agreed to by a narrow vote of 219 to 210 [3 not voting]); Cong. Rec., July 24, 1990, 18838 (Representative Thomas Downey [D-NY] sugar loan rate amendment), 18849 (amendment defeated by a vote of 150 to 271 [1 voting present; 10 not voting]); Cong. Rec., July 25, 1990, 19173 (Representative Byron Dorgan [D-ND]); Cong. Rec., August 1, 1990, 21446 (Dorgan and Tim Johnson [D-SD] amendment), 21448 (amendment was rejected without a recorded vote).

43. Cong. Rec., July 25, 19214 (Ranking Member Ed Madigan).

44. Cong. Rec., July 19, 1990, 18148 (Chairman Leahy); Cong. Rec., August 1, 1990, 21426–27 (Representative Silvio Conte [R-MA] amendment), 21428–29 (Conte amendment), 21432–33 (Representative Armey), 21435–37 (Representative Conte), 21441 (Representative Schumer).

45. Cong. Rec., July 19, 1990, 18148.

46. Cong. Rec., July 19, 1990, 18165–66 (Senator Reid), 18170–71 (Senator David Pryor [D-AR] opposed), 18173 (Senators Baucus and Conrad opposed).

47. Cong. Rec., July 25, 1990, 19197 (Representative Armey), 19194 (Representative Sedgewick [Bill] Green [R-NY]).

48. Cong. Rec., July 25, 1990, 19206 (Representative Major Owens [D-NY]).

49. Cong. Rec., July 25, 1990, 19194–96 (Representatives Larry Combest [R-TX], Armey, Roberts, and Neal Smith [D-IA]), 19200–19202 (Representative Roberts); Cong. Rec., August 1, 1990, 21430–31 (Representative Thomas Huckaby [D-LA]).

50. Cong. Rec., July 25, 1990, 19190–91 (Representative English [D-OK]), 19208 (Representative Charlie Stenholm [D-TX]).

51. Cong. Rec., July 25, 1990, 19205 (Representative Ronald Marlenee [R-MT]), 19208 (Representative Stenholm), 19183, 19189–90 (Representative Glickman), 19209–10 (Representative Dorgan), 19213 (Glickman's substitute defeated 174 to 251 [7 not voting]).

52. Cong. Rec., July 19, 1990, 18151 (Senator Lugar); Cong. Rec., July 27, 1990, 19835–36 (Senator Robert Kasten [R-WI] amendment on conservation compliance provisions), 19839 (Kasten amendment agreed to without a recorded vote after receiving an endorsement of support from Chairman Leahy), 19839–40 (second Kasten amendment [Wetlands Reserve program] agreed to without a recorded vote), 19831 (Senator Frank Lautenberg [D-NJ] amendment on pesticides), 19832 (Senator Grassley).

53. Cong Rec., July 23, 1990, 18742 (Representative Roberts); Cong. Rec., August 1, 1990, 21545 (Representative English), 21544 (Chairman de la Garza [en bloc] amendments).

54. Cong. Rec., July 25, 19217 (Schumer-Armey amendment defeated 159 to 263 [2 voting present; 8 not voting]). See also Cong. Rec., August 1, 1990, 21428–29 (Representative Silvio Conte [R-MA]), 21444 (Conte's amendment to the Huckaby substitute defeated 171 to 250 [1 voting present; 10 not voting]), 21445 (Huckaby substitute [preserving the three-entity rule] agreed to by the House on a vote of 375 to 45 [1 voting present and 11 not voting]), 21445 (Conte amendment as amended by Huckaby agreed to without a recorded vote).

55. Cong. Rec., August 1, 1990, 21550 (food stamp program), 21559 (Representative William Frenzel [R-MN] amendments [en bloc] to the food stamp program), 21566 (Frenzel's amendments defeated by a vote of 83 to 336 [13 not voting]), 21593 (Madigan motion to recommit), 21594 (motion to recommit defeated and the 1990 farm bill was subsequently passed by a vote of 327 to 91 [14 not voting]).

56. "Lean Farm Package."

57. Cong. Rec., October 26, 1990, 34544 (triple base language contained in the Omnibus Budget Reconciliation Act of 1990 [H.R. 5835] read into the Congressional Record); Food, Agriculture, Conservation, and Trade Act of 1990, P.L. No. 101-624 (1990).

58. "Lean Farm Package."

59. Democrats held onto their congressional majorities in the 1990 midterm elections. See Rhodes Cook, "Reflections on 1990 Election," *Public Perspective*, January/February 1991, https://ropercenter.cornell.edu/public-perspective/ppscan/22/22009.pdf; Michael Oreskes, "The 1990 Elections: The Future-Redistricting; Elections Strengthen Hand of Democrats in '91 Redistricting," *New York Times*, November 8, 1990, http://www.nytimes.com/1990/11/08/us/1990-elections-future-redistricting-elections-strengthen-hand-democrats-91.html; Mark P. Petracca, "Midterm Congressional Elections," *Los Angeles Times*, November 17, 1990, http://articles.latimes.com/1990-11-17/local/me-3972_1_house-seats-midterm-elections-post-war-elections; Cohen and Rogers, "Politics of Dealignment."

60. Alvarez and Nagler, "Economics, Issues," 715.

61. Orden, Paarlberg, and Roe, *Policy Reform*, 85–113.

62. Winders, *Politics*, 181; Orden, Paarlberg, and Roe, *Policy Reform*, 118–21.

63. Aldrich and Rohde, "Transition to Republican," 9242–44; Adam Clymer, "The 1994 Elections: Congress the Overview; G.O.P. Celebrates Its Sweep to Power; Clinton Vows to Find Common Ground," *New York Times*, November 10, 1994, http://www.nytimes.com/1994/11/10/us/1994-elections-congress-overview-gop-celebrates-its-sweep-power-clinton-vows.html?pagewanted=all.

64. Winders, *Politics*, 183–91.

65. "Biography," https://www.roberts.senate.gov/public/index.cfm?p=Biography; Kansas Historical Society, Charles Patrick Roberts, *kansaspedia*, last modified August 2013, https://www.kshs.org/kansapedia/charles-patrick-roberts/17016.

66. Schertz and Doering, *Making of*, 7–9.

67. "Plan to Cut Farm Programs Stalls," in CQ *Almanac 1995*, 51st ed., 3-47–3-56 (Washington DC: Congressional Quarterly, 1996), http://library.cqpress.com/cqalmanac/cqal95-1100229; Schertz and Doering, *Making of*, 14.

68. "GOP Throws Down Budget Gauntlet," in CQ *Almanac 1995*, 51st ed., 2-20–2-33 (Washington DC: Congressional Quarterly, 1996), http://library.cqpress.com/cqalmanac/cqal95-1099934; Schertz and Doering, *Making of*, 16, 26–27.

69. Schertz and Doering, *Making of*, 9–10.

70. Schertz and Doering, *Making of*, 16; Winders, *Politics*, 174.

71. Schertz and Doering, *Making of*, 12, 40.

72. Schertz and Doering, *Making of*, 21, 40–42, 56–57.

73. Schertz and Doering, *Making of*, 21, 26–27.

74. Schertz and Doering, *Making of*, 6–36.

75. Schertz and Doering, *Making of*, 35–36, 13–14, 59.

76. Schertz and Doering, *Making of*, 63–64.

77. "Plan to Cut Farm Programs."

78. Schertz and Doering, *Making of*, 56, 73.

79. Cong. Rec., October 26, 1995, 29508 (Representative de la Garza quoting Chairman Roberts).

80. "Plan to Cut Farm Programs."

81. Schertz and Doering, *Making of*, 63.

82. Schertz and Doering, *Making of*, 63.

83. "Plan to Cut Farm Programs"; Schertz and Doering, *Making of*, 63.

84. Schertz and Doering, *Making of*, 66–67 (reporting that Roberts was defeated 22 to 25 when Representatives Emerson, Combest, Saxby Chambliss [R-GA], and William Baker [R-CA] voted against the chair along with all Democrats).

85. Schertz and Doering, *Making of*, 63 (reporting Emerson-Combest defeated 23 to 26), 66–67 (reporting that a total of 18 Democrats, led by Stenholm and Cal Dooley [D-CA], backed it along with only three Republicans).

86. Schertz and Doering, *Making of*, 67 (quoting Harold Volkmer [D-MO]).

87. Schertz and Doering, *Making of*, 67–68.

88. Cong. Rec., October 24, 1995, E2009. See also Cong. Rec., October 26, 1995, 29511 (Representative Thomas Ewing [R-IL]).

89. Cong. Rec., October 26, 1995, 29504-5 (Representative Kika de la Garza [D-TX]), 29506 (Representative Charlie Stenholm [R-TX]), 29508 (Representative David Minge [D-MN]), 29509 (Representative Earl Pomeroy [D-ND]).

90. Cong. Rec., October 26, 1995, 29912–13 (227 to 203 [3 not voting]).

91. "Plan to Cut Farm Programs"; Schertz and Doering, *Making of*, 70.

92. Cong. Rec., October 27, 1995, 30378 (Senator Harkin [D-IA] amendment defeated on point of order by a vote of 31 to 68), 30379–80 (Senator Paul Wellstone [D-MN] amendment defeated on a motion to table by a vote of 64 to 35), 30462 (passed by a vote of 52 to 47).

93. "Plan to Cut Farm Programs"; Schertz and Doering, *Making of*, 71.

94. Schertz and Doering, *Making of*, 71–76, 80–82; "Plan to Cut Farm Programs."

95. Schertz and Doering, *Making of*, 74–76.

96. Cong. Rec., November 17, 1995, 33800–33801 (by a vote of 237 to 189 [7 not voting]), 33641 (by a vote of 52 to 47).

97. Schertz and Doering, *Making of*, 21.

98. Cong. Rec., July 27, 1995, 20652.

99. Schertz and Doering, *Making of,* 42.

100. Schertz and Doering, *Making of,* 52.

101. Schertz and Doering, *Making of,* 49–50.

102. Cong. Rec., December 6, 1995, 35656–57 (President Clinton's veto message); Schertz and Doering, *Making of,* 75, 85, 25.

103. See, e.g., Krishnakumar, "Reconciliation and the Fiscal Constitution," 589.

104. "Plan to Cut Farm Programs."

105. "Plan to Cut Farm Programs"; Orden, Paarlberg, and Roe, "What Is Happening," 2; "Longstanding Farm Laws Rewritten," in CQ *Almanac 1996,* 52nd ed., 3-15–3-26 (Washington DC: Congressional Quarterly, 1997), http://library .cqpress.com/cqalmanac/cqal96-1092001; Schertz and Doering, *Making of,* 92.

106. Schertz and Doering, *Making of,* 108–9.

107. Schertz and Doering, *Making of,* 111; "Longstanding Farm Laws."

108. "Longstanding Farm Laws."

109. Agricultural Market Transition Act of 1996, S. 1541, 104th Cong. (1996); "Longstanding Farm Laws"; Schertz and Doering, *Making of,* 95–96.

110. Cong. Rec., February 1, 1996, 1994 [S672–74].

111. Cong. Rec., February 1, 1996, 1995 [S672–74].

112. See e.g., Cong. Rec., February 1, 1996, 1996 [S674] (Senator Cochran), 1998 [S676–77] (Senator Grassley), 2004 [S682-83] (Majority Leader Bob Dole [R-KS]).

113. Cong. Rec., January 31, 1996, 1700 [S576–77] (Senator Byron Dorgan [D-ND]).

114. Cong. Rec., February 1, 1996, 2001–2 [S679–81] (Senator Tom Daschle [D-SD]).

115. Cong. Rec., February 1, 1996, 2001–2 [S679–81] (Senator Tom Daschle [D-SD]).

116. It would damage farm programs in the public eye by paying farmers even if they didn't farm and leaving in place no real safety net for when prices inevitably fell. See Cong. Rec., February 1, 1996, 1999–2000 [S677].

117. Cong. Rec., February 1, 1996, 2003 [S681].

118. Cong. Rec., February 1, 1996, 2005 [S683] (cloture vote was 53 to 45 [1 not voting]).

119. Cong. Rec., February 1, 1996, 2008 [S685–86] (Senator Leahy), 2045 [S721] (Senator Dole), 2049 [S725] (Senators Dole and Daschle), 2054 [S729–30] (Senator Dorgan).

120. Cong. Rec., February 6, 1996, 2371–72 [S906-7] (cloture motion failed by a vote of 59 to 34 [7 not voting]), 2377 [S913] (unanimous consent to modify Craig amendment); Cong. Rec., February 7, 1996, 2487–88 [S1014–15] (Senator Craig modification), 2474 [S1012–13] (Senator Lugar amendment to add

four titles), 2488 [S1015–16] (Senator Harkin), 2492 [S1018–19] (Senator Rick Santorum [R-PA] amendment on the peanut program), 2503 [S1030] (Senator Judd Gregg [R-NH] amendment to strike the sugar program), 2509 [S1036] (Senator Herb Kohl's [D-WI] amendment passed 50 to 46 [4 not voting]), 2511 [S1037] (Harkin farmer-owned reserve amendment lost 35 to 61 [4 not voting]), 2513–14 [S1040] (sugar amendment defeated 35 to 61 [4 not voting]), 2514 [S1040] (Senator Dorgan amendment), 2518–19 [S1043–44] (Senator Daschle amendment), 2527 [S1052–53] (Dorgan's amendment rejected on a 48 to 48 [4 not voting] vote, while the Daschle alternative farm program amendment rejected on a 33 to 63 vote [4 not voting]).

121. Cong. Rec., February 7, 1996, 2531 [S1056–57] (Senator Baucus [D-MT]), 2534 [S1059] (final passage by a vote of 64 to 32 [4 not voting]).

122. Agriculture Market Transition Act, H.R. Rep. No. 104-462 (1996).

123. "Longstanding Farm Laws."

124. Cong. Rec., February 27, 1996, 2967 [H1328] (farm bill); Cong. Rec., February 28, 1996, 3086–88 [H1403] (rule), 3092 [H1409] (Representative John Boehner [R-OH]), 3097–98 [H1414] (House agreed to the rule by a vote of 244 to 168 [19 not voting]), 3136–44 [H1451–60] (Representative Frank amendment, Representative Steve Chabot [R-OH] amendment defeated 167 to 253 [11 not voting]), 3145[H1460] (Representative Christopher Shays [R-CT] peanut amendment), 3151–52 [H1466–67] (amendment defeated, 209 to 212 [10 not voting]), 3152–65 [H1467–79] (Representative George Miller [D-CA] sugar amendment defeated 208 to 217 [1 voting present; 5 not voting]), 3173–74 [H1487–88] (Solomon amendment agreed to by a vote of 258 to 164 [1 voting present; 8 not voting]).

125. Cong. Rec., February 28, 1996, 3098–99 [H1415–16].

126. Cong. Rec., February 28, 1996, 3099 [H1416].

127. H.R. Rep. No. 104-462, at 2.

128. H.R. Rep. No. 104-462, at 42–43.

129. Cong. Rec.,February 28, 1996, 3100 [H1415–16] , 3112 [H1428–29] (Representative Bill Barrett [R-NE]), 3114 [H1430] (Chairman Roberts). See also, Cong. Rec., October 26, 1995, 29508 (de la Garza quoting Chairman Roberts's argument before the Rules Committee for Freedom to Farm that it will "lock up the baseline for farmers" so that they will not take further cuts in future budget reduction efforts); Cong. Rec., February 1, 1996, 1998–99 [S676] (Senator Grassley).

130. Cong. Rec., February 28, 1996, 3103 [H1419] (Representative Frank Lucas [R-OK]), 3104 [H1420] (Representative Saxby Chambliss [R-GA]); Cong. Rec., February 29, 1996, 3247 [H1539] (Chairman Roberts and Representative Chambliss colloquy).

131. Cong. Rec., February 28, 1996, 3100, 3101-2 [H1415–16].

132. Cong. Rec., February 28, 1996, 3110 [H1426–27] (Representative Charlie Stenholm [D-TX]), 3102 [H1419] (Representative Volkmer [D-MO]), 3107 [H1424] (Representative Earl Pomeroy [D-ND]).

133. Cong. Rec., February 28, 1996, 3136 [H1452] (Representative Barney Frank [D-MA]), 3108 [H1424] (Representative de la Garza).

134. Cong. Rec., February 29, 1996, 3217–19 [H1509–15] (Representative Sherwood Boehlert [R-NY] conservation amendment, opposed by Representative Bob Livingston [R-LA], chair of the Appropriations Committee), 3223 [H1515–16] (Boehlert amendment passed 372 to 37 [22 not voting]); Cong. Rec., February 27, 1996, 3039 [H1397] (de la Garza amendment to restore 1949 act permanent law); Cong. Rec., February 29, 1996, 3248–49 [H1539–40] (Representative Stenholm), 3250 [H1540] (Chairman Roberts opposed), 3255 [H1546] (amendment defeated 163 to 258 [10 not voting]), 3257 [H1548] (Stenholm motion to recommit), 3282–83 [H1572–73] (Roberts point of order sustained, and Stenholm had to offer another motion to recommit), 3284–85 [H1574–75] (motion to recommit defeated 156 to 267 [8 not voting], and 1996 farm bill passed by a vote of 270 to 155 [H1575] [6 not voting]).

135. Schertz and Doering, *Making of*, 117; Cong. Rec., March 27, 1996, 6569–72 [S2997–99] (Senate Chairman Lugar).

136. Cong. Rec., March 28, 1996, 6729 [S3038] (Chairman Lugar), 6733 [S3042] (Senator Leahy); King, "Welfare Reform," 360.

137. Cong. Rec., March 27, 1996, 6572–73 [S2999]; Cong. Rec, March 28, 1996, 6773–74 [S3080–81] (Senator Harkin).

138. Cong. Rec., March 27, 1996, 6573 [S2999] (Senator Daschle); Cong. Rec, March 28, 1996, 6737 [S3046] (Senator Conrad), 6758 [S3066] (Senator Baucus).

139. Cong. Rec., March 27, 1996, 6570–72 [S2997], 6577 [S3003] (Senator Cochran); Cong. Rec, March 28, 1996, 6728–29 [S3038–39] (Chairman Lugar).

140. Cong. Rec., March 28, 1996, 6794 [S3100–3101] (passage by 74 to 26), 7064 [H3147] (Conf. Rep.), 7071 [H3154] (Representative Pomeroy), 7086 [H3168] (farm bill passed by a vote of 318 to 89 [24 not voting]), 7068 [H3150] (Chairman Roberts statement); Federal Agriculture Improvement and Reform Act of 1996, P.L. No. 104-127, 110 Stat. 888 (1996).

141. Schertz and Doering, *Making of*, 121.

142. Schertz and Doering, *Making of*, 121; Winders, *Politics*, 195.

143. See, generally, Winders, *Politics*, chapter 7.

144. "Farm Aid Rift Prompts Veto, Forces Agriculture Bill into Year-End Spending Package," in CQ *Almanac 1998*, 54th ed., 2-3-2-12 (Washington DC: Congressional Quarterly, 1999), http://library.cqpress.com/cqalmanac /cqal98-0000191019; Lauck, "After Deregulation," 30.

145. Scott, "Exported to Death," 87.

146. Winders, *Politics*, 195–96.

147. Cong. Rec., July 15, 1998, s8183 (Daschle's amendment tabled by a vote of 56 to 43 [1 not voting]); Cong. Rec., July 16, 1998, s8283; Cong. Rec., October 2, 1998, H9288 (House consideration began), H9331 (Representative Earl Pomeroy [D-ND]), H9335 (conference report passed by a vote of 333 to 53 [48 not voting]); Cong. Rec., October 5, 1998, s11437 (cloture motion prevailed 93 to 0 to end debate on the conference report), s11569 (conference report passed by a vote of 55 to 43 [2 not voting]); Cong. Rec., October 19, 1998, H11044 (Omnibus Appropriations Bill provided $1.5 billion for 1998 crop losses due to disasters, $875 million for multiyear losses, and $3.057 billion for market loss assistance), H11055–56; Cong. Rec., October 20, 1998, H1591 (omnibus passed 333 to 88 [13 not voting]); Cong. Rec., October 21, 1998, s12809 (the bill was good enough for Daschle's, Dorgan's, and Conrad's votes, but not Baucus's; the Senate agreed to the conference report by a vote of 65 to 29 [6 not voting]).

148. Scott, "Exported to Death," 87.

149. Cong. Rec., July 14, 1998, s8093 (Senators Harkin and Daschle), s8100–8101 (Senator Harkin).

150. Cong. Rec., July 14, 1998, s8107–8 (Senator Paul Wellstone [D-MN]), s8125–26 (Senator Daschle amendment); Cong. Rec., July 15, 1998, s8174 (Senator Harkin), s8180 (Senator Mary Landrieu [D-LA]); Cong. Rec., July 16, 1998, s8284–25 (Senator Conrad).

151. "Farm Aid Rift"; Scott, "Exported to Death," 87.

152. "Farm Aid Rift"; Lauck, "After Deregulation," 30.

153. Cong. Rec., July 14, 1998, s8110, s8116–18; Cong. Rec., July 15, 1998, s8170–71 (Chairman Lugar), s8174 (Senator Cochran).

154. GAO, letter to Secretary of Agriculture Dan Glickman, June 30, 2000 (GAO/RCED-00–177R Market Loss Assistance Payments), https://www.gao .gov/archive/2000/rc00177r.pdf; Cong. Rec., August 2, 1999, s9960 (Senators Daschle, Conrad, Dorgan, and Harkin), s9978 (Senators Daschle and Cochran competing provisions), s9980 (Senator Harkin), s9980–81(colloquy between Senators Harkin and Dorgan), s9988 (Senator Harkin), s9989 (Democratic provisions); Cong. Rec., August 3, 1999, s10067–68 (Senators Conrad and Baucus), s10099 (Senator Daschle).

155. Cong. Rec., August 3, 1999, s10100 (Daschle motion to table the Cochran amendment defeated by a vote of 47 to 51 [2 not voting]), s10102 (Senate tables the Democratic package by a vote of 54 to 44 [2 not voting]); Cong. Rec., August 4, 1999, s10168–69 (Senator Cochran amendment for $7 billion), s10171–72 (Cochran modified his amendment), s10173 (Cochran amendment agreed to by a voice vote), s10174 (Senator Dorgan effort for $9.837 billion), s10176 (Senate agreed to table it by a vote of 55 to 44 [1 not voting]), s10179 (Conrad amendment to increase assistance defeated on a motion

to table 51 to 48 [1 not voting]), s10183 (adoption of the Cochran amendment as amended by a vote of 89 to 8 [3 not voting]). See also Cong. Rec., October 1, 1999, H9217 (Ag Appropriations Subcommittee Chairman Skeen [R-NM]), H9226 (Representative Collin Peterson [D-MN]), H9227 (Representative Rosa DeLauro [D-CT]), H9212 (Representative Marcy Kaptur [D-OH]), H9230 (Representative Kaptur), H9228 (Representative Stenholm), H9230 (Representative Larry Combest [R-TX]), H9230 (Representative Frank Lucas [R-OK]), H9236–37 (Kaptur's motion to recommit defeated by a vote of 187 to 228 [18 not voting]; conference report agreed to by a vote of 240 to 175 [18 not voting]).

156. Cong. Rec., October 12, 1999, s12405 (Senate consideration), s12406 (Senator Cochran), s12410 (Senators Dorgan and Conrad also supported the bill), s12416 (Senator Daschle); Cong. Rec., October 13, 1999, s12472–74 (Senators Baucus and Wellstone), s12458–65, s12419–20 (cloture invoked by a vote of 79 to 20 [1 not voting]), s12502 (Senator Harkin), s12504 (passed Senate by a vote of 74 to 26); "Putting Differences Aside, Lawmakers Clear $69 Billion Agriculture Spending Bill," in CQ Almanac 1999, 55th ed., 2-5-2-16 (Washington DC: Congressional Quarterly, 2000), http://library.cqpress.com/cqalmanac/cqal99-0000201126.

157. Orden, Paarlberg, and Roe, *Policy Reform*, 8–10; Don Paarlberg, "Obituary for a Farm Program," *Choices*, 1st Quarter 1999, http://ageconsearch.umn.edu/record/131675/files/DonPaarlberg.pdf; Cong. Rec., February 1, 1996, 2004 (Senator Dole); Cong. Rec., February 28, 1996, 3100 (Senator Roberts).

158. Orden, Paarlberg, and Roe, *Policy Reform*, 80–82.

159. The bill freed farmers to make planting decisions in order to compete for export markets. Farmers responded mostly by increasing soybean acres and reducing wheat acres, while corn and cotton remained fairly consistent. See appendix 1, fig. 2.

160. Cong. Rec., February 1, 1996, 1998–99 [s676] (Senator Grassley); Orden, Paarlberg, and Roe, *Policy Reform*, 147–48; Winders, *Politics*, 173.

161. See, e.g., Cong. Rec., February 28, 1996, 3100; Cong. Rec., February 29, 1996, 3247 (Chairman Roberts).

162. See, e.g., Paarlberg and Paarlberg, "Agricultural Policy," 159.

163. "Plan to Cut"; Schertz and Doering, *Making of*, 56–57.

164. See, e.g., Orden, Paarlberg, and Roe, *Policy Reform*, 148.

165. Paarlberg and Paarlberg, "Agricultural Policy," 160.

166. See, e.g., Bullock, Hoffman, and Gaddie, "Consolidation of the White Southern," 31–43; Hayes and McKee, "Toward a One-Party South?"

7. Cotton, Ethanol, and Risk Management

1. Chite, "Emergency Funding."

2. Cong. Rec., March 22, 2000, s1557 (Senate Ag Committee Chairman Richard Lugar [R-IN]), s1560 (Senator Roberts), s1585 (Senator Grassley).

3. Glauber, "Growth of," 483.

4. Congressional Budget Office (CBO), *The Economic and Budget Out-look: Fiscal Years 2000–2009* (Washington DC: CBO, 1999), https://www.cbo.gov/sites/default/files/106th-congress-1999-2000/reports/eb0199.pdf.

5. "Farm Aid Bill Clears with Higher Subsidies to Offset Low Crop Prices," in *CQ Almanac 2000*, 56th ed., 4-3-4-5 (Washington DC: Congressional Quarterly, 2001), http://library.cqpress.com/cqalmanac/cqal00-834-24312-1083071.

6. Cong. Rec., March 22, 2000, S1558–59 (Chairman Harkin and Senator Roberts).

7. Cong. Rec., March 22, 2000, S1557 (Chairman Lugar)); "Farm Aid Bill Clears."

8. See, e.g., Cong. Rec., March 22, 2000, S1558 (Senator Tom Harkin [D-IA]), S1557 (Chairman Lugar), S1559 (Senator Roberts); "Farm Aid Bill Clears."

9. Cong. Rec., March 22, 2000, S1576 (Senator Blanche Lincoln [D-AR]), S1578 (Senator Thad Cochran [R-MS]).

10. Cong. Rec., March 22, 2000, S1556, S1559 (Senator Roberts), S1562 (Chairman Lugar and Minority Leader Daschle), S1563 (Senator Chuck Schumer [D-NY]), S1571–72 (Senator Edward Kennedy [D-MA]); Cong. Rec., March 23, 2000, S1628 (Senator Leahy).

11. Cong. Rec., March 23, 2000, S1631–42 (bill text; passed by a vote of 95 to 5).

12. Cong. Rec., May 25, 2000, S4437–38 (Senator Lincoln), S4441 (passed by a vote of 91 to 4 [5 not voting]); "Farm Aid Bill Clears."

13. See, generally, Kelley, "Agricultural Risk Protection Act."

14. Glauber, "Crop Insurance," 1179–95; Glauber, "Growth of," 482–88.

15. "Farm Bill Delayed a Year," in *CQ Almanac 2001*, 57th ed., 3-3-3-8 (Washington DC: Congressional Quarterly, 2002), http://library.cqpress.com/cqalmanac/cqal01-106-6384-328622.

16. J. T. Smith, "Stenholm Says Bipartisanship Sold Farm Bill," *Farm Progress*, January 8, 2004, http://farmprogress.com/story-stenholm-says-bipartisanship-sold-farm-bill-9-1026; Dan Morgan, Sara Cohen, and Gilbert M. Gaul, "Powerful Interests Ally to Restructure Agriculture Subsidies," *Washington Post*, December 22, 2006, http://www.washingtonpost.com/wp-dyn/content/article/2006/12/21/ar2006122101634_pf.html.

17. "Combest, Larry Ed," History, Art, and Archives, U.S. House of Representatives, n.d., http://history.house.gov/People/Detail/11272.

18. Michael deCourcy Hinds, "Man in the News: Charles W. Stenholm; The Texas Congressman behind the Amendment," *New York Times*, June 11, 1992, http://www.nytimes.com/1992/06/12/us/man-in-the-news-charles-w-stenholm-the-texas-congressman-behind-the-amendment.html; "Stenholm, Charles Walter," Biographical Directory of the United States Congress, n.d., http://bioguide.congress.gov/scripts/biodisplay.pl?index=s000851.

19. Hasen, "Bush v. Gore," 377; Chemerinsky, "Bush v. Gore," 1093.

20. U.S. Senate, "The Unforgettable 107th Congress," November 22, 2002, https://www.senate.gov/artandhistory/history/minute/unforgettable_107th _congress.htm.

21. "Arena Profile: Sen. Tom Harkin," *Politico*, 2011, http://www.politico .com/arena/bio/sen_tom_harkin.html; "Harkin, Thomas Richard," Biographical Directory of the United States Congress, n.d., http://bioguide.congress .gov/scripts/biodisplay.pl?index=h000206.

22. "Farm Bill Delayed."

23. Smith, "Stenholm Says"; Dan Morgan, Sara Cohen, and Gilbert M. Gaul, "Powerful Interests Ally to Restructure Agriculture Subsidies," *Washington Post*, December 22, 2006, http://www.washingtonpost.com/wp-dyn /content/article/2006/12/21/ar2006122101634_pf.html.

24. Morgan, Cohen, and Gaul, "Powerful Interests."

25. "Farm Bill Delayed."

26. Cong. Rec., October 3, 2001, H6167–68 ("Farm Security Act of 2001"), H6170 (Chairman Larry Combest [R-TX]).

27. Cong. Rec., October 3, 2001, H6226–27.

28. Cong. Rec., October 3, 2001, H6171.

29. Cong. Rec., October 3, 2001, H6226 (Representative Leonard Boswell [D-IA] amendment), H6235 (Boswell amendment defeated 100 to 323 [1 voting present; 6 not voting]).

30. Cong. Rec., October 3, 2001, H6230 (Representative Nick Smith [R-MI] amendment), H6231 (Chairman Combest), H6232–33 (Representatives Stenholm and Chambliss), H6236–37 (Smith amendment defeated by a vote of 187 to 238 [5 not voting]); Cong. Rec., October 4, 2001, H6327–28 (Representative Dan Miller [R-FL] amendments to reform the sugar program), H6333 (Representative Mark Kirk [R-IL]), H6337 (Representative Peterson), H6340 (Ranking Member Stenholm), H6341 (sugar amendment defeated by a vote of 177 to 239 [14 not voting]).

31. Cong. Rec., October 4, 2001, H6301 (Representative Kind), H6298 (text of amendment).

32. Cong. Rec., October 4, 2001, H6300, H6302 (Representative Holden [D-PA]). See also Cong. Rec., October 3, 2001, H6230 (Representative Nick Smith [R-MI] amendment), H6231 (Combest), H6232–33 (Representatives Stenholm and Chambliss), H6236–37 (Smith amendment defeated by a vote of 187 to 238 [5 not voting]).

33. See, generally, Cong. Rec., October 4, 2001; "Farm Bill Delayed."

34. See, e.g., Cong. Rec., October 4, 2001, H6302 (Representative Frank Lucas [R-OK]), H6311 (Representative Earl Pomeroy [D-ND]).

35. Cong. Rec., October 4, 2001, H6324 (by a vote of 200 to 226 [6 not voting]).

36. CBO, "Cost Estimate: H.R. 2646, Farm Security Act of 2001," September 14, 2001, https://www.cbo.gov/publication/13269.

37. Farm Security Act of 2001, H.R. Rep. No. 107-191, at 6–12 (2001); Cong. Rec., October 3, 2001, H6187–6225; CBO, "Cost Estimate: H.R. 2646" (September 2001). See also CBO, "Cost Estimate: H.R. 2646, Farm Security Act of 2001," August 23, 2001, 2, https://www.cbo.gov/publication/13246.

38. Cong. Rec., October 3, 2001, H6174; Cong. Rec., October 4, 2001, H6295–99.

39. Cong. Rec., October 3, 2001, H6170 (Combest).

40. Cong. Rec., October 3, 2001, H6171.

41. Cong. Rec., October 3, 2001, H6174 (Representative Saxby Chambliss [R-GA]).

42. Cong. Rec., October 3, 2001, H6171–72 (Representative Frank Lucas [R-OK]).

43. Cong. Rec., October 5, 2001, H6407 (substitute amendment was agreed to without a recorded vote), H6410 (final passage by a vote of 291 to 120 [19 not voting]).

44. "Farm Bill Delayed."

45. Cong. Rec., December 10, 2001, S12765–67 (Chairman Harkin).

46. Cong. Rec., November 30, 2001, S12271 (Chairman Harkin); Cong. Rec., December 5, 2001, S12401–2 (Chairman Harkin and Ranking Member Lugar [R-IN] debate cloture).

47. Agriculture, Conservation, and Rural Enhancement Act of 2001, S. Rep. 107-117 (2001).

48. Cong. Rec., December 11, 2001, S12845 (Lugar substitute amendment), S12970.

49. He added that the 1996 farm bill did not expire until 2002. See Cong. Rec., December 5, 2001, S12402.

50. Cong. Rec., December 5, 2001, S12402.

51. Cong. Rec., December 11, 2001, 12848–50.

52. Cong. Rec., December 12, 2001, S12991.

53. Cong. Rec., December 5, 2001, S12402–3.

54. Cong. Rec., December 10, 2001, S12764 (Chairman Harkin); Cong. Rec., December 11, 2001, S12827, S12830 (Senator Lugar).

55. Cong. Rec., December 5, 2001, S12402–3.

56. Cong. Rec., December 5, 2001, S12404 (Senator Conrad), S12406 (Senate agreed to proceed to the bill by a vote of 73 to 26 [1 not voting]); Cong. Rec., December 10, 2001, S12755 (debate began), S12764 (Chairman Harkin); Cong. Rec., December 12, 2001, S12995 (Harkin moved to table the Lugar amendment, and the Senate agreed to table it by a vote of 70 to 30).

57. Cong. Rec., December 13, 2001, s13092 (Majority Leader Daschle and Minority Leader Trent Lott [R-MS]), s13111–12 (the Senate failed to invoke cloture by a vote of 53 to 45 [2 not voting]).

58. Cong. Rec., December 14, 2001, s13255–57.

59. Cong. Rec., December 14, 2001, s13252–54.

60. Cong. Rec., December 14, 2001, 13253–54.

61. Cong. Rec., December 14, 2001, s13262–64 (Senator Byron Dorgan [D-ND]); Cong. Rec., December 17, 2001, s13349 (motion to invoke cloture on the farm bill); Cong. Rec., December 18, 2001, s13423 (motion to invoke cloture again failed on a vote of 54 to 43 [3 not voting]); Cong. Rec., December 19, 2001, s13650–52 (Senator Tim Hutchinson [R-AR]), s13654–55 (Senator Blanche Lincoln), s13655–56 (Senator Roberts).

62. Cong. Rec., December 19, 2001, s13656 (Senate agreed to table the Hutchinson amendment by a vote of 59 to 38 [3 not voting]), s13658 (the cloture vote [third time] by a vote of 54 to 43 [3 not voting]), s13661 (farm bill pulled from the floor); Cong. Rec., December 12, 2001, s13001–3 (Senator Judd Gregg [R-NH] amendment [sugar]), s13020 (tabled by a vote of 71 to 29); Cong. Rec., December 13, 2001, s13093–99 (Senate passed an amendment banning packer ownership of livestock by a vote of 51 to 46 [3 not voting]); Cong. Rec., December 12, 2001, s13110 (Senator John McCain [R-AZ] attacked catfish policy); Cong. Rec., February 12, 2002, s598 (return of packer ban), s608 (packer ban defeated 46 to 53 [1 not voting]).

63. Cong. Rec., February 6, 2002, s402 (Chairman Harkin), s405 (Senator Lugar).

64. Cong. Rec., February 6, 2002, s443–44 (Senators Dorgan and Grassley payment limitations amendment), s451 (Lugar).

65. Cong. Rec., February 6, 2002, s443–44 (Senators Dorgan and Grassley), s451 (Senator Tim Johnson [D-SD]), s455 (Senator Grassley), s457 (Grassley).

66. Cong. Rec., February 7, 2002, s452–53 (Senator Peter Fitzgerald [R-IL]).

67. Cong. Rec., February 7, 2002, s454 (Senator Russ Feingold [D-WI]).

68. See, e.g., Cong. Rec., February 7, 2002, s449 (Senator Chuck Hagel [R-NE] reading into the record a list of Nebraska's top recipients).

69. Cong. Rec., February 7, 2002, s445–47 (Senator Lincoln).

70. Cong. Rec., February 7, 2002, s457 (Senator Lincoln moved to table the amendment, but the motion was overwhelmingly defeated on a vote of 31 to 66 [3 not voting]); Cong. Rec., February 13, 2002, s684 (Senator Cochran).

71. "'02 Farm Bill Revives Subsidies," in CQ Almanac 2002, 58th ed., 4-3–4-7 (Washington DC: Congressional Quarterly, 2003), http://library.cqpress.com/cqalmanac/cqal02-236-10375-664394.

72. Donald C. Smaltz, Final Report of the Independent Counsel in re: Alphonso Michael (Mike) Espy (Washington DC: U.S. Court of Appeals for

the DC Circuit, 2001), http://govinfo.library.unt.edu/oic/smaltz/FnlRpt.htm; Donald C. Smaltz, letter to Albert Gore and Newt Gingrich, February 20, 1998, http://govinfo.library.unt.edu/oic/smaltz/press/reports/finalrpt.htm. See also Associated Press, "Mississippi Farmer Pleads Guilty to USDA Crop Subsidy Fraud," *Post Bulletin*, November 21, 1996, http://www.postbulletin .com/mississippi-farmer-pleads-guilty-to-usda-crop-subsidy-fraud/article _c0b8c6e6-197a-5303-a95d-411ae88008b9.html. The indictment of the farmer, Brook Keith Mitchell, can be found here: http://govinfo.library.unt.edu/oic /smaltz/indict/5m.htm. See also "Billie Sol Estes Investigation," 998–99.

73. Cong. Rec., February 6, 2002, S442–43 (Senator Dick Durbin [D-IL] food stamps amendment passed by a vote of 96 to 1 [3 not voting]); Cong. Rec., February 8, 2002, S534 (Baucus amendment for livestock disaster assistance), S609 (livestock disaster assistance agreed to by a vote of 69 to 30); Cong. Rec., February 11, 2002, S592 (Senator Reid [D-NV]); Cong. Rec., February 13, 2002, S683 (manager's package of amendments), S688 (Senator Ben Nelson [D-NE]), S697 (Senate Majority Leader Daschle), S699 (Senate finally passed the farm bill by a vote of 58 to 40 [2 not voting]).

74. "'02 Farm Bill."

75. Cong. Rec., May 2, 2002, H2031.

76. Cong. Rec., May 2, 2002, H2057 (House passed the conference report by a vote of 280 to 141 [13 not voting]; Cong. Rec., May 8, 2002, S4051 (Senate agreed to the conference report by a vote of 64 to 35 [1 not voting]); Farm Security and Rural Investment Act, P.L. No. 107-171 (2002).

77. CBO, "Cost Estimate: H.R. 2646, Farm Security and Rural Investment Act of 2002," May 6, 2002, https://www.cbo.gov/publication/13637.

78. FAPRI, *2001 U.S. Baseline*; FAPRI, *2002 U.S. Baseline*; FAPRI, *2003 U.S. Baseline*; FAPRI, *2004 U.S. Baseline*; Sumner, "Implications of the US Farm Bill," 99–122.

79. Cong. Rec., May 2, 2002, H2033, H2035 (Representative Tom Latham [R-IA]).

80. Cong. Rec., May 8, 2002, S3980–82, S3983 (Chairman Harkin attacked the 1996 farm bill in response).

81. Cong. Rec., May 8, 2002, S4014–15 (Senator Lugar); Cong. Rec., May 8, 2002, S4018 (Senator Grassley).

82. U.S. Senate, Roll Call Vote 107th Congress, 2nd Session, May 8, 2002, https://www.senate.gov/legislative/LIS/roll_call_lists/roll_call_vote_cfm.cfm ?congress=107&session=2&vote=00103.

83. Morgan, Cohen, and Gaul, "Powerful Interests."

84. Shumaker, "Tearing the Fabric," 548–52; Gillon, "Panel Report," 8.

85. Gillon, "Panel Report," 11–12.

86. Gillon, "Panel Report," 19; Shumaker, "Tearing the Fabric," 565–66.

87. Shumaker, "Tearing the Fabric," 562, 566–68, 576–79; Gillon, "Panel Report," 19–20.

88. Shumaker, "Tearing the Fabric," 580, 589; Schnepf, "Background on," 2.

89. Schnepf, "Background on."

90. Shumaker, "Tearing the Fabric," 579; Schnepf, "Background on," 2; Gillon, "Panel Report," 27.

91. Shumaker, "Tearing the Fabric," 563; Schnepf, "Background on," 3; Gillon, "Panel Report," 31.

92. Gillon, "Panel Report," 27; Shumaker, "Tearing the Fabric," 556–58; Agricultural Appropriations for FY1998, P.L. No. 105-86, https://www.congress.gov/105/plaws/publ86/PLAW-105publ86.pdf.

93. Gillon, "Panel Report," 8; Schnepf, "Background on," 12–13.

94. Shumaker, "Tearing the Fabric," 582–87; Gillon, "Panel Report," 29–30.

95. Gillon, "Panel Report," 36–46; Schnepf, "Background on," 12–13.

96. Shumaker, "Tearing the Fabric," 587–88; Gillon, "Panel Report," 50–52; Schnepf, "Background on," 12–13.

97. Cong. Rec., February 1, 2005, H68–69; Forrest Laws, "Step 2 Helps U.S. Cotton End 05–06 Marketing Year on High Note," *Delta Farm Press*, August 8, 2006, http://deltafarmpress.com/step-2-helps-us-cotton-industry-end-05-06-marketing-year-high-note; Forrest Laws, "Step 2 May Linger through End of July 2006," *Delta Farm Press*, August 25, 2005, http://deltafarmpress.com/step-2-may-linger-through-end-july-2006.

98. Lehrer, *U.S. Farm Bills*; McCarl and Boadu, "Bioenergy," 43; Tyner, "US Ethanol," 646–53.

99. Energy Policy Act of 2005, H.R. 6, 109th Cong. (2005); Cong. Rec., April 20, 2005, H2174 (rule for consideration), H2286; Cong. Rec., April 21, 2005, H2450 (passed the House by a vote of 249 to 183 [3 not voting]); Cong. Rec., June 28, 2005, S7477 (passed Senate by a vote of 85 to 12 [3 not voting]); Cong. Rec., July 27, 2005, H6972–73 (House passed by a vote of 275 to 156 [3 not voting]); Cong. Rec., July 29, 2005, S9374 (passed Senate by a vote of 74 to 26); Energy Policy Act of 2005, P.L. 109-58 (2005).

100. H.R. Rep. No. 109-190 (2005) (Conf. Rep.).

101. Energy Independence and Security Act of 2007, P.L. No. 110-140 (2007); Adam Nagourney, "Democrats Take Control of House; Senate Hangs on Virginia and Montana," *New York Times*, November 8, 2006, http://www.nytimes.com/2006/11/08/us/politics/08elect.html?fta=y&_r=0; William Branigin, "Democrats Take Majority in House; Pelosi Poised to Become Speaker," *Washington Post*, November 8, 2006, http://www.washingtonpost.com/wp-dyn/content/article/2006/11/07/ar2006110700473.html.

102. See, e.g., Lehrer, *U.S. Farm Bills*, 87–98; Morgan, "Farm Bill and Beyond," 20.

103. Cong. Rec., December 13, 2007, s15421; Keith Good, "Bigger Impact on Agriculture: Farm Bill or Energy Bill?," *FarmPolicy.com*, December 24, 2007, http://farmpolicy.com/2007/12/24/bigger-impact-on-agriculture-farm-bill-or -energy-bill/#more-576; Dan Morgan, "Analysis from Washington," *FarmPolicy.com*, February 2008, http://farmpolicy.com/2008/02/.

104. CBO, "Commodity Credit Corporation Account Plus Other Accounts Comparable to the USDA Baseline: March 2007 CBO Baseline," February 23, 2007, https://www.cbo.gov/sites/default/files/51317-2007-03-ccc.pdf.

105. Lehrer, *U.S. Farm Bills*, 111; Morgan, "Farm Bill and Beyond," 20.

106. "About Me," Congressman Collin C. Peterson, n.d., https://collinpeterson .house.gov/about-me/full-biography; "Peterson, Collin Clark," Biographical Directory of the United States Congress, n.d., http://bioguide.congress.gov /scripts/biodisplay.pl?index=p000258.

107. Dan Morgan, Sara Cohen, and Gilbert M. Gaul, "Harvesting Cash," *Washington Post*, 2006, http://www.washingtonpost.com/wp-srv/nation /interactives/farmaid/; Keith Good, "Johanns Comments on House Farm Bill," *FarmPolicy.com*, July 13, 2007, http://farmpolicy.com/2007/07/13/johanns -comments-on-the-house-farm-bill/#more-393.

108. Keith Good, "Conservation and Direct Payment Considerations," *FarmPolicy.com*, July 10, 2007, http://farmpolicy.com/2007/07/10/conservation -direct-payment-considerations/#more-390; Keith Good, "Political Dynamics," *FarmPolicy.com*, July 17, 2007, http://farmpolicy.com/2007/07/17/political -dynamics/#more-396.

109. Dan Morgan, "Analysis from Washington," *FarmPolicy.com*, May 12, 2007, http://farmpolicy.com/2007/05/page/2/; Keith Good, "Energy and Agriculture," *FarmPolicy.com*, May 15, 2007, http://farmpolicy.com/2007/05/15 /energy-and-agriculture/#more-293.

110. Keith Good, "Farm Bill: Budget Remains a Key Issue," *FarmPolicy.com*, May 7, 2007, http://farmpolicy.com/2007/05/07/farm-bill-budget-remains -key-issue/#more-275; Keith Good, "Budget," *FarmPolicy.com*, May 16, 2007, http://farmpolicy.com/2007/05/16/budget/#more-296.

111. Dan Morgan, "House Panel Votes to Extend Controversial Farm Subsidies, Signaling Battle to Come," *Washington Post*, June 20, 2007, http://www .washingtonpost.com/wp-dyn/content/article/2007/06/19/ar2007061902068 _pf.html; Keith Good, "House Ag Committee: Farm Bill Reauthorization Starts Today," *FarmPolicy.com*, May 22, 2007, http://farmpolicy.com/2007/05/22/house -ag-committee-farm-bill-reauthorization-starts-today/#more-310; Keith Good, "*Washington Post*: Harvesting Cash Series Continues," *FarmPolicy.com*, June 20, 2007, http://farmpolicy.com/2007/06/20/washington-post-%e2%80%9charvesting -cash%e2%80%9d-series-continues-2/#more-361; Dan Morgan, "Analysis from Washington," *FarmPolicy.com*, June 23, 2007, http://farmpolicy.com/2007/06/.

112. Keith Good, "House Ag Committee, Day One," *FarmPolicy.com*, July 18, 2007, http://farmpolicy.com/2007/07/18/house-ag-committee-day-one/#more-397.

113. Keith Good, "House Ag Committee, Day Two," *FarmPolicy.com*, July 19, 2007, http://farmpolicy.com/2007/07/19/house-ag-committee-day-two/#more-398.

114. Cong. Rec., July 26, 2007, H8687–88.

115. Keith Good, "House Ag Committee, Day Three," *FarmPolicy.com*, July 20, 2007, http://farmpolicy.com/2007/07/20/house-ag-committee-day-three/#more-399; Dan Morgan, "Analysis from Washington," *FarmPolicy.com*, July 24, 2007, http://farmpolicy.com/2007/07/.

116. Cong. Rec., July 26, 2007, H8676 (the rule), H8686 (the rule agreed to by a vote of 222 to 202 [8 not voting]), H8679 (Ranking Member on House Ag, Representative Bob Goodlatte [R-VA]), H8688 (Representative Goodlatte), H8690 (Representative Frank Lucas [R-OK]), H8692 (Representative Randy Neugebauer [R-TX]), H8693 (Representative Michael Conaway [R-TX]).

117. Cong. Rec., July 27, 2007, H8764 (House Speaker Nancy Pelosi [D-CA]), H8765 (Majority Leader Steny Hoyer [D-MD]).

118. Cong. Rec., July 27, 2007, H8756 (Boehner amendment and Representative Jeff Flake [R-AZ]), H8757 (Chairman Peterson), H8776 (Boehner amendment defeated by a vote of 153 to 271 [13 not voting]).

119. Cong. Rec., July 27, 2007, H8771–72 (Cooper amendment), H8772–73 (Representative Earl Pomeroy [D-ND]), H8774 (debate), H8778 (Cooper's amendment was defeated 175 to 250 [1 voting present; 11 not voting]), H8761 (Representative Danny Davis [D-IL] amendment to reform the sugar program), H8776–77 (amendment defeated 144 to 282 [11 not voting]).

120. Cong. Rec., July 26, 2007, H8701–16 (Kind amendment).

121. Cong. Rec., July 26, 2007, 8717 (Representative Stephanie Herseth-Sandlin [D-SD]), H8719 (Representative Lucas [R-OK]).

122. Cong. Rec., July 26, 2007, H8730 (Kind amendment defeated by a vote of 117 to 309 [11 not voting]; Keith Good, "Tax Provision Passes, Kind Amd. Fails-Final Vote Today," *FarmPolicy.com*, July 27, 2007, http://farmpolicy.com/2007/07/27/tax-provision-passes-kind-amd-fails-final-vote-today/#more-405; Cong. Rec., July 27, 2007, H8769 (Representative Mark Udall [D-CO] cotton amendment), H8777 (Udall's amendment was defeated 175 to 251 [11 not voting]).

123. Cong. Rec., July 27, 2007, H8786 (Goodlatte offered the motion to recommit and strike the tax provisions), H8786–87 (Chairman Rangel and Representative James McCrery [R-LA]), H8787 (motion to recommit defeated by a vote of 198 to 223 [11 not voting]), H8788 (farm bill passed by a vote of 231 to 191 [10 not voting]).

124. Dan Morgan, "Analysis from Washington," *FarmPolicy.com*, July 2007, http://farmpolicy.com/2007/07/.

125. Good, "Tax Provision Passes"; Keith Good, "Farm Bill Debate," *Farm-Policy.com*, July 11, 2007, http://farmpolicy.com/2007/07/11/farm-bill-debate/#more-391; Keith Good, "Farm Bill: Timing Could Be an Issue," *FarmPolicy.com*, June 28, 2007, http://farmpolicy.com/2007/06/28/farm-bill-timing-could-be-an-issue/#more-375.

126. Keith Good, "Details Emerge in Senate Farm Bill Debate," *FarmPolicy.com*, September 12, 2007, http://farmpolicy.com/2007/09/12/details-emerge-in-senate-farm-bill-debate/#more-464.

127. Keith Good, "Additional Reaction to House Bill While Attention Shifts to the Senate," *Farmpolicy.com*, July 29, 2007, http://farmpolicy.com/2007/07/29/additional-reaction-to-house-bill-while-attention-shifts-to-the-senate/#more-408; Keith Good, "Harkin: CSP Will Be Part of the 2007 Farm Bill, 'or There Won't Be a Bill,'" *Farmpolicy.com*, August 15, 2007, http://farmpolicy.com/2007/08/15/harkin-csp-will-be-part-of-the-2007-farm-bill-or-there-wont-be-a-bill/#more-427.

128. Keith Good, "Harkin Previews Farm Bill," *FarmPolicy.com*, August 1, 2007, http://farmpolicy.com/2007/08/01/harkin-previews-farm-bill/#more-412.

129. Keith Good, "Harkin: Title I Farm Bill Draft," *FarmPolicy.com*, August 24, 2007, http://farmpolicy.com/2007/08/24/harkin-title-i-farm-bill-draft/#more-434; Keith Good, "Sen. Conrad: Farm Bill 'At a Critical Moment,'" *FarmPolicy.com*, August 29, 2007, http://farmpolicy.com/2007/08/29/sen-conrad-farm-bill-%e2%80%9cat-a-critical-moment%e2%80%9d/#more-439.

130. Keith Good, "Senate Takes Turn at Farm Bill Development," *FarmPolicy.com*, September 4, 2007, http://farmpolicy.com/2007/09/04/senate-takes-turn-at-farm-bill-development/#more-449.

131. Keith Good, "Farm Bill: Senate Debate Taking Shape," *FarmPolicy.com*, September 8, 2007, http://farmpolicy.com/2007/09/08/farm-bill-senate-debate-taking-shape/#more-457.

132. Keith Good, "Details Emerge in Senate Farm Bill Debate," *FarmPolicy.com*, September 12, 2007, http://farmpolicy.com/2007/09/12/details-emerge-in-senate-farm-bill-debate/#more-464.

133. Keith Good, "Harkin: Ideas Formulating on Farm Bill," *FarmPolicy.com*, September 19, 2007, http://farmpolicy.com/2007/09/19/harkin-ideas-formulating-on-farm-bill/#more-477; Dan Morgan, "Analysis from Washington," *FarmPolicy.com*, September 2007, http://farmpolicy.com/2007/09/.

134. Keith Good, "Senate Farm Bill: Financing Issues Remain," *FarmPolicy.com*, October 2, 2007, http://farmpolicy.com/2007/10/02/senate-farm-bill-financing-issues-remain/#more-496; Keith Good, "Farm Bill Mark-Up

Pushed Back," *FarmPolicy.com*, October 4, 2007, http://farmpolicy.com/2007 /10/04/farm-bill-mark-up-pushed-back/#more-498.

135. Keith Good, "U.S. Notifies WTO on Domestic Support," *FarmPolicy. com*, October 5, 2007, http://farmpolicy.com/2007/10/05/us-notifies-wto-on -domestic-support/#more-499.

136. Keith Good, "As Senate Debate Approaches, Farm Bill Ideas Scrutinized," *FarmPolicy.com*, October 15, 2007, http://farmpolicy.com/2007/10 /15/as-senate-debate-approaches-farm-bill-ideas-scrutinized/#more-509; Dean Kleckner, "Today's Harvest of Shame," *New York Times*, October 15, 2007, http://www.nytimes.com/2007/10/15/opinion/15kleckner.html?_r=2& th=&emc=th&pagewanted=print&oref=slogin&; Keith Good, "WTO Ruling: U.S. Action on Cotton Subsidies Insufficient," *FarmPolicy.com*, October 16, 2007, http://farmpolicy.com/2007/10/16/wto-ruling-us-action-on-cotton -subsidies-insufficient/#more-511; Keith Good, "Diverse Issues Impacting Farm Policy Debate," *FarmPolicy.com*, October 19, 2007, http://farmpolicy .com/2007/10/19/diverse-issues-impacting-farm-policy-debate/#more-514.

137. Food and Energy Security Act of 2007, S. Rep. No. 110-220 (2007); Keith Good, "Senate Farm Bill Plan Emerges," *FarmPolicy.com*, October 18, 2007, http://farmpolicy.com/2007/10/18/senate-farm-bill-plan-emerges /#more-513; Keith Good, "Senate Markup Set for Wednesday: Background Developments," *FarmPolicy.com*, October 21, 2007, http://farmpolicy.com /2007/10/21/senate-markup-set-for-wednesday-background-developments /#more-515; Keith Good, "Details Still Emerging as Sen. Ag. Comm. Takes Up Farm Bill," *FarmPolicy.com*, October 24, 2007, http://farmpolicy.com /2007/10/24/details-still-emerging-as-sen-ag-comm-takes-up-farm-bill /#more-518; David Herszenhorn, "A Bid to Overhaul a Farm Bill Yields Subtle Changes," *New York Times*, October 24, 2007, http://www.nytimes.com /2007/10/24/washington/24farm.html?_r=2&th=&oref=slogin&emc=th& pagewanted=print&oref=slogin.

138. Keith Good, "Sen. Ag. Comm. Markup–Day One," *FarmPolicy.com*, October 25, 2007, http://farmpolicy.com/2007/10/25/sen-ag-comm-markup -day-one/#more-519; Keith Good, "Senate Ag. Comm.-Markup Complete," *FarmPolicy.com*, October 26, 2007, http://farmpolicy.com/2007/10/26/senate -ag-comm-markup-complete/#more-520.

139. Dan Morgan, "Analysis from Washington," *FarmPolicy.com*, October 2007, http://farmpolicy.com/2007/10/; Keith Good, "Senate Farm Bill: Reaction to Comm. Markup Continues," *FarmPolicy.com*, October 27, 2007, http://farmpolicy .com/2007/10/27/senate-farm-bill-reaction-to-comm-markup-continues/#more -522; Keith Good, "Debate Continues as Senate Farm Bill Awaits Floor Action," *FarmPolicy.com*, October 30, 2007, http://farmpolicy.com/2007/10/30/debate -continues-as-senate-farm-bill-awaits-floor-action/#more-525.

140. Keith Good, "Senate Farm Bill: Payment Limits and Other Issues," *FarmPolicy.com*, October 23, 2007, http://farmpolicy.com/2007/10/23/senate -farm-bill-payment-limits-other-issues/#more-517.

141. Cong. Rec., November 5, 2007, s13743–47, s13750, s13751–52 (Ranking Member Chambliss), s13756 (Senator Conrad).

142. Cong. Rec., November 5, 2007, s13746–47.

143. Cong. Rec., November 5, 2007, s13752.

144. Cong. Rec., November 5, 2007, s13752.

145. See, e.g., Malcolm and Aillery, "Growing Crops," 10–15; Tyner, "US Ethanol," 646–53.

146. Cong. Rec., November 6, 2007, s13940 (Senate Majority Leader Harry Reid [d-nv]), s13947–48 (Senator Reid), s13948–49 (Minority Leader Mitch McConnell [r-ky]), s13951 (Senator Reid), s13981 (Senator Reid); Cong. Rec., November 14, 2007, s14354–55 (Reid and McConnell), s14363–64 (Chairman Harkin), s14388–89 (Reid filed the cloture motion); Cong. Rec., November 15, 2007, s14435 (Chairman Harkin); Cong. Rec., November 16, 2007, s14592 (cloture failed 55 to 42 [3 not voting]); Cong. Rec., December 7, 2007, s15013 (Senators Harkin and Chambliss on amendment deal). See also Keith Good, "Ethanol-Energy Bill," *FarmPolicy.com*, November 28, 2007, http://farmpolicy .com/2007/11/28/ethanol-%e2%80%93-energy-bill/#more-553; Keith Good, "Farm Payments Draw Focus, While Market Prices and Income Appear Strong," *FarmPolicy.com*, December 4, 2007, http://farmpolicy.com/2007/12 /04/farm-payments-draw-focus-while-market-prices-income-appear-strong /#more-557; Keith Good, "Deal Made on Senate Farm Bill," *FarmPolicy.com*, December 7, 2007, http://farmpolicy.com/2007/12/07/deal-made-on-senate -farm-bill/#more-560.

147. Cong. Rec., December 7, 2007, s15014 (Dorgan-Grassley amendment followed by Senator Klobuchar's amendment on adjusted gross income eligibility requirement); Cong. Rec., December 11, 2007, s15091–92 (Senator Lugar amendment), s15107 (Senator Lugar; Senator Brown [d-oh] amendment on crop insurance); Cong. Rec., December 12, 2007, s15179–15203 (Dorgan-Grassley amendment), s15207–12 (Senators Lincoln and Jeff Sessions [r-al]), s15212 (Senator Dorgan).

148. Cong. Rec., November 5, 2007, s13761 (Senator Grassley); Cong. Rec., November 6, 2007, s13977 (Senator Grassley).

149. Cong. Rec., December 13, 2007, s15414 (Senator Chambliss), s15416–17 (Senator Lincoln). See also Cong. Rec., November 6, 2007, s13965 (Senator Domenici [r-nm] amendment to double the rfs); Cong. Rec., December 13, 2007, s15421 (the Senate going back and forth between considering the 2007 energy bill and the farm bill).

150. Morgan, Cohen, and Gaul, "Harvesting Cash."

151. Cong. Rec., November 6, 2007, s13955 (Senator Ken Salazar [d-co]), s13976 (Senator Grassley); Cong. Rec., December 12, 2007, s15210–11 (Senator Chambliss).

152. See, e.g., Food and Energy Security Act of 2007, S. 2302, 110th Cong. (2007), https://www.congress.gov/110/bills/s2302/bills-110s2302pcs.xml; Food and Energy Security Act of 2007, S. Rep. No. 110-220 (2007), https://www .congress.gov/congressional-report/110th-congress/senate-report/220/1?q =%7b%22search%22%3a%5b%22food+conservation+energy%22%5d%7d; Johnson et al., "2008 Farm Bill," 65–68.

153. Cong. Rec., November 6, 2007, s13972–73.

154. Cong. Rec., December 12, 2007, s15207–8, s15210–11 (Senator Chambliss), s1521 (Senator Lincoln); Cong. Rec., December 13, 2007, s15385 (Senator Cochran).

155. Cong. Rec., December 13, 2007, s15384–85 (Senator Tim Johnson [d-sd]), s15380 (Senator Reid), s15385 (Dorgan-Grassley amendment defeated by a vote of 56 to 43), s15412–13 (Klobuchar amendment), s15415 (Senator Durbin and Chairman Harkin), s15417 (Senator Klobuchar), s15418 (Klobuchar amendment failed by a vote of 48 to 47 [5 not voting]).

156. Cong. Rec., December 13, 2007, s15404–5 (Senator Brown), s15411–12 (Chairman Harkin supported the amendment), s15405 (Senator Roberts), s15420 (Brown amendment defeated by a vote of 32 to 63 [5 not voting]).

157. Cong. Rec., December 13, 2007, s15448 (Reid), s15449 (Senator Conrad), s15450 (cloture motion passed by a vote of 78 to 12 [10 not voting]); Cong. Rec., December 14, 2007, s15639 (final passage by a vote of 79 to 14 [7 not voting]).

158. Keith Good, "Farm Bill: 'Showdown Is Brewing,'" *FarmPolicy.com*, January 6, 2008, http://farmpolicy.com/2008/01/14/farm-bill-%e2%80 %9cshowdown-is-brewing%e2%80%9d/#more-591; Keith Good, "Farm Bill Funding Level Still an Issue," *FarmPolicy.com*, January 15, 2008, http:// farmpolicy.com/2008/02/15/farm-bill-funding-level-still-an-issue/#more -624; Keith Good, "Farm Bill: Senate Offers Perspective on Spending Level," *FarmPolicy.com*, January 16, 2008, http://farmpolicy.com/2008/02/16/farm -bill-senate-offers-perspective-on-spending-level/#more-625; Keith Good, "Pres. Bush: If No Farm Bill by April 18, Then One Year Extension," *Farm-Policy.com*, March 14, 2008, http://farmpolicy.com/2008/03/14/pres-bush-if -no-farm-bill-by-april-18-then-one-year-extension/#more-656.

159. Dan Morgan, "Analysis from Washington," *FarmPolicy.com*, April 2008, http://farmpolicy.com/2008/04/page/2/.

160. See, e.g., Keith Good, "Senate, House Farm Bill Negotiators Reach Compromise," *FarmPolicy.com*, April 26, 2008, http://farmpolicy.com/2008 /04/26/senate-house-farm-bill-negotiators-reach-compromise/#more-741;

Keith Good, "Farm Bill; Commodity—Food Prices; Pres. Bush—Food Aid," *FarmPolicy.com*, May 2, 2008, http://farmpolicy.com/2008/05/02/farm-bill -commodity-food-prices-pres-bush-food-aid/#more-758; Keith Good, "Farm Bill: Congress Moves Closer to Completion, Executive Branch States Priorities; Biofuels; Post Series on C-SPAN," *FarmPolicy.com*, May 3; 2008, http://farmpolicy.com/2008/05/03/farm-bill-congress-moves-closer-to -completion-executive-branch-states-priorities-biofuels-post-series-on-c -span/#more-759; Keith Good, "Agreement on Farm Bill—Details Expected Today; Biofuels Hearing (Senate); Doha," *FarmPolicy.com*, May 8, 2008, http://farmpolicy.com/2008/05/08/agreement-on-farm-bill-details-expected -today-biofuels-hearing-senate-doha/#more-766; Keith Good, "Farm Bill Conference Agreement Announced, Pres. Bush Promises Veto," *FarmPolicy. com*, May 9, 2008, http://farmpolicy.com/2008/05/09/farm-bill-conference -agreement-announced-pres-bush-promises-veto/#more-767; Keith Good, "House Vote on Farm Bill Conference Report Expected Today," *FarmPolicy. com*, May 14, 2008, http://farmpolicy.com/2008/05/14/house-vote-on-farm -bill-conference-report-expected-today/#more-779; David Herszenhorn, "Tentative Deal Reached in Congress on Farm Bill," *New York Times*, April 26, 2008, http://www.nytimes.com/2008/04/26/washington/26farm.html?_r =2&sq=farmers&st=nyt&scp=4&pagewanted=print&oref=slogin&; David Herszenhorn, "Bush Says Pain from the Economy Defies Easy Fix," *New York Times*, April 30, 2008, http://www.nytimes.com/2008/04/30/washington /30bush.html?_r=1&ref=washington&pagewanted=print&oref=slogin.

161. Cong. Rec., May 14, 2008, H3801 (Chairman Peterson); Dan Morgan, "Analysis from Washington," *FarmPolicy.com*, May 2008, http://farmpolicy .com/2008/05/.

162. Cong. Rec., May 14, 2008, H3784 (conference report consideration), H3801 (Chairman Peterson), H3822 (House passed it by a vote of 318 to 106 [10 not voting]), S4152 (conference report considered by the Senate; Chairman Harkin); Cong. Rec., May 15, 2008, S4239 (passed the Senate by a vote of 81 to 15 [4 not voting]).

163. Cong. Rec., May 21, 2008, H4402 (Chairman Peterson explained), H4411 (House voted to override the veto by a vote of 316 to 108 [11 not voting]; Cong. Rec., May 22, 2008, S4743 (veto override in the Senate), S4749 (Senate voted to override the veto 82 to 13 [4 not voting]).

164. Cong. Rec., May 22, 2008, H4654 (Chairman Peterson explained), H4649–50 (Peterson); Cong. Rec., June 18, 2008, H5535 (veto message), H5536–37 (House voted 317 to 109 [8 not voting], S5741 (Senate voted 80 to 14 [6 not voting]); Cong. Rec., June 5, 2008, S5182–83 (Senator Coburn [R-OK]), S5187 (Chairman Harkin), S5188 (passed the Senate by a vote of 77 to 15 [8 not voting]).

165. Keith Good, "Awkwardly, the 2008 Farm Bill Debate Wraps Up," *FarmPolicy.com*, May 23, 2008, http://farmpolicy.com/2008/05/23/awkwardly-the-2008-farm-bill-debate-wraps-up/#more-796; Keith Good, "UN Conference Wraps Up—Biofuels; Doha Talks; Farm Bill; Crop Concerns; Climate Change Bill," June 6, 2008, http://farmpolicy.com/2008/06/06/un-conference-wraps-up-biofuels-doha-talks-farm-bill-crop-concerns-climate-change-bill/#more-805; "Work Begins on New Farm Bill; Conference Put Off until 2008," in CQ *Almanac 2007*, 63rd ed., ed. Jan Austin, 3-3-3-7 (Washington DC: Congressional Quarterly, 2008), http://library.cqpress.com/cqalmanac/cqal07-1006-44912-2047970.

8. Old Fights Plague the Agricultural Act

1. Rhodes Cook, "From Republican 'Lock' to Republican 'Lockout'?" *Sabato's Crystal Ball*, November 13, 2008, http://www.centerforpolitics.org/crystalball/articles/frc2008111301/. Obama won 364 electoral votes and carried 53 percent of the popular vote, with a margin of 8.25 million votes.

2. Carey Gillam, "U.S. Farm Sector Cautiously Welcomes Obama Win," Reuters, November 6, 2008, http://www.reuters.com/article/sppage014-n05324826-oistl-idusn0532482620081106?sp=true; Herron and Lewis, "Economic Crisis," 1–2; Jickling, "Causes of"; Ed Hornick et.al., "McCain, Obama Headed to Washington for Bailout Talks," CNN.com, September 24, 2008, http://www.cnn.com/2008/politics/09/24/campaign.wrap/index.html?eref=onion.

3. Tyler Atkinson, David Luttrell, and Harvey Rosenblum, "How Bad Was It?: The Costs and Consequences of the 2007–09 Financial Crisis," *Staff Papers, Dallas Fed*, no. 20 (July 2013): 2.

4. USDA, Food and Nutrition Service, "Supplemental Nutrition Assistance Program," last modified March 2018, http://www.fns.usda.gov/pd/supplemental-nutrition-assistance-program-snap; CBO, "Cost Estimate: H.R. 2419, Food, Conservation, and Energy Act of 2008," May 13, 2008, https://www.cbo.gov/publication/41696; CBO, Data and Technical Information, "Food Stamp Program—March 2008 Baseline," March 1, 2008, https://www.cbo.gov/sites/default/files/recurringdata/51312-2008-03-foodstamps.pdf; CBO, Data and Technical Information, "Food Stamp Program—January 2009 Baseline," January 7, 2009, https://www.cbo.gov/sites/default/files/recurringdata/51312-2009-03-snap.pdf; CBO, Data and Technical Information, "Supplemental Nutrition Assistance Program—March 2011 Baseline," 2011, https://www.cbo.gov/sites/default/files/recurringdata/51312-2011-03-snap.pdf.

5. "Federal Budget Deficit Totals $1.4 Trillion in Fiscal Year 2009," CBO (blog), November 6, 2009, https://www.cbo.gov/publication/24992; "The Federal Budget Deficit for 2010," CBO (blog), October 7, 2010, https://www.cbo.gov/publication/25107.

6. Keith Good, "ERS Farm Financial Forecast, President-Elect Obama on Farm Spending, Food Stamps and France," *Farmpolicy.com*, November 26, 2008, http://farmpolicy.com/2008/11/26/ers-farm-financial-forecast-president-elect-obama-on-farm-spending-food-stamps-and-france/#more-940.

7. See, e.g., Jacobson, "Republican Resurgence," 1; Carson and Pettigrew, "Strategic Politicians," 26–36; Aldrich et al., "Blame, Responsibility," 9242–44.

8. "Biography," Congressman Frank Lucas, n.d., https://lucas.house.gov/about-me/full-biography; "Lucas, Frank D.," Biographical Directory of the United States Congress, n.d., http://bioguide.congress.gov/scripts/biodisplay.pl?index=l000491; "Arena Profile: Rep. Frank Lucas," *Politico*, 2012, http://www.politico.com/arena/bio/rep_frank_lucas.html.

9. Chris Clayton, "Ripple Effects," *DTNProgressiveFarmer*, August 26, 2009, http://www.dtnprogressivefarmer.com/dtnag/common/link.do;jsessionid=dba62eef2db6784b2e43946f2bbc2cf6.agfreejvm2?symbolicName=/ag/blogs/template1&blogHandle=policy&blogEntryId=8a82c0bc231572590123570995b5032a&showCommentsOverride=false; Tom Laskawy, "Blanche Lincoln as Ag Chair? Say It Ain't So," *New Republic*, September 8, 2009, http://www.newrepublic.com/blog/the-vine/blanche-lincoln-ag-chair-say-it-aint-so; Alec MacGillis, "$1.5 Billion Farm-Aid Proposal Assailed as Relief for Sen. Blanche Lincoln," *Washington Post*, August 22, 2010, http://www.washingtonpost.com/wp-dyn/content/article/2010/08/21/ar2010082102517.html.

10. Keith Good, "Farm Bill; and Political Notes—Election Impact on Farm Policy and Biofuels," *FarmPolicy.com*, November 4, 2010, http://farmpolicy.com/2010/11/04/farm-bill-and-political-notes-election-impact-on-farm-policy-and-biofuels/#more-3383; Keith Good, "Farm Bill; Ag Economy; Climate; Trade; Animal Agriculture; and Food Safety," *FarmPolicy.com*, November 9, 2010, http://farmpolicy.com/2010/11/09/farm-bill-ag-economy-climate-trade-animal-agriculture-and-food-safety/#more-3394; Keith Good, "Farm Bill; Food Safety; Ag Economy; Climate; and Animal Ag," *FarmPolicy.com*, November 19, 2010, http://farmpolicy.com/2010/11/19/farm-bill-food-safety-ag-economy-climate-and-animal-ag/#more-3525.

11. Keith Good, "Farm Bill; Food Safety; Ag Economy; Climate; Biofuels; USDA Issue; and Animal Ag," *FarmPolicy.com*, November 22, 2010, http://farmpolicy.com/2010/11/22/farm-bill-food-safety-ag-economy-climate-biofuels-usda-issue-and-animal-ag/#more-3536; Keith Good, "Farm Bill; Biofuels; Ag Economy; and Animal Agriculture," *FarmPolicy.com*, November 23, 2010, http://farmpolicy.com/2010/11/23/farm-bill-biofuels-ag-economy-and-animal-agriculture/#more-3546; Liza Mundy, "The Secret History of Women in the Senate," *Politico*, January 2015, http://www.politico.com/magazine/story/2015/01/senate-women-secret-history-113908.html#.vw2yjel_nyd.

12. "Biography," Debbie Stabenow, U.S. Senator for Michigan, n.d., https://www.stabenow.senate.gov/about/biography; "Stabenow, Deborah Ann," Biographical Directory of the United States Congress, n.d., http://bioguide.congress.gov/scripts/biodisplay.pl?index=s000770.

13. USDA, ERS, "Farm Income and Wealth Statistics," https://www.ers.usda.gov/data-products/farm-income-and-wealth-statistics/data-files-us-and-state-level-farm-income-and-wealth-statistics/.

14. Schnepf, "Brazil's WTO Case," 16–23.

15. Schnepf, "Brazil's WTO Case," 24–27.

16. Agriculture Reform, Food and Jobs Act of 2012, S. Rep. No. 112-203, at 39 (2012); Keith Good, "Farm Bill; Budget Issues; and Rural Data," FarmPolicy.com, February 18, 2011, http://farmpolicy.com/2011/02/18/farm-bill-issues-budget-issues-and-rural-data/#more-4038.

17. Charles Abbott, "Farm Cuts May Be Smaller than Expected: Lawmaker," Reuters, July 27, 2011, http://www.reuters.com/article/us-usa-agriculture-idustre76q6e220110727; Keith Good, "Farm Bill; Debt Talks; Ag Economy; Trade; and Regulations," FarmPolicy.com, July 28, 2011, http://farmpolicy.com/2011/07/28/farm-bill-debt-talks-ag-economy-trade-and-regulations/#more-5217.

18. House Committee on the Budget, "The Path to Prosperity: Restoring America's Promise," Fiscal Year 2012 Budget Resolution, April 15, 2011, 4, http://budget.house.gov/uploadedfiles/pathtoprosperityfy2012.pdf (hereafter cited as House FY2012 Budget).

19. See, e.g., Keith Good, FarmPolicy.com, March 2011; Keith Good, "Budget Issues; Farm Bill; Ag Economy; and Biofuels," FarmPolicy.com, April 11, 2011, http://farmpolicy.com/2011/04/11/budget-issues-farm-bill-ag-economy-and-biofuels/#more-4385; Austin and Levit, "Debt Limit," 17–18; Colleen Murray, "As US Reaches Debt Limit, Geithner Implements Additional Extraordinary Measures to Allow Continued Funding of Government Obligations," Treasury Notes (blog), May 16, 2011, http://www.treasury.gov/connect/blog/Pages/Geithner-Implements-Additional-Extraordinary-Measures-to-Allow-Continued-Funding-of-Government-Obligations.aspx.

20. Keith Good, "Farm Bill (Budget Issues); Ag Economy; and Regulations," FarmPolicy.com, August 1, 2011, http://farmpolicy.com/2011/08/02/farm-bill-budget-issues-ag-economy-and-regulations/#more-5277.

21. See, generally, New York Times, Bipartisan Deficit Committee (September 8, 2011).

22. Heniff, Rybicki, and Mahan, "Budget Control Act," 34; Lawrence, "Profiles in Negotiation," 4.

23. Heniff, Rybicki, and Mahan, "Budget Control Act," 34.

24. See, e.g., Keith Good, "Farm Bill; Ag Economy; CFTC Issue; Food Safety; Regulations; and Trade," *FarmPolicy.com*, September 1, 2011, http://farmpolicy .com/2011/09/01/farm-bill-ag-economy-cftc-issue-food-safety-regulations-and -trade/#more-5454; Keith Good, "Farm Bill Update," *FarmPolicy.com*, September 5, 2011, http://farmpolicy.com/2011/09/05/farm-bill-update-4/#more-5464; Keith Good, "Farm Bill; Regulations; Trade; and Ag Economy," *FarmPolicy. com*, September 7, 2011, http://farmpolicy.com/2011/09/07/farm-bill-regulations -trade-and-the-ag-economy/#more-5475; Keith Good, "Farm Bill; Ag Economy; Regulations; Food Safety; and Trade," *FarmPolicy.com*, September 13, 2011, http://farmpolicy.com/2011/09/13/farm-bill-ag-economy-regulations -food-safety-and-trade/#more-5526; Charles Abbott, "Obama Wants to End $5 Billion a Year US Farm Subsidy," Reuters, September 19, 2011, http://www .reuters.com/article/2011/09/19/us-debt-obama-farm-idustre78j00120110919 ?feedType=rss&feedName=everything&virtualBrandChannel=11563.

25. Lawrence, "Profiles in Negotiation," 4.

26. See, e.g., Keith Good, "FarmPolicy.com Interview, Senate Ag Committee Chairwoman Debbie Stabenow (d-Mich.), Farm Bill, Supercommittee, and Research," *FarmPolicy.com*, September 22, 2011, http://farmpolicy .com/wp-content/uploads/2011/09/transcript11sep22interviewChairStabenow .pdf; Keith Good, "Farm Bill; Ag Economy; and Trade," *FarmPolicy.com*, October 14, 2011, http://farmpolicy.com/2011/10/14/farm-bill-ag-economy -and-trade-4/#more-5824; Keith Good, "Farm Bill; Ag Economy; and Senate Procedural Issue," *FarmPolicy.com*, October 17, 2011, http://farmpolicy .com/2011/10/17/farm-bill-ag-economy-and-senate-procedural-issue/#more -5830; Keith Good, "Farm Bill; Ag Economy; and Regulations," *FarmPolicy.com*, October 18, 2011, http://farmpolicy.com/2011/10/18/5840/#more -5840; Keith Good, "Farm Bill; Ag Economy; Regulations; and Trade," *FarmPolicy.com*, October 26, 2011, http://farmpolicy.com/2011/10/26/farm-bill -ag-economy-regulations-and-trade-3/#more-5937; David Rogers, "Deal Close on Cut in Farm Subsidies," *Politico*, October 12, 2011, http://www .politico.com/story/2011/10/deal-close-on-cut-in-farm-subsidies-065834; Charles Abbott, "U.S. Farm Subsidy Reform May Be Tied to Budget Cuts," Reuters, October 13, 2011, http://www.reuters.com/article/2011/10/13/us-usa -agriculture-idustre79c70l20111013?feedType=rss&feedName=everything& virtualBrandChannel=11563; Alan Bjerga and Derek Wallbank, "Farm Cuts Should Be Held to $23 Billion, Super Committee Told," *Bloomberg*, October 17, 2011, http://www.bloomberg.com/news/articles/2011-10-17/farm-cuts -should-be-capped-at-23-billion-supercommittee-told.

27. David Rogers, "Farm Lobby's Power Withers as Subsidies Face Cuts," *Politico*, September 20, 2011, http://www.politico.com/story/2011/09/farm -lobbys-power-withers-063988.

28. Orden and Zulauf, "Political Economy," 4.

29. Charles Abbott, "US Farm Law Overhaul Would Boost Some Crop Subsidy Rates," Reuters, November 14, 2011, http://af.reuters.com/article /commoditiesNews/idafnle7ad1bb20111114?sp=true; Keith Good, "Farm Bill; Budget Issues; Ag Economy; and Trade," FarmPolicy.com, November 14, 2011, http://farmpolicy.com/2011/11/14/farm-bill-budget-issues-ag-economy -and-trade/#more-6095; Keith Good, "Farm Bill; Supercommittee; and Land Values," FarmPolicy.com, November 17, 2011, http://farmpolicy.com/2011/11 /17/farm-bill-supercommittee-and-land-values/#more-6125; David Rogers, "Farm Bill: Corn Belt vs. Great Plains Dems," Politico, November 16, 2011, http://www.politico.com/news/stories/1111/68552_Page2.html; David Rogers, "Rare Deal: $182B Approps Bill Passes," Politico, November 17, 2011, http:// www.politico.com/news/stories/1111/68630_Page2.html; William Neuman, "Farmers Facing Loss of Subsidy May Get a New One," New York Times, October 17, 2011, http://www.nytimes.com/2011/10/18/business/when-one -farm-subsidy-ends-another-may-rise-to-replace-it.html.

30. See, e.g., Harwood et al., "Managing Risk in Farming"; Newton and Coppess, "Rethinking Revenue."

31. Keith Good, "Farm Bill and Supercommittee Issues," FarmPolicy. com, October 28, 2011, http://farmpolicy.com/2011/10/28/farm-bill-and -supercommittee-issues/#more-5969; David Rogers, "Farm Bill: Corn Belt vs. Great Plains Dems," Politico, November 16, 2011, http://www.politico.com /news/stories/1111/68552_Page2.html.

32. Keith Good, "Farm Bill and Supercommittee Issues," FarmPolicy. com, November 18, 2011, http://farmpolicy.com/2011/11/18/farm-bill-and -supercommittee-issues-2/#more-6132; Keith Good, "Supercommittee Falters, Farm Bill Implications," FarmPolicy.com, November 22, 2011, http:// farmpolicy.com/2011/11/22/supercommittee-falters-farm-bill-implications /#more-6161.

33. Monke, "Budget Issues Shaping a 2012 Farm Bill"; Charles Abbott, "US Farmers Get Record $10 Bln from Crop Insurance," Reuters, February 27, 2012, http://www.reuters.com/article/2012/02/27/usa-agriculture -idusl2e8drbpf20120227; Keith Good, "Budget and Farm Bill; Ag Economy; Biofuels; and Regulations," FarmPolicy.com, January 3, 2012, http:// farmpolicy.com/2012/01/03/budget-and-farm-bill-ag-economy-biofuels-and -regulations/#more-6392; Keith Good, "Budget Issues—Senate Farm Bill Hearing; Policy Issues; Ag Economy; and China," FarmPolicy.com, February 16, 2012, http://farmpolicy.com/2012/02/16/budget-issues-senate-farm -bill-hearing-policy-issues-ag-economy-and-china/#more-6643.

34. Ron Hays, "National Cotton Council President Mark Lange Examines Farm Bill's Progress, Prospects," Oklahoma Farm Report, April

2, 2012, http://oklahomafarmreport.com/wire/news/2012/04/04515
_LangeOnFarmBill04022102_172016.php (reporting comments of the head
of the National Cotton Council of America); Sara Wyant, "Southern Ag Stakes
Out Policy Territory in Upcoming Farm Bill Debate," *Agri-Pulse*, April 4,
2012, http://www.agri-pulse.com/Southern-ag-stakes-out-policy-territory-in
-uncoming-farm-bill-debate-04042012.asp. See also Ken Anderson, "NCGA
Lobbying against Target Price Option," *Brownfield*, January 2, 2012, http://
brownfieldagnews.com/2012/01/02/ncga-lobbying-against-target-price-option
/?utm_source=feedburner&utm_medium=twitter&utm_campaign=Feed%3a
+BrownfieldAgNews+%28brownfield+Ag+News%29; Keith Good, "Budget;
Farm Bill and Policy; Regulations; MF Global; and the Ag Economy," *Farm-
Policy.com*, February 3, 2012, http://farmpolicy.com/2012/02/03/budget-farm
-bill-and-policy-regulations-mf-global-and-the-ag-economy/#more-6564;
Keith Good, "Budget Developments; Farm Bill; and Policy Issues; Ag Economy; and China," *FarmPolicy.com*, February 17, 2012, http://farmpolicy.com
/2012/02/17/budget-developments-farm-bill-and-policy-issues-ag-economy
-and-china/#more-6650; Keith Good, "Farm Bill; Ag Economy; Trade; and
Regulations," *FarmPolicy.com*, March 15, 2012, http://farmpolicy.com/2012/03
/15/farm-bill-ag-economy-trade-and-regulations-4/#more-6912.

35. Keith Good, "Farm Bill and Policy Issues; Regulations; and Trade,"
FarmPolicy.com, February 28, 2012, http://farmpolicy.com/2012/02/28/quick
-take-senate-agricultre-committee-hearing-conservation/#more-6760; Keith
Good, "Farm Bill and Policy Issues; Regulations; Trade; and Ag Economy,"
FarmPolicy.com, March 1, 2012, http://farmpolicy.com/2012/03/01/farm-bill
-and-policy-issues-regulations-trade-and-ag-economy/#more-6802; Keith
Good, "Farm Bill; Budget; Ag Economy; and Trade," *FarmPolicy.com*, March
20, 2012, http://farmpolicy.com/2012/03/20/farm-bill-budget-ag-economy
-and-trade/#more-6936.

36. House Committee on the Budget, "The Path to Prosperity: A Blueprint
for American Renewal," Fiscal Year 2013 Budget Resolution, March 20, 2012,
http://budget.house.gov/uploadedfiles/blueprint_final_3192012.pdf (hereafter cited as House FY2013 Budget); Charles Abbott, "House Agriculture Plan
Would Slash Crop, Insurance Payouts," Reuters, March 20, 2012, https://www
.reuters.com/article/us-usa-budget-farm/house-agriculture-plan-would-slash
-crop-insurance-payouts-idUSBRE82J0QT20120320; Keith Good, "Farm
Bill; Budget Committee Chairman Paul Ryan Releases House GOP Budget
Outline," *FarmPolicy.com*, March 21, 2012, http://farmpolicy.com/2012/03
/21/farm-bill-budget-committee-chairman-paul-ryan-releases-house-gop
-budget-outline/#more-6948.

37. David Rogers, "Ryan Plan Triggers Budget Wars Anew," *Politico*, March
20, 2012, http://www.politico.com/story/2012/03/ryan-plan-triggers-budget

-wars-anew-074224; David Rogers, "Paul Ryan Budget Goes There Again," *Politico*, March 22, 2012, http://www.politico.com/story/2012/03/ryan-budget -goes-there-again-074377; David Rogers, "Paul Ryan Plan Vote a Devil's Bargain?" *Politico*, March 29, 2012, http://www.politico.com/story/2012/03/ryan -plan-vote-a-devils-bargain-074662; Keith Good, "Farm Bill and Ag Economy," *FarmPolicy.com*, March 23, 2012, http://farmpolicy.com/2012/03/23/farm -bill-and-the-ag-economy/#more-6964; Keith Good, "Farm Bill; Ag Economy; and Regulations," *FarmPolicy.com*, March 30, 2012, http://farmpolicy .com/2012/03/30/farm-bill-ag-economy-and-regulations-3/#more-7009; Collin Peterson, "FY2013 Republican Government: Implications for Agriculture," House Committee on Agriculture, press release, March 28, 2012, http://democrats.agriculture.house.gov/inside/Pubs/fy2013%20republican %20budget%20implications%20for%20agriculture.pdf.

38. David Lawder, "House Panels Turn Budget Axe to Automatic Cuts," Reuters, April 12, 2012, http://www.reuters.com/article/2012/04/12/us-usa -budget-cuts-idusbre83b1qi20120412; Jonathan Weisman, "House Republicans to Tackle Ambitions Budget," *New York Times*, April 13, 2012, http:// www.nytimes.com/2012/04/14/us/politics/house-republicans-to-tackle-federal -budget.html?_r=3&emc=tnt&tntemail0=yHouse Republicans.

39. Spar, "Budget 'Sequestration,'" 8–9; GAO, "Crop Insurance: Savings."

40. David Rogers, "Republicans to Slash Food Stamps," *Politico*, April 16, 2012, http://www.politico.com/story/2012/04/republicans-to-slash-food -stamps-075190; Keith Good, "Farm Bill; Ag Economy; Budget Issues; and Oil-Gas Prices," *FarmPolicy.com*, April 18, 2012, http://farmpolicy.com/2012/04/18 /farm-bill-ag-economy-budget-issues-and-oil-gas-prices/#more-7110; Emily Stephenson, "House Panel Okays $33 Billion in Food Stamp Cuts," Reuters, April 18, 2012, https://www.reuters.com/article/us-usa-agriculture-stamps /house-panel-okays-33-billion-in-food-stamp-cuts-idUSBRE83H16320120418.

41. House FY2013 Budget, 25.

42. Compare CBO, "Supplemental Nutrition Assistance Program: CBO Baseline," March 2012, https://www.cbo.gov/sites/default/files/51312-2012 -03-snap.pdf; CBO, "March 2012 Baseline for Farm Programs," March 2012, https://www.cbo.gov/sites/default/files/51317-2012-03-usda.pdf.

43. See, e.g., Schertz and Doering, *Making of*, 8–9 (quoting Secretary Butz, who compared food stamps and farm program spending, saying the line for food stamps "represents welfare and it keeps increasing"; about the line for farm programs he said, "Look at the decline in this line! Now, it is welfare, too. But it is for our kind of people!").

44. Keith Good, "Senate Farm Bill Draft; Chairman Lucas; Ag Economy; and Regulatory Issues," *FarmPolicy.com*, April 23, 2012, http://farmpolicy

.com/2012/04/23/senate-farm-bill-draft-chairman-lucas-ag-economy-and
-regulatory-issues/#more-7158.

45. CBO, "Cost Estimate: S.3240, Agriculture Reform, Food and Jobs Act
of 2012," May 24, 2012, 7, https://www.cbo.gov/publication/43273; Aussenberg and Perl, "Changing the Treatment," 2–8.

46. Keith Good, "Farm Bill; Food Safety; Ag Economy; and Regulations,"
FarmPolicy.com, April 5, 2012, http://farmpolicy.com/2012/04/05/farm-bill
-food-safety-ag-economy-and-regulations/#more-7040; David Rogers,
"Farm Bill: Corn Belt vs. Great Plains Dems," *Politico*, November 16, 2011,
http://www.politico.com/news/stories/1111/68552_Page2.htmlFarm bill; Gary
Schnitkey, "Risk Implications of Commodity Programs in the 2012 Farm
Bill," *farmdoc daily*, April 10, 2012, http://farmdocdaily.illinois.edu/2012/04
/risk-implications-of-commodity.html; Gary Schnitkey, "Graphical Illus-
trations of Proposed Farm Revenue Programs and Crop Insurance," *farmdoc daily*, April 12, 2012, http://farmdocdaily.illinois.edu/2012/04/graphical
-illustrations-of-pro.html; Sara Wyant, "Revenue Programs and Crop Insurance: What Works?," *Agri-Pulse*, April 11, 2012, http://www.agri-pulse.com
/revenue-programs-and-crop-insurance-what-works-04112012.asp.

47. *Hearing on Risk Management and Commodities in the 2012 Farm Bill
Before the U.S. Senate Committee on Agriculture, Nutrition, and Forestry*,
112th Cong. 8–9 (2012) (testimony of Travis Henry Satterfield; testimony of
Jimbo Grissom), http://www.agriculture.senate.gov/download/testimony
/satterfield-testimony-3-15-12, http://www.agriculture.senate.gov/download
/testimony/grissom-testimony-3-15-12.

48. Schnepf, "Status of"; Campiche, "Details of," 570; Gorter, "2012 US
Farm Bill."

49. Sara Wyant, "Senate Ag Panel Sets Farm Bill Markup for Next Week,"
Agri-Pulse, April 18, 2012, http://www.agri-pulse.com/Senate-ag-panel-sets
-farm-bill-markup-for-next-week-04182012.asp; David Rogers, "Delay Sought
on Farm Bill," *Politico*, April 23, 2012, http://www.politico.com/story/2012
/04/delay-sought-on-farm-bill-075515.

50. Keith Good, "Farm Bill; and the Ag Economy," *FarmPolicy.com*, April
17, 2012, http://farmpolicy.com/2012/04/17/farm-bill-and-the-ag-economy
-3/#more-7107.

51. Senate Committee on Agriculture, Nutrition, and Forestry, Committee
Print, April 20, 2012, http://www.ag.senate.gov/download/?id=8cc7b60d-3758
-41b6-be4c-1703321d2b39; Rogers, "Delay Sought"; David Rogers, "Action
on Farm Bill Postponed," April 24, 2012, http://www.politico.com/story/2012
/04/panel-takes-up-new-farm-bill-075562; Keith Good, "Farm Bill; Budget;
IFPRI Report; Ag Economy; Immigration; and Regulations," *FarmPolicy.*

com, April 24, 2012, http://farmpolicy.com/2012/04/24/farm-bill-budget-ifpri
-report-ag-economy-immigration-and-regulations/#more-7168; Keith Good,
"Farm Bill (Mark-up Today); BSE; House Ag Comm.; Budget Issues; and
Food Prices," *FarmPolicy.com*, April 26, 2012, http://farmpolicy.com/2012/04
/26/farm-bill-mark-up-today-bse-house-ag-comm-animal-ag-budget-issues
-and-food-prices/#more-7192. See also Charles Abbott, "Senate Bill Pro-
poses 21st Century US Farm Supports," Reuters, April 20, 2012, http://articles
.chicagotribune.com/2012-04-20/business/sns-rt-usa-agriculturel2e8fkcli
-20120420_1_farm-subsidies-farm-bill-money-for-conservation-programs.

52. See, e.g., David Rogers, "Regional Battle Opens over Major Farm Bill,"
Politico, April 26, 2012, http://www.politico.com/story/2012/04/landmark
-farm-bill-clears-senate-panel-075665regional battle.

53. Senate Committee on Agriculture, Nutrition, and Forestry, "Business Meeting: Farm Bill Markup," April 26, 2012, http://www.ag.senate.gov
/hearings/business-meeting-farm-bill-markup; Senate Committee on Agriculture, Manager's Amendment to the Committee Print, 2012, http://www.ag
.senate.gov/download/?id=6ec94a6e-fdc0-42b5-a615-ed0dbeaf659d; Senate
Committee on Agriculture, Red-Line Summary of Manager's Amendment
to the 2012 Farm Bill Committee Print, April 2012, 19–20, http://www.ag
.senate.gov/download/?id=f2d32515-a59b-466a-8831-ac3c84e11c29.

54. S. Rep. 112-203, at 46; Senate Committee on Agriculture, Nutrition,
and Forestry, "Baucus Amendment #12," n.d., http://www.agriculture.senate
.gov/download/?id=7574bda0-4df4-4cd1-82cc-c0b8cedb8d61; Senate Committee on Agriculture, "Business Meeting."

55. It passed by a vote of 16 to 5. See S. Rep. 112-203, at 46; David Rogers, "Regional Battle Opens over Major Farm Bill," *Politico*, April 26, 2012,
http://www.politico.com/story/2012/04/landmark-farm-bill-clears-senate
-panel-075665regional battle; Keith Good, "Farm Bill; Appropriations; and
Regulations," *FarmPolicy.com*, April 27, 2012, http://farmpolicy.com/2012/04
/27/farm-bill-appropriations-and-regulations/#more-7203; Sarah Gonza-
lez, "Senate Ag Committee Passes Bipartisan Farm Bill with Regional Differences," *Agri-Pulse*, April 26, 2012, http://www.agri-pulse.com/Senate-Ag
-Committee-passes-bipartisan-Farm-Bill-regional-differences-04262012
.asp; Paul Hollis, "Peanut Growers Call for 'Producer Choice' Crop Program," *Southeast Farm Press*, April 27, 2012, http://southeastfarmpress.com
/government/peanut-growers-call-producer-choice-crop-program.

56. See, e.g., David Bennett, "Will House Agriculture Committee Rec-
tify Senate Farm Bill Flaws?," *Delta Farm Press*, April 30, 2012, http://
deltafarmpress.com/print/government/will-house-agriculture-committee
-rectify-senate-farm-bill-flaws; Keith Good, "Farm Bill; Trade; Ag Economy; MF Global; Biofuels; and Political Notes," *FarmPolicy.com*, May 4, 2012,

http://farmpolicy.com/2012/05/04/farm-bill-trade-ag-economy-mf-global
-biofuels-and-political-notes/#more-7270; Sara Wyant, "Stabenow and Roberts on Farm Bill: Don't Kick the Can down the Road," *Agri-Pulse*, May 16,
2012, http://www.agri-pulse.com/Stabenow-roberts-dont-kick-can-05162012
.asp#.t7pb44t3meE.twitter; David Rogers, "Senate Agriculture Committee
Files Farm Bill," *Politico*, May 24, 2012, http://www.politico.com/news/stories
/0512/76751.html; David Rogers, "Farm Bill Savings Would Be $23.6 Bil-
lion, CBO Says," *Politico*, May 25, 2012, http://www.politico.com/story/2012
/05/cbo-farm-bill-would-save-236b-076770. See also S. Rep. 112-203, at 4.

57. Keith Good, "Farm Bill; Budget; Ag Economy; Regulations; and Trade,"
FarmPolicy.com, May 30, 2012, http://farmpolicy.com/2012/05/30/farm-bill
-budget-ag-economy-regulations-and-trade/#more-7401.

58. Cong. Rec., June 5, 2012, S3715, S3716 (Senator Roberts); Cong. Rec.,
June 6, 2012, S3753 (Senator Roberts). See also Cong. Rec., June 7, 2012,
S3808–16 (debate over budget resolution).

59. Cong. Rec., June 5, 2012, S3715 (Chairwoman Stabenow); Cong. Rec.,
June 6, 2012, S3750 (Stabenow).

60. Cong. Rec., June 5, 2012, S3715–16; Cong. Rec., June 6, 2012, S3749–50
(Chairwoman Stabenow).

61. Cong. Rec., June 5, 2012, S3715–16.

62. Cong. Rec., June 5, 2012, S3716; Cong. Rec., June 6, 2012, S3753.

63. Cong. Rec., June 6, 2012, 3755, S3763 (Senator Dick Durbin [D-IL]);
Cong. Rec., June 7, 2012, S3807 (Senator Ben Nelson [D-NE]).

64. Cong. Rec., June 7, 2012, S3805–6 (Senator Conrad), S3829 (Senator
Hoeven [R-ND]); Cong. Rec., June 20, 2012, S4356–58 (Senator Baucus).

65. Cong. Rec., June 7, 2012, S3826 (Senator Saxby Chambliss [R-GA]).

66. Cong. Rec., June 7, 2012, S3824–25.

67. Cong. Rec., June 6, 2012, S3753.

68. Cong. Rec., June 19, 2012, S4269–70 (Grassley amendment), S4273–74
(passed by a vote of 75 to 24 [1 not voting]. See also Cong. Rec., June 20, 2012,
S4337–38 (Senator Tom Coburn's amendment to apply the AGI eligibility test
to the conservation programs agreed to by a vote of 63 to 36 [1 not voting]).

69. Cong. Rec., June 6, 2012, S3763 (Senator Durbin); Cong. Rec., June
20, 2012, S4354–55 (Senator Thune alternative defeated by a vote of 44 to 55
[1 not voting]), S4355–56 (Durbin/Coburn amendment agreed to by a vote
of 66 to 33 [1 not voting]).

70. Keith Good, "Farm Bill; EPA Issues; and Trade," *FarmPolicy.com*, June
21, 2012, http://farmpolicy.com/2012/06/21/farm-bill-epa-issues-and-trade
/#more-7583; Charles Abbott, "Analysis: U.S. Budget Axe May Spare Costly
Crop Insurance," Reuters, June 20, 2012, http://www.reuters.com/article/2012
/06/20/us-usa-agriculture-insurers-idusbre85j1hv20120620.

71. Cong. Rec., June 20, 2012, s4353–54 (Chambliss amendment; by a vote of 52 to 47 [1 not voting]); David Rogers, "House Agriculture Committee Pulls Back on Farm Bill Markup," *Politico*, June 20, 2012, http://www.politico.com/story/2012/06/cantor-hits-pause-on-farm-bill-action-077651.

72. Cong. Rec., June 13, 2012, s4124–25 (sugar program phaseout tabled by a vote of 50 to 46 [4 not voting]), s4126 (Senator Coburn); Cong. Rec., June 20, 2012, s4349–50 (Senator Pat Toomey [R-PA] sugar amendment narrowly defeated by a vote of 46 to 53 [1 not voting]).

73. Cong. Rec., June 13, 2012, s4124 (Senator Rand Paul [R-KY] amendment to cut SNAP tabled by a vote of 65 to 33 [2 not voting]).

74. Cong. Rec., June 19, 2012, s4270–71 (Senator Jeff Sessions [R-AL] amendment), s4272 (Sessions amendment defeated by a vote of 43 to 56 [1 not voting] and his other amendment to eliminate bonus payments to states was defeated by a vote of 41 to 58 [1 not voting]).

75. Cong. Rec., June 6, 2012, 3754–55 (Senator Kirsten Gillibrand [D-NY]).

76. Cong. Rec., June 6, 2012, 3754–55 (Gillibrand amendment); Cong. Rec., June 19, 2012, s4277 (Gillibrand amendment).

77. Cong. Rec., June 19, 2012, s4278 (Gillibrand amendment defeated by a vote of 33 to 66 [1 not voting]).

78. Cong. Rec., June 5, 2012, s3714 (Majority Leader Harry Reid [D-NV], motion to invoke cloture on the motion to proceed); Cong. Rec., June 7, 2012, s3803 (debating the motion), s3808 (cloture agreed to by a vote of 90 to 8 [2 not voting]); Senate Committee on Agriculture, Nutrition, and Forestry, "90 Senators Vote to Proceed to 2012 Farm Bill," press release, June 7, 2012, http://www.ag.senate.gov/newsroom/press/release/90-senators-vote-to-proceed-to-2012-farm-bill; David Rogers, "Farm Bill on the Move in Senate," *Politico*, June 4, 2012, http://www.politico.com/news/stories/0612/77035.html; David Rogers, "Farm Bill Confidence Builds in Senate," *Politico*, June 14, 2012, http://www.politico.com/story/2012/06/confidence-builds-on-farm-bill-077431; Keith Good, "Farm Bill; Animal Agriculture; Ag Economy; and MF Global," *FarmPolicy.com*, June 5, 2012, http://farmpolicy.com/2012/06/05/farm-bill-animal-agriculture-ag-economy-and-mf-global/#more-7439; Keith Good, "Farm Bill; Animal Agriculture; and Agricultural Economy," *FarmPolicy.com*, June 8, 2012, http://farmpolicy.com/2012/06/08/farm-bill-animal-agriculture-and-agricultural-economy/#more-7473; Keith Good, "Farm Bill Issues; National Journal—Chairman Lucas; and Animal Agriculture," *FarmPolicy.com*, June 11, 2012, http://farmpolicy.com/2012/06/11/farm-bill-issues-national-journal-chairman-lucas-and-animal-agriculture/#more-7486; Ed O'Keefe, "Senate Begins Debate on Slimmed Down Farm Bill," *Washington Post*, June 6, 2012, http://www.washingtonpost.com/politics/senate-begins-debate-on-slimmed-down-farm-bill/2012/06/06/gjqao9shjv_story.html.

79. Cong. Rec., June 13, 2012, s4124–26 (Majority Leader Reid filled the amendment tree; Senator Coburn).

80. Lawrence, "Profiles in Negotiation," 6–7; David Rogers, "Big Breakthrough on Farm Bill," *Politico*, June 18, 2012, http://www.politico.com /story/2012/06/farm-bill-backers-scramble-for-consent-077550; Ron Nixon, "Stack of Farm Proposals Is Coming Up for Votes," *New York Times*, June 19, 2012, http://www.nytimes.com/2012/06/20/us/politics/amendments -trimmed-senate-moves-ahead-on-farm-bill.html?_r=1&emc=tnt&tntemail0 =y; Ron Nixon, "Senate Weighs Bill Overhauling Agriculture Programs," *New York Times*, June 20, 2012, http://www.nytimes.com/2012/06/21/us /politics/senate-debates-new-farm-bill.html?_r=1&emc=tnt&tntemail0= y; Keith Good, "Senate Moves Forward on Farm Bill; EPA; Ag Economy; and Trade," *FarmPolicy.com*, June 19, 2012, http://farmpolicy.com/2012 /06/19/senate-moves-forward-on-farm-bill-epa-ag-economy-and-trade /#more-7563; Paul Kane, "Farm Bill Politics Mutes Partisanship in Senate," *Washington Post*, June 20, 2012, http://www.washingtonpost.com/politics /farm-bill-politics-mutes-partisanship-in-senate/2012/06/20/gjqas7v7pV _print.html.

81. Cong. Rec., June 19, 2012, s4266–67 (Senate proceeds), s4268–69 (Senators Stabenow and Roberts); Meredith Shiner, "Senate Agrees on Way Forward on Farm Bill," *Roll Call*, June 18, 2012, http://www.rollcall.com/news /Senate-Agrees-on-Way-Forward-on-Farm-Bill-215476-1.html?pos=hftxt.

82. Cong. Rec., June 20, 2012, s4356 (Senators Reid and McConnell); Cong. Rec., June 21, 2012, s4396–97 (Chairwoman Stabenow), s4397–98 (Senator Roberts and final passage by a vote of 64 to 35; Majority Leader Reid and McConnell).

83. See, e.g., Cong. Rec., June 21, 2012, s4398 (Senator Chambliss); David Rogers, *Politico*, June 22, 2012; Alan Bjerga, "Civil War Redux in U.S. Farm Bill Debate," *Bloomberg*, June 28, 2012, http://www.bloomberg.com/news /articles/2012-06-28/civil-war-redux-in-u-s-farm-bill-debate.

84. Jim Wiesemeyer, "Winners and Losers in Senate Farm Bill," *AgWeb. com*, June 21, 2012, http://www.agweb.com/article/winners_and_losers_in _senate_farm_bill/.

85. David Rogers, "Senate, House Farm Bills Follow Different Paths," *Politico*, July 8, 2012, http://www.politico.com/news/stories/0712/78220.html.

86. David Rogers, "House Farm Bill Cuts Deeper," *Politico*, July 5, 2012, http://www.politico.com/story/2012/07/house-farm-bill-cuts-deeper-078156; Charles Abbott, "U.S. House Farm Bill Cuts $35 Billion, Boost Crop Supports," Reuters, July 5, 2012, http://www.reuters.com/article/2012/07/05/us -usa-agriculture-idusbre86418820120705; Keith Good, "Farm Bill (House Draft); Ag Economy; and Animal Agriculture," *FarmPolicy.com*, July 6, 2012,

http://farmpolicy.com/2012/07/06/farm-bill-house-draft-ag-economy-and
-animal-agriculture/#more-7651.

87. David Rogers, "GOP Fights over Major Food Stamps," *Politico*, July 3,
2012, http://www.politico.com/story/2012/07/gop-fights-over-major-food
-stamp-crackdown-078096; Keith, Good, "Farm Bill; Budget Issues; Ag
Economy; EPA; Trade; and Biofuels," *FarmPolicy.com*, July 3, 2012, http://
farmpolicy.com/2012/07/03/farm-bill-budget-issues-ag-economy-epa-trade
-and-biofuels/#more-7640.

88. Rosenbaum and Dean, "House Agriculture Committee."

89. Monica Davey, "Searing Sun and Drought Shrivel Corn in Midwest,"
New York Times, July 4, 2012, http://www.nytimes.com/2012/07/05/us/for
-midwest-corn-crop-the-pressure-rises-like-the-heat.html?_r=1&emc=tnt&
tntemail0=y; USDA, "US Drought and Your Food Costs," https://www.usda
.gov/media/blog/2012/08/10/us-drought-and-your-food-costs; P. Westcott and
M. Jewison, USDA, ERS, "Weather Effects on Expected Corn and Soybean
Yields," last modified July 26, 2013, https://www.ers.usda.gov/publications
/pub-details/?pubid=36652 ; Joshua Zumbrun and Mark Drajem, "Worst-
in-Generation Drought Dims U.S. Farm Economy Hopes," *Bloomberg*, July
16, 2012, http://www.bloomberg.com/news/articles/2012-07-16/worst-in
-generation-drought-dims-u-s-farm-economy-hopes.

90. The vote to report the bill was 35 to 11 and bipartisan. See David Rogers,
"House Panel Approves Farm Bill," *Politico*, July 11, 2012, http://www.politico
.com/news/stories/0712/78394.html; Daniel Looker, "House Committee
Approves Farm Bill," *Agriculture.com*, July 11, 2012, http://www.agriculture
.com/news/policy/house-committee-approves-farm-bill_4-ar25150; Keith
Good, "Farm Bill; Ag Economy; Animal Agriculture; and Biofuels," *Farm-
Policy.com*, July 12, 2012, http://farmpolicy.com/2012/07/12/farm-bill-ag
-economy-animal-agriculture-and-biofuels/#more-7681.

91. David Rogers, "Boehner: No Decision on Farm Bill Vote," *Politico*, July
12, 2012, http://www.politico.com/story/2012/07/boehner-no-decision-on
-farm-bill-vote-078444; Keith Good, "Drought Concerns Grow as Farm Bill
Awaits House Floor Action," *FarmPolicy.com*, July 16, 2012, http://farmpolicy
.com/2012/07/16/drought-concerns-grow-as-farm-bill-awaits-house-floor
-action/#more-7738.

92. Peter Whoriskey and Michael A. Fletcher, "Drought in U.S. Reaching Levels Not Seen in 50 Years, Pushing up Corn Prices," *Washington Post*,
July 16, 2012, http://www.washingtonpost.com/business/economy/drought
-in-us-reaching-levels-not-seen-in-50-years-pushing-up-corn-prices/2012
/07/16/gjqa0lsopW_story.html; Peter Baker, "Drought Puts Food at Risk, U.S.
Warns," *New York Times*, July 18, 2012, http://www.nytimes.com/2012/07/19
/us/drought-puts-food-at-risk-us-warns.html?_r=1&emc=tnt&tntemail0=y;

John Eligon, "Widespread Drought Is Likely to Worsen," *New York Times*, July 19, 2012, http://www.nytimes.com/2012/07/20/science/earth/severe-drought -expected-to-worsen-across-the-nation.html?ref=todayspaper.

93. Erik Wasson, "No Farm Bill Action Scheduled for Next Week," *The Hill*, July 20, 2012, http://thehill.com/policy/finance/239205-no-farm-bill -scheudled-for-next-week; David Rogers, "Tongues Wag after Boehner Farm Bill Remark," *Politico*, July 24, 2012, http://www.politico.com/story/2012/07 /tongues-wag-after-boehner-farm-bill-remark-078919; David Rogers, "Farm Bill Causes All-Out Scramble," *Politico*, July 25, 2012, http://www.politico .com/news/stories/0712/78959.html; David Rogers, "A New Twist in Farm Bill Drama," *Politico*, July 26, 2012, http://www.politico.com/news/stories /0712/79023.html; David Rogers, "House Pitches Shorter Farm Bill," *Politico*, July 27, 2012, http://www.politico.com/news/stories/0712/79070.html ?hp=14; David Rogers, "Ag Panels Look at Disaster Relief," *Politico*, July 30, 2012, http://www.politico.com/story/2012/07/ag-panels-look-at-disaster-relief -079173; Ron Nixon, "Major Bill Delayed, House Works on Short-Term Farm Measure," *New York Times*, July 26, 2012, http://thecaucus.blogs.nytimes.com /2012/07/26/house-works-on-a-short-term-farm-bill/.

94. Erik Wasson, "GOP Scrambles for Farm Bill Votes," *The Hill*, July 31, 2012, http://thehill.com/homenews/house/241177-gop-scrambles-for-farm -bill-votes; Erik Wasson, "Obama Calls on Congress to Pass Stalled Farm Bill," *The Hill*, August 7, 2012, http://thehill.com/policy/finance/242621-obama -calls-on-congress-to-pass-stalled-farm-bill-; David Rogers, "House GOP Gives Boost to Drought Bill," *Politico*, August 1, 2012, http://www.politico .com/story/2012/08/house-gop-gives-boost-to-drought-aid-079302; David Rogers, "House Passes Drought, Disaster Aid," *Politico*, August 2, 2012, http:// www.politico.com/story/2012/08/house-passes-drought-disaster-aid-079335; Jennifer Steinhauer, "Pile of Bills Is Left Behind as Congress Goes to Campaign," *New York Times*, August 2, 2012, http://www.nytimes.com/2012/08 /03/us/politics/house-passes-short-term-farm-relief-bill.html?_r=1; Brian Sullivan, "Worst Drought Covers Nearly One-Fourth of Contiguous U.S.," *Bloomberg*, August 9, 2012, https://www.bloomberg.com/news/articles/2012 -08-09/worst-drought-covers-nearly-one-fourth-of-contiguous-u-s-.

95. David Rogers, "GOP Infighting as Farm Bill Suffers," *Politico*, September 11, 2012, http://www.politico.com/story/2012/09/gop-infighting-as -farm-bill-suffers-081066; Daniel Looker, "Farm Bill Is Still Doable, Stabenow Says," *Agriculture.com*, September 11, 2012, http://www.agriculture.com /news/policy/farm-bill-is-still-doable-stabenow-says_4-ar26304; Jennifer Steinhauer, "Deal on Farm Bill Appears Unlikely," *New York Times*, September 12, 2012, http://www.nytimes.com/2012/09/13/us/congressional-deal-on -a-farm-bill-appears-unlikely.html?ref=todayspaper.

96. David Rogers, "Farm Bill Finale: Milk, Mayhem," *Politico*, September 17, 2012, http://www.politico.com/story/2012/09/farm-bill-finale-milk-mayhem -081313; David Rogers, "John Boehner Vague on Farm Bill Direction," *Politico*, September 20, 2012, http://www.politico.com/news/stories/0912/81449 .html?hp=l18; Charles Abbott, "US Farm Law to Expire, House Republicans Split on Food Stamps," Reuters, September 21, 2012, http://in.reuters.com /article/2012/09/20/usa-congress-farmbill-idinl1e8kk9cw20120920; Erik Wasson, "Dems Hit House Leaders over Farm Bill Expiration," *The Hill*, October 1, 2012, http://thehill.com/policy/finance/259447-dems-look-to-hit-house -leaders-as-farm-bill-expires; Molly K. Hooper, "Farm Bill Backers: We Have the Votes," *The Hill*, September 22, 2012, http://thehill.com/homenews /house/251129-farm-bill?utm_campaign=thehill&utm_source=twitterfeed &utm_medium=twitter.

97. Keith Good, "Farm Bill (Political Notes); and the Ag Economy," *Farm-Policy.com*, November 7, 2012, http://farmpolicy.com/2012/11/07/farm-bill -political-notes-and-the-ag-econom/#more-10422.

98. Charles Abbott, "Crop Insurance a Post-election Target, Farm Bill Elusive," Reuters, November 7, 2012, http://www.reuters.com/article/2012 /11/07/us-usa-election-agriculture-idusbre8a62h520121107; David Rogers, "Farm Bill Still Has a Shot," *Politico*, November 15, 2012, http://www.politico .com/story/2012/11/farm-bill-still-has-a-shot-083934; David Rogers, "Five-Year Farm Bill Gets New Life," *Politico*, November 29, 2012, http://www .politico.com/story/2012/11/five-year-farm-bill-gets-new-life-084420; David Rogers, "'Great Progress' on 5-Year Farm Bill," *Politico*, December 6, 2012, http://www.politico.com/story/2012/12/great-progress-on-5-year-farm-bill -084685; Phillip Brasher, "Lucas Suggests Path Forward for Farm Bill," *Roll Call*, November 30, 2012, http://www.rollcall.com/news/lucas_suggests_path _forward_for_farm_bill-219565-1.html; Jonathan Weisman, "Initial Deficit Cuts Are Sticking Point in Negotiations," *New York Times*, December 3, 2012, http://www.nytimes.com/2012/12/04/us/politics/in-fiscal-cliff-talks-first -step-is-the-hardest.html?emc=tnt&tntemail0=y; Ron Nixon, "Stalled Farm Bill Is Pushed for Its Savings," *New York Times*, December 5, 2012, http:// www.nytimes.com/2012/12/06/us/politics/stalled-farm-bill-could-help-with -deficit-reduction.html.

99. David Rogers, "Farm Bill Talks Stall; Milk Prices at Risk," *Politico*, December 11, 2012, http://www.politico.com/story/2012/12/farm-bill-talks -stumble-milk-prices-at-risk-084938; David Rogers, "Huge Divide Remains on Farm Bill," *Politico*, December 12, 2012, http://www.politico.com/story /2012/12/huge-divide-remains-on-farm-bill-085021; David Rogers, "Tom Vilsack Warns Congress on Farm Bill," *Politico*, December 13, 2012, http:// www.politico.com/story/2012/12/vilsack-warns-congress-on-farm-bill

-085063; David Rogers, "Boehner's Stand Dims Farm Bill Hopes," *Politico*, December 18, 2012, http://www.politico.com/story/2012/12/boehners -stand-dims-farm-bill-hopes-085280; David Rogers, "With Obama's Urging, Congress Tries for Farm Bill Extension to Avoid Dairy Disaster," *Politico*, December 28, 2012, http://www.politico.com/story/2012/12/congress -tries-to-avoid-dairy-disaster-085567; David Rogers, "3 Bills Take Aim at Milk Prices," *Politico*, December 30, 2012, http://www.politico.com/story /2012/12/3-bills-take-aim-at-milk-prices-085583; Ron Nixon, "With Farm Bill Stalled, Consumers May Face Soaring Milk Prices," *New York Times*, December 20, 2012, http://www.nytimes.com/2012/12/21/us/milk-prices -could-double-as-farm-bill-stalls.html?emc=tnt&tntemail0=y&_r=0; Jonathan Weisman and Jennifer Steinhauer, "Senators Return with 5 Days Left and No Clear Fiscal Path," *New York Times* December 26, 2012, http:// www.nytimes.com/2012/12/27/us/politics/little-sense-of-fiscal-urgency-as -senators-prepare-to-return.html?ref=todayspaper; Keith Good, "Budget and the Farm Bill," *FarmPolicy.com*, December 24, 2012, http://farmpolicy .com/2012/12/24/budget-and-the-farm-bill/#more-11003; Keith Good, "Senate Passes Fiscal Cliff Deal-Farm Bill Concerns Noted," *FarmPolicy.com*, January 1, 2013, http://farmpolicy.com/2013/01/01/senate-passes-fiscal-cliff -deal-farm-bill-concerns-noted/#more-11103.

100. Alan Bjerga, "Fix on 'Dairy Cliff' Sends Farm Bill Back to Square One," *Bloomberg*, January 2, 2013, http://www.bloomberg.com/news/articles /2013-01-02/fix-on-dairy-cliff-sends-farm-bill-back-to-square-one; Keith Good, "Farm Bill Extension—Reactions, Details," *FarmPolicy.com*, January 4, 2013, http://farmpolicy.com/2013/01/04/farm-bill-extension-reactions -details/#more-11137.

101. Sara Wyant, "Cochran Selected as Senate Agriculture Committee Ranking Member," *Agri-Pulse*, January 3, 2013, http://www.agri-pulse.com /Cochran-selected-as-Senate-Agriculture-Committee-ranking-member -01032013.asp; David Rogers, "Cochran Wins Senate Ag Post," *Politico*, January 3, 2013, http://www.politico.com/story/2013/01/cochran-wins-senate -ag-post-085741.

102. David Rogers, "Senate Democrats Reach Sequester Deal," *Politico*, February 14, 2013, http://www.politico.com/story/2013/02/senate-democrats-reach -sequester-deal-087667; Charles Abbott, "Republicans Seek Farm Subsidy Cuts, Mull Food Stamps to Cash," Reuters, March 12, 2013, http://www.reuters .com/article/2013/03/12/us-usa-fiscal-agriculture-idusbre92b0rx20130312. See also House Committee on the Budget, "The Path to Prosperity: A Responsible, Balanced Budget," Fiscal Year 2014 Budget Resolution, March 2013, http://budget.house.gov/uploadedfiles/fy14budget.pdf (hereafter cited as FY 2014 House Budget).

103. Daniel Looker, "Farm Bill in Works, Still Difficult," *Agriculture.com*, April 8, 2013, http://www.agriculture.com/news/policy/farm-bill-in-wks-still-difficult_4-ar30929; Ellyn Ferguson, "Cochran Brings Southern Perspective to Senate Agriculture Committee," *Roll Call*, April 20, 2013, http://www.rollcall.com/news/cochran_brings_southern_perspective_to_senate_agriculture_committee-224199-1.html?pos=hftxt.

104. David Rogers, "Frank Lucas to Move on Farm Bill," *Politico*, April 18, 2013, http://www.politico.com/story/2013/04/farm-bill-frank-lucas-090289; David Rogers, "New Farm Bill Leans on Food Stamps," *Politico*, May 6, 2013, http://www.politico.com/story/2013/05/frank-lucas-retools-farm-bill-090984; Charles Abbott, "Lawmakers Plan to Start Drafting Farm Bill Next Week," Reuters, May 7, 2013, http://www.reuters.com/article/2013/05/07/us-usa-agriculture-farm-bill-idusbre9460pn20130507.

105. Lawrence, "Profiles in Negotiation," 9; Senate Committee on Agriculture, Nutrition, and Forestry, Committee Print of the Agricultural Reform, Food and Jobs Act of 2013, http://www.ag.senate.gov/download/?id=8413a981-3431-486a-a62f-626a1d6c2374; Agriculture Reform, Food and Jobs Act of 2013, S. Rep. 113-88 (2013), https://www.agriculture.senate.gov/download/?id=7dea2eaa-3195-4d72-abcd-825b12e437e1.

106. S. Rep. 113-88, 154; Keith Good, "Farm Bill; and Budget Issues," *FarmPolicy.com*, May 16, 2013, http://farmpolicy.com/2013/05/16/farm-bill-and-the-ag-economy-thursday/#more-12496.

107. See, e.g., David Rogers, "House Panel Approves Farm Bill," *Politico*, May 15, 2013, http://www.politico.com/story/2013/05/farm-bill-advances-91436.html; Daniel Looker, "Senate Committee Advances Farm Bill," *Agriculture.com*, May 13, 2013, http://www.agriculture.com/news/policy/senate-committee-advces-farm-bill_4-ar31524.

108. Keith Good, "Farm Bill; and the Ag Economy," *FarmPolicy.com*, May 15, 2013, http://farmpolicy.com/2013/05/15/farm-bill-and-the-ag-economy-wednesday/#more-12473 (quoting Senator John Thune [R-SD]).

109. The bill was again reported by a strong bipartisan vote, 36 to 10. See David Rogers, "House Panel Approves Farm Bill," *Politico*, May 15, 2013, http://www.politico.com/story/2013/05/farm-bill-advances-91436.html; Mary Claire Jalonick, "House Panel Oks Farm Bill with Food Stamp Cuts," Associated Press, May 16, 2013, http://bigstory.ap.org/article/house-panel-set-ok-cut-food-stamps; Keith Good, "Farm Bill; and the Ag Economy," *FarmPolicy.com*, May 16, 2013, http://farmpolicy.com/2013/05/16/farm-bill-and-the-ag-economy-thursday/#more-12496.

110. Daniel Looker, "House Progresses on Farm Bill," *Agriculture.com*, May 16, 2013, http://www.agriculture.com/news/policy/house-progresses-on-farm-bill_4-ar31564; Ron Nixon, "House Agriculture Committee Approves

Farm Bill," *New York Times*, May 16, 2013, http://thecaucus.blogs.nytimes
.com/2013/05/16/house-agriculture-committee-approves-farm-bill/?_r=0.

111. David Rogers, "House Panel Approves Farm Bill," *Politico*, May 15,
2013, http://www.politico.com/story/2013/05/farm-bill-advances-91436.html;
Looker, "House Progresses."

112. Cong. Rec., May 20, 2013, s3589 (Majority Leader Reid); Cong. Rec.,
May 22, 2013, s3717–21 (budget debate); Cong. Rec., May 23, 2013, s3798–3805;
Cong. Rec., June 4, 2013, s3925–26; David Rogers, "Senate Debates Farm
Bill," *Politico*, May 20, 2013, http://www.politico.com/story/2013/05/senate
-debates-farm-bill-091636.

113. Cong. Rec., May 20, 2013, s3591–92; Cong. Rec., May 22, 2013, s3729–30;
S. Rep. 113-88, at 101–6; National Association of Conservation Districts,
"Conservation Compliance Coalition Praises Agreement in Senate Farm
Bill," press release, May 14, 2013, http://www.nacdnet.org/news/newsroom
/2013/conservation-compliance-coalition-praises-agreement-in-senate-farm
-bill; Mikkel Pates, "ND Senators Decry Crop Insurance, Conservation Compliance Link," *Agweek*, May 16, 2013, http://www.agweek.com/event/article
/id/20915/; Daniel Looker, "Senate Committee Advances Farm Bill," *Agri-
culture.com*, May 13, 2013, http://www.agriculture.com/news/policy/senate
-committee-advces-farm-bill_4-ar31524.

114. Cong. Rec., May 20, 2013, s3594–95 (Senator McCain); Cong. Rec.,
May 23, 2013, s3805 (Senators Feinstein and McCain, amendment to prohibit premium subsidy for tobacco insurance), s3816–17 (Feinstein-McCain
amendment defeated by a vote of 44 to 52 [4 not voting]); Cong. Rec., May 21,
2013, s3636–37 (Gillibrand crop insurance amendment), s3647–48 (amendment defeated by a vote of 26 to 70 [4 not voting]). See also David Rogers,
"Crop Insurance Debate Grows in Senate," *Politico*, May 23, 2013, http://www
.politico.com/story/2013/05/crop-insurance-debate-grows-in-senate-091851.

115. Cong. Rec., May 23, 2013, s3817–19 (debate on the Durbin-Coburn
crop insurance–AGI amendment), s3820 (amendment passed by a vote of
59 to 33 [8 not voting]).

116. Ron Nixon, "Record Taxpayer Cost Is Seen for Crop Insurance,"
New York Times, January 15, 2013, http://www.nytimes.com/2013/01/16
/us/politics/record-taxpayer-cost-is-seen-for-crop-insurance.html?_r=0;
USDA, ERS, "Crop Insurance Indemnities Rise with Drought," Septem-
ber 1, 2016, https://www.ers.usda.gov/data-products/chart-gallery/gallery
/chart-detail/?chartId=77081; Bruce Sherrick and Gary Schnitkey, "Crop
Insurance Program Losses in Perspective," *farmdoc daily*, March 30, 2013,
Dept. of Agricultural and Consumer Economics, University of Illinois at
Urbana-Champaign, http://farmdocdaily.illinois.edu/2013/05/crop-insurance
-losses-perspective.html.

117. Cong. Rec., May 20, 2013, S3592 (Senator Cochran); Cong. Rec., May 22, 2013, S3725 (Senator Chambliss).

118. Cong. Rec., May 23, 2013, S3825.

119. Cong. Rec., June 6, 2013, S3980 (Senator Thune).

120. Cong. Rec., May 23, 2013, S3825 (Senator Thune).

121. Cong. Rec., May 20, 2013 S3593.

122. Cong. Rec., June 3, 2013, S3896 (Chairwoman Stabenow); David Rogers, "Crunch Time for Senate Farm Bill," *Politico*, June 5, 2013, http://www.politico.com/story/2013/06/senate-farm-bill-stall-will-be-felt-across-capitol-092311.

123. Cong. Rec., May 23, 2013, S3816 (Majority Leader Reid); Cong. Rec., June 4, 2013, S3924 (Chairwoman Stabenow), S3933–41 (Senators Coburn and Mary Landrieu [D-LA] and Chairwoman Stabenow); Erik Wasson, "Agreement on Farm Bill Floor Amendments Eludes Senate," *The Hill*, June 1, 2013, http://thehill.com/homenews/senate/302859-agreement-on-farm-bill-amendments-eludes-senate; David Rogers, "Stabenow: Farm Bill Cloture Near," *Politico*, June 4, 2013, http://www.politico.com/story/2013/06/debbie-stabenow-farm-bill-092207; Keith Good, "Farm Bill; Appropriations; Wheat; and the Ag Economy," *FarmPolicy.com*, June 5, 2013, http://farmpolicy.com/2013/06/05/farm-bill-appropriations-wheat-and-the-ag-economy-wednesday/#more-12802.

124. Cong. Rec., June 4, 2013, S3949 (Majority Leader Reid moved to invoke cloture on the bill); Cong. Rec., June 6, 2013, S3970 (Majority Leader Reid), S3975 (Chairwoman Stabenow), S3976 (the Senate agreed to invoke cloture and end debate on the farm bill by a vote of 75 to 22 [2 not voting]), S3979–80 (Senator Thune).

125. Cong. Rec., May 21, 2013, S3641–42 (Roberts amendment on eliminating categorical eligibility for SNAP benefits), S3647 (amendment defeated by a vote of 40 to 58 [2 not voting]). See also Keith Good, "Farm Bill; Ag Economy; Immigration; and, CFTC," *FarmPolicy.com*, May 22, 2013, http://farmpolicy.com/2013/05/22/farm-bill-ag-economy-immigration-and-cftc-wednesday/#more-12583.

126. Cong. Rec., May 22, 2013, S3727 (Senator Shaheen [D-NH] amendment to reform the sugar program), S3746 (amendment defeated by a vote of 45 to 54 [1 not voting]); Keith Good, "Farm Bill; and the Ag Economy," *FarmPolicy.com*, May 23, 2013, http://farmpolicy.com/2013/05/23/farm-bill-and-the-ag-economy-thursday-2/#more-12623.

127. Cong. Rec., June 10, 2013, S4043–44, S4052 (by a vote of 66 to 27 [7 not voting]); David Rogers, "Farm Bill Passes the Senate," *Politico*, June 10, 2013, http://www.politico.com/story/2013/06/farm-bill-passes-senate-with-bipartisan-majority-092539; Ron Nixon, "Senate Passes Farm Bill; House

Vote Is Less Sure," *New York Times*, June 10, 2013, http://www.nytimes.com /2013/06/11/us/politics/senate-passes-farm-bill-house-vote-is-less-sure.html ?emc=tnt&tntemail0=y&_r=1; Keith Good, "Farm Bill; Ag Economy; Smithfield; and Immigration," *FarmPolicy.com*, June 11, 2013, http://farmpolicy.com /2013/06/11/farm-bill-ag-economy-smithfield-and-immigration-tuesday/ #more-12851.

128. Cong. Rec., June 18, 2013, H3708, H3719–20 (rule approved by a vote of 232 to 193 [9 not voting]); Cong. Rec., June 19, 2013, H3787–3850 (text of the bill); David Rogers, "Nancy Pelosi: 'Not Likely' to Back Farm Bill," *Politico*, June 18 2013, http://www.politico.com/story/2013/06/nancy-pelosi -house-dems-farm-bill-092975; Keith Good, "Farm Bill; Appropriations; and Immigration," *FarmPolicy.com*, June 19, 2013, http://farmpolicy.com/2013/06 /19/farm-bill-appropriations-and-immigration-wednesday/#more-12910.

129. Cong. Rec., June 18, 2013, H3721–22.

130. Cong. Rec., June 18, 2013, H3722.

131. Cong. Rec., June 19, 2013, H3907–8 (Representative Ron Kind [D-WI] amendment), H3850–51 (Representative Jim McGovern [D-MA] amendment), H3878 (McGovern amendment defeated 188 to 234 [12 not voting]); Cong. Rec., June 20, 2013, H3946 (Kind amendment defeated on a vote of 208 to 217 [9 not voting]), H3933 (Representative Joseph Pitts [R-PA] amendment on sugar program reform), H3962 (amendment was defeated by a vote of 206 to 221 [7 not voting]). See also David Rogers, "Farm Bill Advances in House," *Politico*, June 19, 2013, http://www.politico.com/story/2013/06/farm -bill-093056; Keith Good, "Farm Bill," *FarmPolicy.com*, June 20, 2013, http:// farmpolicy.com/2013/06/20/farm-bill-thursday/#more-12928.

132. Cong. Rec., June 19, 2013, H3855–56 (Representative Bob Gibbs [R-OH] amendment; withdrawn).

133. Cong. Rec., June 20, 2013, H3953–54 (Fortenberry amendment), H3963–64 (amendment agreed to by a vote of 230 to 194 [10 not voting]).

134. Cong. Rec., June 19, 2013, H3857 (Foxx amendment to cap spending on the farm programs at 110 percent of the CBO predicted levels), H3878–79 (Foxx amendment agreed to by a vote of 267 to 156 [11 not voting]).

135. Cong. Rec., June 20, 2013, H3957–59 (Southerland amendment).

136. See, e.g., Cong. Rec., June 20, 2013, H3959 (Representative Gwen Moore [D-WI] argued that it contained a "perverse incentive for States to end SNAP benefits for people").

137. Cong. Rec., June 20, 2013, H3960 (Ranking Member Peterson).

138. Cong. Rec., June 20, 2013, H3959 (Majority Leader Cantor pointed to the 1996 reforms for welfare and work requirements).

139. Cong. Rec., June 20, 2013, H3959 (Majority Leader Cantor), H3964 (Southerland amendment agreed to by a vote of 227 to 198 [9 not voting]).

140. Cong. Rec., June 20, 2013, H3967 (Chairman Lucas; the House voted against passage by a vote of 195 to 234 [6 not voting]).

141. David Rogers, "Uncertain Future for Farm Policy," *Politico*, June 20, 2013, http://www.politico.com/story/2013/06/house-rejects-farm-bill-093118; Mary Clare Jalonick, "House Rejects Farm Bill, 62 Republicans Vote No," Associated Press, June 20, 2013, http://bigstory.ap.org/article/house-vote -farm-bill-cuts-crop-insurance; Charles Abbott, "UPDATE2-U.S. House Deals Shock Defeat to Republican Farm Bill," Reuters, June 20, 2013, http://www .reuters.com/article/2013/06/20/usa-agriculture-idusl2n0ewlcl20130620; Keith Good, "Farm Bill," *FarmPolicy.com*, June 21, 2013, http://farmpolicy .com/2013/06/21/farm-bill-friday/#more-12958.

142. Agriculture and Consumer Protection Act of 1973, H.R. Rep. No. 93-337 (1973).

143. Lawrence, "Profiles in Negotiation," 10; Hamilton, "2014 Farm Bill," 8.

144. David Rogers, "How the Farm Bill Failed," *Politico*, June 23, 2013, http://www.politico.com/story/2013/06/how-the-farm-bill-failed-093209.

145. Charles Abbott, "Senate Leader Reid Says No to Farm Law Extension," Reuters, June 24, 2013, http://www.reuters.com/article/2013/06/24/us-usa -agriculture-farm-bill-idusbre95n14m20130624; Ron Nixon, "Farm Bill Defeat Shows Agriculture's Waning Power," *New York Times*, July 2, 2013, http://www .nytimes.com/2013/07/03/us/politics/farm-bill-defeat-shows-agricultures -waning-power.html?_r=1&; Robert Costa, "Cantor Scolds Committee Chairmen," *National Review*, July 8, 2013, http://www.nationalreview.com/corner /352923/cantor-scolds-committee-chairmen-robert-costa.

146. Cong. Rec., July 11, 2013, H4382 (Representative Jim McGovern [D-MA]); Keith Good, "Farm Bill; Ag Economy; and Immigration," *FarmPolicy. com*, July 10, 2013, http://farmpolicy.com/2013/07/10/farm-bill-ag-economy -and-immigration-wednesday-3/#more-13152; Keith Good, "Farm Bill; Smithfield Hearing; GMO Wheat; and Immigration," *FarmPolicy.com*, July 11, 2013, http://farmpolicy.com/2013/07/11/farm-bill-smithfield-hearing-gmo-wheat -and-immigration-thursday/#more-13177.

147. The bill also added a full repeal of the 1949 act's place as permanent law. Cong. Rec., July 11, 2013, H4376 (Representative Pete Sessions [R-TX]).

148. Cong. Rec., July 11, 2013, H4376 (Representative Sessions).

149. Cong. Rec., July 11, 2013, H4376 (Representative Sessions), H4381 (Representative Sessions), H4390-91 (Chairman Lucas), H4458-59 (Ranking Member Peterson).

150. Cong. Rec., July 11, 2013, H4458.

151. Cong. Rec., July 11, 2013, H4470.

152. Cong. Rec., July 11, 2013, H4382-83 (Representative Jim McGovern [D-MA]).

153. Cong. Rec., July 11, 2013, H4377–78 (Representative John Lewis [D-GA]), 4381 (Representative Emmanuel Cleaver [D-MO]), H4382 (Representative Steny Hoyer [D-MD]), H4384 (Representative Louise Slaughter [D-NY]), H4394 (bill read into the record), H4394–4458 (rule passed the House by a vote of 223 to 195 [16 not voting]).

154. Cong. Rec., Jul. 11, 2013, H4385 (Minority Leader Nancy Pelosi [D-CA]).

155. See, e.g., Cong. Rec., July 11, 2013, H4384–86 (Representative McGovern).

156. Cong. Rec., July 11, 2013, H4458–59, H4468.

157. Cong. Rec., July 11, 2013, H4469.

158. Cong. Rec., July 11, 2013, H4474 (by a vote of 216 to 208 [11 not voting]); David Rogers, "Farm Bill 2013: House Narrowly Passes Pared-Back Version," Politico, July 11, 2013, http://www.politico.com/story/2013/07/farm-bill-2013-house-passes-94031.html; Keith Good, "Farm Bill; and the Ag Economy," FarmPolicy.com, July 12, 2013, http://farmpolicy.com/2013/07/12/farm-bill-and-the-ag-economy-friday/#more-13209.

159. David Rogers, "Senate Pushes House on Farm Bill," Politico, July 15, 2013, http://www.politico.com/story/2013/07/senate-house-farm-bill-conference-094236; David Rogers, "Frank Lucas: Food Stamps Still in Play," Politico, July 16, 2013, http://www.politico.com/story/2013/07/frank-lucas-food-stamps-farm-bill-094304; David Rogers, "Ag Panel Heads Meet on Farm Bill," Politico, July 18, 2013, http://www.politico.com/story/2013/07/agriculture-panel-farm-bill-094448; Keith Good, "Farm Bill; and the Ag Economy," FarmPolicy.com, July 16, 2013, http://farmpolicy.com/2013/07/16/farm-bill-and-the-ag-economy-tuesday-2/#more-13244.

160. House Republicans planned to pass it after the August recess. See David Rogers, "Cantor Takes Fresh Cut at Food Stamps," Politico, August 1, 2013, http://www.politico.com/story/2013/08/food-stamps-bill-house-gop-095084; Keith Good, "Farm Bill; RFS; Ag Economy; and Immigration," FarmPolicy.com, August 7, 2013, http://farmpolicy.com/2013/08/07/.

161. David Rogers, "Vilsack Opposes Farm Bill Extension," Politico, September 10, 2013, http://www.politico.com/story/2013/09/tom-vilsack-farm-bill-extension-096560; Keith Good, "Farm Bill; Budget; Ag Economy; Biofuels; and EPA," FarmPolicy.com, September 11, 2013, http://farmpolicy.com/2013/09/11/farm-bill-budget-ag-economy-biofuels-and-epa-wednesday/#more-13629.

162. David Rogers, "House GOP Seeks Cuts in Food Stamps," Politico, September 16, 2013, http://www.politico.com/story/2013/09/house-gop-seeks-tighter-food-stamp-rules-096873; David Rogers, "SNAP Showdown Set in House," Politico, September 18, 2013, http://www.politico.com/story/2013/09/snap-food-stamps-showdown-set-in-house-097021.

163. Cong. Rec., September 19, 2013, H5694 (the House nutrition bill), H5706 (Chairman Lucas).

164. Cong. Rec., September 19, 2013, H5709.

165. Cong. Rec., September 19, 2013, H5710.

166. FY2013 House Budget, 25; FY2014 House Budget, 4, 7–8.

167. Cong. Rec., September 19, 2013, H5706 (Representative Marcia Fudge [D-OH]).

168. Cong. Rec., September 19, 2013, H5721 (by a vote of 217 to 210 [6 not voting]); Ron Nixon, "House Republicans Pass Deep Cuts in Food Stamps," *New York Times*, September 19, 2013, http://www.nytimes.com/2013/09/20 /us/politics/house-passes-bill-cutting-40-billion-from-food-stamps.html; David Rogers, "House Approves Plan to Cut Food Stamps," *Politico*, September 19, 2013, http://www.politico.com/story/2013/09/food-stamp-cuts -house-097090; Keith Good, "Farm Bill; and the Ag Economy," *FarmPolicy. com*, September 20, 2013, http://farmpolicy.com/2013/09/20/farm-bill-and -the-ag-economy-friday-2/#more-13706. See also Todd Kuethe and Jonathan Coppess, "Mapping the Fate of the Farm Bill," *farmdoc daily*, December 5, 2013, http://farmdocdaily.illinois.edu/2013/12/mapping-fate-of-farm -bill.html; Todd Kuethe and Jonathan Coppess, "Mapping the Farm Bill: Voting in the House of Representatives," *farmdoc daily* (4):70, April 17, 2014, http://farmdocdaily.illinois.edu/2014/04/mapping-the-farm-bill-voting-in -the-house-of-representatives.html.

169. Ron Nixon, "Time Short, House Says It Seeks a New Farm Bill," *New York Times*, September 24, 2013, http://www.nytimes.com/2013/09/25/us /politics/time-short-house-says-it-seeks-new-farm-bill.html?partner=rss& emc=rss&smid=tw-thecaucus&_r=0; David Rogers, "Farm Bill Advances in House," *Politico*, September 26, 2013, http://www.politico.com/story/2013/09 /farm-bill-advances-house-097435; Keith Good, "Budget Issues; Farm Bill; and the Ag Economy," *FarmPolicy.com*, October 1, 2013, http://farmpolicy .com/2013/10/01/.

170. David Rogers, "House-Senate Farm Bill Talks OK'd," *Politico*, October 11, 2013, http://www.politico.com/story/2013/10/house-senate-farm -bill-talks-098198; David Rogers, "Farm Bill Conferees Huddle," *Politico*, October 16, 2013, http://www.politico.com/story/2013/10/farm-bill -2013-098403; David Rogers, "Farm Bill Negotiators Resume Talks This Week," *Politico*, November 5, 2013, http://www.politico.com/story/2013 /11/farm-bill-negotiators-talks-099415; Keith Good, "Sunday Update: Farm Bill—House Names Conferees," *FarmPolicy.com*, October 13, 2013, http://farmpolicy.com/2013/10/13/sunday-update-farm-bill-house-names -conferees/#more-13918; Jonathan Weisman and Ashley Parker, "Republicans Back Down, Ending Crisis over Shutdown and Debt Limit," *New York*

Times, October 16, 2013, http://www.nytimes.com/2013/10/17/us/congress
-budget-debate.html?_r=0; Erik Wasson and Russell Berman, "Boehner
to Ryan: Hands off Farm Bill," *The Hill*, November 14, 2013, http://thehill
.com/policy/finance/190256-boehner-no-farm-bill-savings-in-budget-deal.
See also Lisa Mascaro, Michael A. Memoli, and Brian Bennett, "Government Crisis Is Averted—For Now," *Los Angeles Times*, October 16, 2013,
http://www.latimes.com/nation/la-na-government-shutdown-20131017
-story.html#page=1.

171. Douglas Elmendorf, letter to Frank D. Lucas, September 16, 2013,
https://www.cbo.gov/publication/44583; Douglas Elmendorf, letter to Frank
D. Lucas, January 28, 2014, https://www.cbo.gov/publication/45049.

172. Lawrence, "Profiles in Negotiation," 13–14; Hamilton, "2014 Farm
Bill," 34–35.

173. David Rogers, "Farm Bill Talks Stumble," *Politico*, November 21, 2013,
https://www.politico.com/story/2013/11/farm-bill-update-100217; Charles
Abbott, "Food Stamps, Subsidies Knotty Issues as U.S. Farm Talks Drag
on," Reuters, November 22, 2013, http://in.reuters.com/article/2013/11/21/usa
-agriculture-negotiations-idinl2n0j61bm20131121; Jerry Hagstrom, "How
Cheap Corn and Wheat Could Cost the GOP," *National Journal*, January 5,
2014, http://www.nationaljournal.com/outside-influences/how-cheap-corn
-and-wheat-could-cost-the-gop-20140105. See also Todd Kuethe and Jonathan Coppess, "Mapping the Fate of the Farm Bill: A Closer Look at SNAP,"
farmdoc daily, January 10, 2014, http://farmdocdaily.illinois.edu/2014/01
/mapping-fate-of-farm-bill-2.html; Carl Zulauf and Gary Schnitkey, "Farm
Bill Conference Issues," *farmdoc daily*, Dept. of Agricultural and Consumer
Economics, University of Illinois at Urbana-Champaign, August 22, 2013,
http://farmdocdaily.illinois.edu/2013/08/farm-bill-conference-issues.html;
Carl Zulauf, "Big Picture View of the Crop Safety Net Debate," *farmdoc daily*,
Dept. of Agricultural and Consumer Economics, University of Illinois at
Urbana-Champaign, November 21, 2013, http://farmdocdaily.illinois.edu
/2013/11/big-picture-view-of-the-crop-s.html.

174. Campiche, "Details of," 570; Gorter, "2012 US Farm Bill."

175. David Rogers, "Big Trades Advance Farm Bill Talks," *Politico*, December 4, 2013, http://www.politico.com/story/2013/12/farm-bill-talks-progress
-100670; Carl Zulauf and Jonathan Coppess, "2013 Farm Bill Update—
November 2013," *farmdoc daily*, November 14, 2013, http://farmdocdaily
.illinois.edu/2013/11/2013-farm-bill-update-november.html.

176. See, e.g., Hamilton, "2014 Farm Bill," 23–24.

177. Schnepf, "U.S. Peanut Program"; Schnepf, "Farm Safety-Net."

178. David Rogers, "Stabenow Optimistic on Farm Bill," *Politico*, January 7, 2014, http://www.politico.com/story/2014/01/debbie-stabenow

-farm-bill-101846; David Rogers, "Farm Bill in Trouble," *Politico*, January 9, 2014, http://www.politico.com/story/2014/01/farm-bill-trouble-frank-lucas-101980.

179. Tom Steever, "Grassley Remains Confident of Farm Bill Passage," *Brownfield*, December 3, 2013, http://brownfieldagnews.com/2013/12/03/grassley-remains-confident-farm-bill-passage/; Charles Abbott, "UPDATE1—Boehner Blasts Lack of Farm Bill Progress, Supports Extension," Reuters, December 5, 2013, http://www.reuters.com/article/2013/12/05/usa-agriculture-boehner-idusl2n0jk1r020131205; David Rogers, "Reformers Could Be Crucial to Farm Bill Passage," *Politico*, December 9, 2013, http://www.politico.com/story/2013/12/reformers-crucial-farm-bill-passage-agriculture-100903; David Rogers, "No Farm Bill in 2013," *Politico*, December 10, 2013, http://www.politico.com/story/2013/12/no-farm-bill-in-2013-agriculture-debbie-stabenow-frank-lucas-100966.

180. David Rogers, "Farm Bill Talks in Final Stretch," *Politico*, December 13, 2013, http://www.politico.com/story/2013/12/farm-bill-update-frank-lucas-101128; Jonathan Coppess, "Reviewing the USDA Proposal to Limit Farm Program Payment Eligibility," *farmdocdaily* (5):64, April 8, 2015, http://farmdocdaily.illinois.edu/2015/04/reviewing-usda-proposal-to-limit-payment-eligibility.html.

181. Keith Good, "Farm Bill; Budget; Ag Economy; and, Political Notes," *FarmPolicy.com*, January 15, 2014, http://farmpolicy.com/2014/01/15/farm-bill-budget-ag-economy-and-political-notes-wednesday-2/#more-14425; David Rogers, Haggling over Farm Bill Final Points," *Politico*, January 23, 2014, http://www.politico.com/story/2014/01/farm-bill-2014-update-102536.

182. Lawrence, "Profiles in Negotiation," 13.

183. David Rogers, "New Farm Bill Readied for Final Debate," *Politico*, January 26, 2014, http://www.politico.com/story/2014/01/new-house-farm-bill-102626; David Rogers, "Farm Bill Agreement Heading to Floor," *Politico*, January 27, 2014, http://www.politico.com/story/2014/01/beef-lobby-loses-push-to-repeal-labeling-rules-102654; David Rogers, "House Poised for Vote on Farm Deal," *Politico*, January 28, 2014, http://www.politico.com/story/2014/01/house-farm-bill-deal-agriculture-102757.

184. Cong. Rec., January 27, 2014, H1269–1425; Cong. Rec., January 29, 2014, H1485.

185. Cong. Rec., January 29, H1485.

186. David Rogers, "House Clears Farm Bill," *Politico*, January 29, 2014, http://www.politico.com/story/2014/01/house-farm-bill-102806; Lawrence, "Profiles in Negotiation," 13–14; Hamilton, "2014 Farm Bill," 34–35.

187. Cong. Rec., January 29, 2014, H1486 (Representative Jim McGovern [D-MA]), H1500 (by a vote of 251 to 166 [14 not voting]).

188. Cong. Rec., January 30, 2014, s619; Cong. Rec., February 3, 2014, s666–67, s690 (Senate invoked cloture by a vote of 72 to 22 [6 not voting]); David Rogers, "Senate Advances Farm Bill," *Politico*, February 3, 2014, http://www.politico.com/story/2014/02/farm-bill-update-103049.

189. Cong. Rec., February 3, 2014, s667.

190. Cong. Rec., February 3, 2014, 673; Cong. Rec., February 4, 2014, s726.

191. Cong. Rec., February 4, 2014, s736 (by a vote of 68 to 32); David Rogers, "Congress Approves 5-Year Farm Bill," *Politico*, February 4, 2014, http://www.politico.com/story/2014/02/congress-approves-farm-bill-103114; Keith Good, "Farm Bill Passes the Senate," *FarmPolicy.com*, February 5, 2014, http://farmpolicy.com/2014/02/05/farm-bill-passes-the-senate/#more-14594.

192. Michael D. Shear, "In Signing Farm Bill, Obama Extols Rural Growth," *New York Times*, February 7, 2014, http://www.nytimes.com/2014/02/08/us/politics/farm-bill.html; Keith Good, "Farm Bill; Ag Economy; Labeling Issues; Food Safety; and Immigration," *FarmPolicy.com*, February 8, 2014, http://farmpolicy.com/2014/02/08/weekend-update-president-obama-signs-2014-farm-bill-into-law/#more-14619.

193. House FY2012 Budget, 4, 11, 25; House FY2013 Budget, 7–8, 37.

194. See, e.g., Rosenbaum, "Relationship Between."

195. Rosenbaum, "Ryan Budget Would Slash."

196. CBO, "Cost Estimate: Agricultural Reconciliation Act of 2012," April 18, 2012, https://www.cbo.gov/publication/43185; CBO, "Cost Estimate: H.R. 6083, Federal Agriculture Reform and Risk Management Act of 2012," July 26, 2012, https://www.cbo.gov/publication/43486; Douglas Elmendorf, letter to Bob Goodlatte, June 7, 2013, https://www.cbo.gov/publication/44325; Federal Agricultural Reform and Risk Management Act of 2013, H.R. 1947, 113th Cong. (2013), https://www.congress.gov/bill/113th-congress/house-bill/1947.

197. Elmendorf to Lucas, September 16, 2013.

198. Schertz and Doering, *Making of*, 8–9 (quoting Secretary Butz).

199. See, e.g., Schnepf, "Measuring Equity"; Schnepf, "Farm Safety-Net."

200. CBO's June 2017 Baseline for Farm Programs, June 29, 2017, https://www.cbo.gov/sites/default/files/recurringdata/51317-2017-06-usda.pdf.

201. CBO, "Supplemental Nutrition Assistance Program: CBO's January 2017 Baseline," January 2017, https://www.cbo.gov/sites/default/files/recurringdata/51312-2017-01-snap.pdf.

202. For example, USDA has estimated that roughly $53 billion of the program was redeemed at supermarkets and stores and that for every dollar in SNAP benefits, between $0.17 and $0.47 of new spending is on food products. See, e.g., USDA, ERS, "Economic Linkages: Supplemental Nutrition Assistance Program (SNAP) Linkages with the General Economy," last modified October 4, 2017, https://www.ers.usda.gov/topics/food-nutrition

-assistance/supplemental-nutrition-assistance-program-snap/economic
-linkages/; Canning, *Food Dollar Series*. USDA also estimates that in general, $0.16 of every dollar spent on food goes directly to the farmer; farm bill food assistance programs were estimated to spend over $70 billion per year. See Canning, *Food Dollar Series*; Elmendorf to Lucas, January 28, 2014.

203. Johnson and Monke, "What Is a Farm Bill"; Johnson and Monke, "What Is the Farm Bill."

204. Gary Schnitkey, Jonathan Coppess, Nick Paulson, and Carl Zulauf, "Perspectives on Commodity Program Choices under the 2014 Farm Bill," *farmdoc daily* (5):111, June 16, 2015, Department of Agricultural and Consumer Economics, University of Illinois at Urbana-Champaign, http://farmdocdaily .illinois.edu/2015/06/perspectives-on-commodity-program-choices.html; Carl Zulauf, Gary Schnitkey, Jonathan Coppess, and Nick Paulson, "2014 Farm Bill Crop Program Election, Part II," *farmdoc daily* (5):113, June 18, 2015, Department of Agricultural and Consumer Economics, University of Illinois at Urbana-Champaign, http://farmdocdaily.illinois.edu/2015/06/2014 -farm-bill-crop-program-election-part2.html; Jonathan Coppess, "Evaluating Commodity Program Choices in the New Farm Bill," *farmdoc daily* (4):21, February 6, 2014, Dept. of Agricultural and Consumer Economics, University of Illinois at Urbana-Champaign, http://farmdocdaily.illinois .edu/2014/02/evaluating-commodity-program-choices-in-new-farm-bill .html; Jonathan Coppess and Nick Paulson, "Agriculture Risk Cover and Price Loss Coverage in the 2014 Farm Bill," *farmdoc daily* (4):32, February 20, 2014, Dept. of Agricultural and Consumer Economics, University of Illinois at Urbana-Champaign, http://farmdocdaily.illinois.edu/2014/02 /arc-and-plc-in-2014-farm-bill.html; Gary Schnitkey, Jonathan Coppess, Nick Paulson, and Carl Zulauf, "Overview of Commodity Program Decisions from the 2014 Farm Bill," *farmdoc daily*, (4):223, November 18, 2014, Dept. of Agricultural and Consumer Economics, University of Illinois at Urbana-Champaign, http://farmdocdaily.illinois.edu/2014/11/overview-of -commodity-program-decision-2014-farm-bill.html.

205. J. Coppess, N. Paulson, C. Zulauf, and G. Schnitkey. "Farm Bill Round 1: Dairy, Cotton and the President's Budget," *farmdoc daily* (8):25, Department of Agricultural and Consumer Economics, University of Illinois at Urbana-Champaign, February 14, 2018, http://farmdocdaily.illinois .edu/2018/02/farm-bill-round-1-dairy-cotton-and-the-budget.html.

9. Trying to Reason with the Fault Lines

1. Cochrane, *Curse*, 9–10 (from a speech he gave to New York dairy farmers in 1954).

2. Cochrane, *Curse*, 19–24 (taken from a 1958 essay).

3. Cochrane, *Curse*, 27, 41.

4. See, e.g., Schnepf, "U.S. Peanut Program."

5. See, e.g., Cochrane, *Curse*, 68.

6. 7 U.S.C. §1501 et seq.

7. GAO, "Crop Insurance: Reducing."

8. Stubbs, "Conservation Compliance," 10–11.

9. See, e.g., USDA, Natural Resource Conservation Service, Conservation Effects Assessment Project (CEAP), National and Regional Assessments, https://www.nrcs.usda.gov/wps/portal/nrcs/main/national/technical/nra/ceap/.

10. See, e.g., USDA, Farm Service Agency, Economic and Policy Analysis, Natural Resource Analysis, https://www.fsa.usda.gov/programs-and-services/economic-and-policy-analysis/natural-resources-analysis/index.

11. Claassen, Cattaneo, and Johansson, "Cost-Effective Design," 741–42.

12. Agricultural Act of 2014, H.R. Rep. No. 113-333, at 410–13 (2014) (Conf. Rep.); Jonathan Coppess, "Dead Zones and Drinking Water, Part 1: RCPP and Review," *farmdoc daily* (6):37, February 25, 2016. Department of Agricultural and Consumer Economics, University of Illinois at Urbana-Champaign, http://farmdocdaily.illinois.edu/2016/02/dead-zones-drinking-water-part1.html.

13. Angelo, "Corn, Carbon and Conservation," 601.

14. See, e.g., Eubanks, "Rotten System," 245–48; Craig Cox, "Data Show Farmers Must Do More to Protect the Environment, Public Health," *AgMag*, October 13, 2016, http://www.ewg.org/agmag/2016/10/new-ewg-database-details-30-billion-spent-us-farm-conservation-programs.

15. Harwood et al., "Managing Risk in Farming," 12–13.

16. Waldron, "Legislation, Authority," 2198.

17. Waldron, "Legislation, Authority," 2198, 2208; Eskridge, Frickey, and Garrett, *Legislation*, 103–6; Eskridge, and Ferejohn, "Article I," 523.

18. Eskridge, Frickey, and Garrett, *Legislation*, 3.

19. Eskridge, Frickey, and Garrett, *Legislation*, 70, 79–81.

20. Waldron, "Legislation, Authority," 2210, 2213, 2202.

21. Waldron, "Dignity of Legislation," 641. See also Waldron, "Legislation, Authority," 2202.

22. Waldron, "Legislation, Authority," 2204.

23. Eskridge, Frickey, and Garrett, *Legislation*, 72–74.

24. Eskridge, Frickey, and Garrett, *Legislation*, 87–89; Waldron "Dignity of Legislation," 640.

25. Eskridge, Frickey, and Garrett, *Legislation*, 98.

26. Eskridge and Ferejohn, "Article I," 560.

27. Waldron, "Dignity of Legislation," 640, 660–62, 664.

28. Eskridge, Frickey, and Garrett, *Legislation*, 95, 99.

29. See, e.g., Binder and Lee, "Making Deals in Congress," 54–55.

30. In one sense this might fit in James Madison's vision expressed in the *Federalist* that "ambition must be made to counteract ambition" in the system. See, e.g., Eskridge and Ferejohn, "Article I," 560–61 (quoting Madison).

31. See, e.g., Eskridge, Frickey, and Garrett, *Legislation*, 188–89, 191–95.

32. See, e.g., Waldron, "Legislation, Authority," 2212.

33. See, e.g., Binder and Lee, "Making Deals in Congress," 54–55.

34. See, e.g., Binder and Lee, "Making Deals in Congress," 54.

35. See, e.g., Waldron, "Legislation, Authority," 2214.

36. Eskridge, Frickey, and Garrett, *Legislation*, 70, 79–81.

37. See, e.g., Waldron, "Dignity of Legislation," 634.

38. See, e.g., FY2012 and FY2013 House Budgets, chapter 9.

39. See, e.g., Daniel, "Crossroads," 433–36.

40. See, e.g., Joan C. Williams, "What So Many People Don't Get about the U.S. Working Class," *Harvard Business Review*, November 10, 2016, https://hbr.org/2016/11/what-so-many-people-dont-get-about-the-u-s-working-class; Joan C. Williams, "The Dumb Politics of Elite Condescension," *New York Times*, May 27, 2017, https://www.nytimes.com/2017/05/27/opinion/sunday/the-dumb-politics-of-elite-condescension.html?mcubz=0&_r=0.

41. House FY2013 Budget, 25.

42. Perkins, *Crisis*, 146 (quoting Secretary Wallace).

BIBLIOGRAPHY

Aldrich, John H., Bradford H. Bishop, Rebecca S. Hatch, D. Sunshine Hilly-
    gus, and David Rohde. "Blame, Responsibility, and the Tea Party in
    the 2010 Midterm Elections." *Political Behavior* (2013). DOI 10.1007/
    s11109-013-9242-4.
Aldrich, John H., and David W. Rohde. "The Transition to Republican Rule
    in the House: Implications for Theories of Congressional Politics." *Polit-
    ical Science Quarterly* 112, no. 4 (Winter 1997–98): 541–67.
Alvarez, R. Michael, and Jonathan Nagler. "Economics, Issues and the Perot
    Candidacy: Voter Choice in the 1992 Presidential Election." *American
    Journal of Political Science* 39, no. 3 (August 1995): 714–44.
Angelo, Mary Jane. "Corn, Carbon, and Conservation: Rethinking U.S. Agri-
    cultural Policy in a Changing Global Environment." *George Mason Law
    Review* 17 (2010): 593–660.
Ash, Mark, George Douvelis, Jaime Castaneda, and Nancy Morgan. *Oilseeds:
    Background for 1995 Farm Legislation.* Agricultural Economic Report
    No. (AER-715). Washington DC: U.S. Department of Agriculture, Eco-
    nomic Research Service, 1995. https://www.ers.usda.gov/publications
    /pub-details/?pubid=40666.
Atkinson, Tyler, David Luttrell, and Harvey Rosenblum. "How Bad Was It?:
    The Costs and Consequences of the 2007–09 Financial Crisis." *Staff
    Papers, Dallas Fed*, no. 20 (July 2013).
Aussenberg, Randi Alison, and Libby Perl. "Changing the Treatment of LIHEAP
    Receipt in the Calculation of SNAP Benefits." Rep. No. R42591. Washing-
    ton DC: Congressional Research Service, 2013. https://www.everycrsreport
    .com/files/20130917_R42591_432a64502391af3ffd09ba9562017
    aba556e894b.pdf.
Austin, D. Andrew, and Mindy R. Levit, "The Debt Limit: History and Recent
    Increases." Rep. No. RL31967. Washington DC: Congressional Research
    Service, 2013.

Babcock, Bruce. "Welfare Effects of Title One Programs." In *The Economic Welfare and Trade Relations Implications of the 2014 Farm Bill*, edited by Vincent H. Smith, 25–30. Bingley, UK: Emerald Group, 2015.

Barton, Weldon V. "Coalition Building in the United States House of Representatives: Agricultural Legislation in 1973." In *Cases in Public Policy-Making*, edited by James E. Anderson, 141–61. New York: Praeger, 1976.

Becnel, Thomas A. *Senator Ellender of Louisiana, A Biography*. Baton Rouge: Louisiana State Press, 1996.

Benedict, Murray R. *Farm Policies of the United States, 1790–1950: A Study of Their Origins and Development*. New York: Twentieth Century Fund, 1953.

Berry, Wendell. "The Pleasures of Eating." In *The Art of the Commonplace: The Agrarian Essays of Wendell Berry*. Berkeley CA: Counterpoint, 2002.

Binder, Sarah A., and Frances E. Lee. "Making Deals in Congress." In *Negotiating Agreement in Politics*, edited by Jane Mansbridge and Cathie Jo Martin, 54–72. Washington DC: American Political Science Association, 2013.

Blanpied, Nancy A. *Farm Policy: The Politics of Soil, Surpluses, and Subsidies*. Washington DC: Congressional Quarterly, 1984.

Bodnick, Marc A. "'Going Public' Reconsidered: Reagan's 1981 Tax and Budget Cuts, and Revisionist Theories of Presidential Power." *Congress and the Presidency* 17, no. 1 (1990): 13–28.

Boehm, William T. "Agricultural Policy: Some Hard Choices Ahead." *Southern Journal of Agricultural Economics* (July 1981): 1–9.

Bonnen, James T. "Implications for Agricultural Policy." *American Journal of Agricultural Economics* 55, no. 3 (August 1973): 391–98.

Bowers, Douglas E., Wayne D. Rasmussen, and Gladys L. Baker. "History of Agricultural Price-Support and Adjustment Programs, 1933–84." *Agriculture Information Bulletin*, no. 485 (1984).

Breimyer, Harold F. "Agricultural Philosophies and Policies in the New Deal." *Minnesota Law Review* 68 (1983): 333–52.

Briscoe, Dolph, IV. "LBJ and Grassroots Federalism: Congressman Bob Poage, Race, and Change in Texas." *Presidential Studies Quarterly* 45, no. 4 (2015): 814.

Bullock, Charles S., III, Donna R. Hoffman, and Ronald Keith Gaddie. "The Consolidation of the White Southern Congressional Vote." *Political Research Quarterly* 58, no. 2 (June 2005): 231–43.

Burford, Roger L. "The Federal Cotton Programs and Farm Labor Force Adjustments." *Southern Economic Journal* 33, no. 2 (October 1966): 223–36.

Campbell, Christiana McFadyen. *The Farm Bureau and the New Deal: A Study of the Making of National Farm Policy, 1933–40*. Urbana: University of Illinois Press, 1962.

Campiche, Judy. "Details of the Proposed Stacked Income Protection Plan (stax) Program for Cotton Producers and Potential Strategies for Extension Education." *Journal of Agricultural and Applied Economics* 43, no. 3 (August 2013): 569–75. http://ageconsearch.umn.edu/bitstream/155445/2/jaae453ip17.pdf.

Canning, Patrick. *A Revised and Expanded Food Dollar Series: A Better Understanding of Our Food Costs.* Economic Research Report No. ERR-114. Washington DC: U.S. Department of Agriculture, Economic Research Service, February 2011. https://www.ers.usda.gov/publications/pub-details/?pubid=44827.

Caro, Robert. *Master of the Senate.* New York: Vintage Books, 2003.

———. *The Passage of Power.* New York: Vintage Books, 1990.

Carson, Jamie L., and Stephen Pettigrew. "Strategic Politicians, Partisan Roll Calls, and the Tea Party: Evaluating the 2010 Midterm Elections." *Electoral Studies* 32 (2013): 26–36.

Chemerinsky, Erwin. "Bush v. Gore Was Not Justiciable." *Notre Dame Law Review* 76, no. 4 (2000–2001): 1093–1112.

Chen, James. "The Story of Wickard v. Filburn: Agriculture, Aggregation, and Commerce." Abstract. In *Constitutional Law Stories*, 2nd ed., edited by Michael C. Dorf. New York: Foundation Press, 2008. http://ssrn.com/abstract=1268162.

Chite, Ralph M. "Emergency Funding for Agriculture: A Brief History of Supplemental Appropriations, FY1989-FY20009." Washington DC: Congressional Research Service, 2008.

Christenson, Reo M. *The Brannan Plan: Farm Politics and Policy.* Ann Arbor: University of Michigan Press, 1959.

Claassen, Roger, Andrea Cattaneo, and Robert Johansson. "Cost-Effective Design of Agri-Environmental Payment Programs: U.S. Experience in Theory and in Practice." *Ecological Economics* 65, no. 4 (2008): 737–52.

Claffey, Barbara A., and Thomas A. Stucker. "Food Stamp Program." In *Food Policy and Farm Programs*, edited by Don F. Hadwiger and Ross B. Talbot, 40–53. New York: Proceedings of the Academy of Political Science, 1982.

Cochrane, Willard W. *The Curse of American Agricultural Abundance.* Lincoln: University of Nebraska Press, 2003.

———. *The Development of American Agriculture: A Historical Analysis.* 2nd ed. Minneapolis: University of Minnesota Press, 1993.

Cochrane, Willard W., and Mary E. Ryan. *American Farm Policy 1948–1973.* Minneapolis: University of Minnesota Press, 1976.

Cohen, Joshua, and Joel Rogers. "The Politics of Dealignment." Madison: Center on Wisconsin Strategy, 1990. http://cows.org/joel/pdf/a_048.pdf.

Collin, Andrea Winkjer, and Richard E. Collin. *Mr. Wheat: A Biography of U.S. Senate Milton R. Young*. Bismark ND: Smoky Water Press, 2010.

Conrad, David Eugene. *The Forgotten Farmers: The Story of Sharecroppers in the New Deal*. Urbana: University of Illinois Press, 1965.

Dallek, Robert. *An Unfinished Life: John F. Kennedy 1917–1963*. New York: Back Bay Books, 2003.

Daniel, Pete. "The Crossroads of Change: Cotton, Tobacco, and Rice Cultures in the Twentieth-Century South." *Journal of Southern History* 50, no. 3 (August 1984): 429–56.

Davis, Chester C. "The Development of Agricultural Policy since the End of the World War." In *Yearbook of Agriculture 1940: Farmers in a Changing World*, by U.S. Dept. of Agriculture. Washington DC: U.S. Government Printing Office, 1940.

Dean, Virgil W. "Charles F. Brannan and the Rise and Fall of Truman's 'Fair Deal' for Farmers." *Agricultural History* 69, no. 1 (Winter 1995): 28–53.

——. "The Farm Policy Debate of 1949–50: Plains State Reaction to the Brannan Plan." *Great Plains Quarterly*, paper no. 773 (1993): 33–46.

——. "Why Not the Brannan Plan?" *Agricultural History* 70, no. 2 (Spring 1996): 268–82.

Dimitri, Carolyn, Anne Effland, and Neilson Conklin. "The 20th Century Transformation of U.S. Agriculture and Farm Policy." Economic Information Bulletin No. 3. Washington DC: U.S. Department of Agriculture, Economic Research Service, 2005.

Dixon, Deborah P., and Holly M. Hapke. "Cultivating Discourse: The Social Construction of Agricultural Legislation." *Annals of the Association of American Geographers* 93, no. 1 (2003): 42–164.

Egan, Timothy. *The Worst Hard Time: The Untold Story of Those Who Survived the Great American Dust Bowl*. Boston: Houghton Mifflin, 2006.

Eskridge, William N., Jr., and John Ferejohn. "The Article I, Section 7 Game." *Georgetown Law Journal* 80, no. 3 (1991–92): 523–64.

Eskridge, William N., Jr., Philip P. Frickey, and Elizabeth Garrett. *Legislation and Statutory Interpretation*. 2nd ed. New York: Foundation Press, 2006.

Eubanks, William S., II. "A Rotten System: Subsidizing Environmental Degradation and Poor Public Health with Our Nation's Tax Dollars." *Stanford Environmental Law Journal* 28, no. 2 (2009): 213–310.

Ferejohn, John. "Logrolling in an Institutional Context: A Case Study of Food Stamp Legislation." In *Congress and Policy Change*, edited by Gerald C. Wright, Leroy N. Reiselbach, and Lawrence C. Dodd, 220–63. New York: Agathon Press, 1986.

Finegold, Kenneth. "From Agrarianism to Adjustment: The Political Origins of New Deal Agricultural Policy." *Politics and Society* 11, no. 1 (1982): 1–27.

Fite, Gilbert C. *Cotton Fields No More: Southern Agriculture, 1865–1980.* Lexington: University Press of Kentucky, 2015.

——. "Farmer Opinion and the Agricultural Adjustment Act of 1933." *Mississippi Valley Historical Review* 48, no. 4 (1962): 656–73.

——. *George N. Peek and the Fight for Farm Parity.* Norman: University of Oklahoma Press, 1954.

Food and Agricultural Policy Research Institute (FAPRI). *2001 U.S. Baseline Briefing Book.* Columbia: University of Missouri, 2001. https://www.fapri.missouri.edu/wp-content/uploads/2015/03/FAPRI-MU-TDr-01-01.pdf.

——. *2002 U.S. Baseline Briefing Book.* Columbia: University of Missouri, 2002. https://www.fapri.missouri.edu/wp-content/uploads/2015/03/FAPRI-MU-TDr-02-02.pdf.

——. *2003 U.S. Baseline Briefing Book.* Columbia: University of Missouri, 2003. https://www.fapri.missouri.edu/wp-content/uploads/2015/03/FAPRI-MU-TDr-04-03.pdf.

——. *2004 U.S. Baseline Briefing Book.* Columbia: University of Missouri, 2004. https://www.fapri.missouri.edu/wp-content/uploads/2015/03/FAPRI-MU-TDr-01-04.pdf.

Forsythe, James L. "Clifford Hope of Kansas: Practical Congressman and Agrarian Idealist." *Agricultural History* 51, no. 2 (1977): 406–20.

Fortier, John C., and Norman J. Ornstein. "Presidential Succession and Congressional Leaders." *Catholic University Law Review* 53, no. 4 (2003–2004): 993–1014.

Frischnecht, Reed L. "The Commodity Credit Corporation: A Case Study of a Government Corporation." *Western Political Quarterly* 6, no. 3 (September 1953): 559–69.

Gillon, William A. "The Panel Report in the U.S.-Brazil Cotton Dispute: WTO Subsidy Rules Confront U.S. Agriculture." *Drake Journal of Agricultural Law* 10, no. 1 (2005): 7–56.

Glade, Edward H., Jr., Leslie Meyer, and Stephen MacDonald. *Cotton: Background for 1995 Farm Legislation.* Agricultural Economic Report No. (AER-706). Washington DC: U.S. Department of Agriculture, Economic Research Service, 1995. https://www.ers.usda.gov/publications/pub-details/?pubid=40611.

Glauber, Joseph W. "Crop Insurance Reconsidered." *American Journal of Agricultural Economy* 86, no. 5 (2004): 1179–95.

——. "The Growth of the Federal Crop Insurance Program, 1990–2011." *American Journal of Agricultural Economy* 95, no. 2 (2013): 482–88.

Goodwin, Barry K., and Vincent H. Smith. "The 2014 Farm Bill—An Economic Welfare Disaster or Triumph?" In *The Economic Welfare and*

Trade Relations Implications of the 2014 Farm Bill, edited by Vincent H. Smith, 1–10. Bingley, UK: Emerald Group, 2015.

Gorter, Harry de. "The 2012 US Farm Bill and Cotton Subsidies." Issue Paper No. 46. Geneva: International Centre for Trade and Sustainable Development, 2012. http://www.ictsd.org/downloads/2012/12/us-farm-bill -2012-and-cotton-subsidies.pdf.

Guither, Harold D. "Tough Choices: Writing the Food Security Act of 1985." Washington DC: American Enterprise Institute for Public Policy Research, 1986.

Hackbart-Dean, Pamela. "The Greatest Civics Lesson in Our History: Herman Talmadge and Watergate from a Twenty-Five-Year Perspective." *Georgia Historical Quarterly* 83, no. 2 (1999): 314–21.

Hadwiger, Don F., and Ross Talbot. *Pressures and Protests: The Kennedy Farm Program and the Wheat Referendum of 1963.* Ames: Iowa State University Press, 1965.

Hallberg, M. C. *Policy for American Agriculture: Choices and Consequences.* Ames: Iowa State University Press, 1992.

Hamilton, Neil D. "The 2014 Farm Bill: Lessons in Patience, Politics, and Persuasion." *Drake Journal of Agricultural Law* 19, no. 1(2014), 1–37.

Hansen, John Mark. *Gaining Access: Congress and the Farm Lobby, 1919–1981.* Chicago: University of Chicago Press, 1991.

Hardin, Charles M. "Farm Price Policy and the Farm Vote." *Journal of Farm Economics* 37, no. 4 (November 1955): 601–24.

Harkin, Ruth R., and Thomas R. Harkin. "'Roosevelt to Reagan' Commodity Programs and the Agriculture and Food Act of 1981." *Drake Law Review* 31, no. 3 (1981): 499–518.

Harris, Douglas B. "Legislative Parties and Leadership Choice: Confrontation or Accommodation in the 1989 Gingrich-Madigan Whip Race." *American Politics Research* 34, no. 2 (March 2006): 189–222.

Harwood, Joy, Richard Heifner, Keith Coble, Janet Perry, and Agapi Somwaru. "Managing Risk in Farming: Concepts, Research, and Analysis." Agricultural Economic Report No. 774. Washington DC: U.S. Department of Agriculture, Economic Research Service, March 1999.

Hasen, Richard L. "Bush v. Gore and the Future of Equal Protection Law in Elections." *Florida Study University Law Review* 29, no. 2 (2001–2): 377–406.

Hayes, Danny, and Seth C. McKee. "Toward a One-Party South?" *American Politics Research* 36, no. 1 (2008). http://apr.sagepub.com/content/early/2007/10/03/1532673x07307278.short.

Heacock, Walter J. "William B. Bankhead and the New Deal." *Journal of Southern History* 2, no. 3 (August 1955): 347–59.

Heinz, John P. "The Political Impasse in Farm Support Legislation." *Yale Law Journal* 71, no. 5 (April 1962): 952–78.

Heniff, Bill, Jr., Elizabeth Rybicki, and Shannon M. Mahan. "The Budget Control Act of 2011." Rep. No. R41965. Washington DC: Congressional Research Service, 2011.

Herron, Michael C., and Jeffrey B. Lewis. "Economic Crisis, Iraq, and Race: A Study of the 2008 Presidential Election." *Election Law Journal* 9, no. 1 (2010): 41–62.

Hershey, Marjorie Randon. "Richard Lugar and the New Politics of 'Civil' Engagement." *Indiana Magazine of History* 108, no. 3 (September 2012), 274–79.

Hoffman, Linwood, Grace V. Chomo, and Sara Schwartz. *Wheat: Background for 1995 Farm Legislation.* Agricultural Economic Report No. (AER-712). Washington DC: U.S. Department of Agriculture, Economic Research Service, 1995. https://www.ers.usda.gov/publications/pub-details/?pubid=40644.

Hoffsommer, Harold. "The AAA and the Cropper." *Social Forces* 13, no. 4 (May 1935): 494–502.

Hollis, Daniel W. "'Cotton Ed Smith': Showman or Statesman?" *South Carolina Historical Magazine* 71, no. 4 (October 1970): 235–56.

Hunter, Dan. "Reason Is Too Large: Analogy and Precedent in Law." *Emory Law Journal* 50, no. 4 (2001): 1197–1264.

Hurt, R. Douglas. *Problems of Plenty: The American Farmer in the Twentieth Century.* Chicago: Ivan R. Dee, 2002.

Infanger, Craig L., William C. Bailey, and David R. Dyer. "Agricultural Policy in Austerity: The Making of the 1981 Farm Bill." *American Journal of Agricultural Economics* 65, no. 1 (February 1983): 1–9.

Jacobson, Gary C. "The Republican Resurgence in 2010." *Political Science Quarterly* 126, no. 1 (2011): 27–52.

Jensen, Helen H. "Food Insecurity and the Food Stamp Program." *American Journal of Agricultural Economics* 84, no. 5 (2002): 1215–28.

Jickling, Mark. "Causes of the Financial Crisis." Rep. No. R40173. Washington DC: Congressional Research Service, 2010.

Johnson, James P. "Legislative Consideration of the Agriculture and Consumer Protection Act of 1973." *North Dakota Law Review* 50, no. 2 (1973–74): 279–98.

Johnson, Renee, et al. "The 2008 Farm Bill: Major Provisions and Legislative Action." Rep. No. RL34696. Washington DC: Congressional Research Service, 2008.

Johnson, Renee, and Jim Monke, "What Is the Farm Bill?" Rep. No. RS22131. Washington DC: Congressional Research Service, 2014.

———. "What Is the Farm Bill?" Rep. No. RS22131. Washington DC: Congressional Research Service, 2017.

Kelley, Christopher R. "The Agricultural Risk Protection Act of 2000: Federal Crop Insurance, the Non-insured Crop Disaster Assistance Program, and the Domestic Commodity and Other Farm Programs." *Drake Journal of Agricultural Law* 6, no. 1 (2001): 141–74.

Kile, Orville Merton. *The Farm Bureau through Three Decades.* Baltimore: Waverly Press, 1948.

Kirby, Jack Temple. "The Southern Exodus, 1910–1960: A Primer for Historians." *Journal of Southern History* 49, no. 4 (November 1983): 585–600.

———. "The Transformation of Southern Plantations c. 1920–1960." *Agricultural History* 57, no. 3 (July 1983): 257–76.

Kirkendall, Richard S. "The New Deal and Agriculture." In *The New Deal*, vol. 1, edited by John Braeman, Robert Hamlett, and David Brody. Columbus: Ohio State University Press, 1975.

Krishnakumar, Anita S. "Reconciliation and the Fiscal Constitution: The Anatomy of the 1995–96 Budget 'Train Wreck.'" *Harvard Journal on Legislation* 35, no. 2 (1998): 589–622.

King, Ronald F. "Welfare Reform: Block Grants, Expenditure Caps, and the Paradox of the Food Stamp Program." *Political Science Quarterly* 114, no. 3 (1999): 359–85.

Landers, Patti S. "The Food Stamp Program: History, Nutrition Education, and Impact." *Journal of the American Dietetic Association* 107, no. 11 (2007): 1945–51.

Lauck, Jon. "After Deregulation: Constructing Agricultural Policy in the Age of 'Freedom to Farm.'" *Drake Journal of Agricultural Law* 5, no. 1 (Spring 2000): 3–56.

Lawrence, Jill. "Profiles in Negotiation: The 2014 Farm and Food Stamp Deal." Washington DC: Center for Effective Public Management at Brookings, October 2015. http://www.brookings.edu/research/papers/2015/10/23 -farm-bill-negotiation-lawrence.

Lehrer, Nadine. *U.S. Farm Bills and Policy Reforms: Ideological Conflicts over World Trade, Renewable Energy and Sustainable Agriculture.* Amherst NY: Cambria Press, 2010.

Lerner, Mitchell. "Robert Harold Duke. LBJ and Grassroots Federalism: Congressman Bob Poage, Race, and Change in Texas." *American Historical Review* 120, no. 3 (June 2015): 1054–55.

Levins, Richard A., and Willard W. Cochrane. "The Treadmill Revisited." *Land Economics* 72, no. 4 (November 1996): 550–53.

Lin, William, Peter Riley, and Sam Evans. *Feed Grains: Background for 1995 Farm Legislation.* Agricultural Economic Report No. (AER-714). Washington DC: U.S. Department of Agriculture, Economic Research Service, 1995. https://www.ers.usda.gov/publications/pub-details/?pubid=40658.

Lord, Russell. *The Wallaces of Iowa*. Boston: Riverside Press, 1947.

Lynch, Megan S. "The Budget Reconciliation Process: Timing of Legislative Action." Rep. No. RL30458. Washington DC: Congressional Research Service, February 23, 2016.

Lyons, Michael S., and Marcia Whicker Taylor. "Farm Politics in Transition: The House Agriculture Committee." *Agricultural History* 55, no. 2 (April 1981): 128–46.

Malcolm, Scott, and Marcel Aillery. "Growing Crops for Biofuels Has Spillover Effects." *Amber Waves* 7, no, 1 (March 2009): 10–15. https://www.ers.usda.gov/amber-waves/2009/march/growing-crops-for-biofuels-has-spillover-effects/.

Malone, Linda A. "A Historical Essay on the Conservation Provisions of the 1985 Farm Bill: Sodbusting, Swampbusting, and the Conservation Reserve." *University of Kansas Law Review* 24, no. 3 (1985–86): 577–98.

Mann, Susan A. "Sharecropping in the Cotton South: A Case of Uneven Development in Agriculture." *Rural Sociology* 49, no. 4 (1984): 412–29.

Matusow, Allen J. *Farm Policies and Politics in the Truman Years*. Cambridge MA: Harvard University Press, 1967. Reprint, New York: Atheneum, 1970.

May, Irvin, Jr. "Marvin Jones: Agrarian and Politician." *Agricultural History* 51, no. 2 (1977): 421–40.

Mayer, Leo V. "Farm Exports and Soil Conservation." In *Food Policy and Farm Programs*, edited by Don F. Hadwiger and Ross B. Talbot, 99–111. New York: Proceedings of the Academy of Political Science, 1982.

McCarl, Bruce A., and Fred O. Boadu. "Bioenergy and U.S. Renewable Fuels Standards: Law, Economic, Policy/Climate Change and Implementation Concerns." *Drake Journal of Agricultural Law* 14, no. 1 (2009): 43–74.

McConnell, Grant. *The Decline of Agrarian Democracy*. Berkeley: University of California Press, 1959.

McCoy, Donald R. "Senator George S. McGill and the Election of 1938." *Kansas History* 4, no. 1 (Spring 1981): 2. https://www.kshs.org/publicat/history/1981spring_mccoy.pdf.

McCullough, David. *Truman*. New York: Simon & Schuster, 1992.

McGranahan, David A., Paul W. Brown, Lisa A. Shulte, and John C. Tyndall. "A Historical Primer on the U.S. Farm Bill: Supply Management and Conservation Policy." *Journal of Soil and Water Conservation* 68, no. 3 (2013): 67A–73A.

Miller, James A., and James D. Range. "Reconciling an Irreconcilable Budget: The New Politics of the Budget Process." *Harvard Journal on Legislation* 20, no. 1 (1983): 4–30.

Minchin, Timothy J. "'An Historic Upset': Herman Talmadge's 1980 Senate Defeat and the End of a Political Dynasty." *Georgia Historical Quarterly* 99, no. 3 (2015): 156–97.

Monke, Jim. "Budget Issues Shaping a 2012 Farm Bill." Rep. No. R40532. Washington DC: Congressional Research Service, 2012.

———. "Budget Issues That Shaped the 2014 Farm Bill." Rep. No. R42484. Washington DC: Congressional Research Service, 2014.

Morgan, Dan. "The Farm Bill and Beyond." Economic Policy Paper Series 2010. Washington DC: German Marshall Fund of the United States, 2010.

Morgan, Robert J. *Governing Soil Conservation: Thirty Years of the New Decentralization*. Baltimore: Johns Hopkins University Press, 1965.

Nelson, Lawrence J. *King Cotton's Advocate: Oscar G. Johnston and the New Deal*. Knoxville: University of Tennessee Press, 1999.

———. "Oscar Johnston, the New Deal, and the Cotton Subsidy Payments Controversy, 1936–1937." *Journal of Southern History* 40, no. 3 (August 1974): 399–416.

Newton, John, and Jonathan Coppess. "Rethinking Revenue: Policy Design Options for Farm Bill Commodity Programs." Discussion draft paper presented at the 2016 AAEA Annual Meeting, 2016. On file with author.

Nordin, Dennis Sven, and Roy Vernon Scott. *From Prairie Farmer to Entrepreneur: The Transformation of Midwestern Agriculture*. Bloomington: Indiana University Press, 2005.

Norris, Pippa. "1986 US Elections: National Issues or Pluralistic Diversity?" *Political Quarterly* 58, no. 2 (April 1987): 194–207.

Olson, Allen H. "Federal Farm Programs—Past Present and Future—Will We Learn from Our Mistakes?" *Great Plains Natural Resources Journal* 6, no. 1 (2001): 1–29.

Orden, David. "Exchange Rate Effects on Agricultural Trade and Trade Relations." *Journal of Agricultural and Applied Economics* 34, no. 2 (August 2002): 303–12.

Orden, David, David Blandford, and Timothy Josling. "Determinants of Farm Policies in the United States, 1996–2008." Agricultural Distortions Working Paper 81. Washington DC: World Bank Development Research Group, May 2009. http://documents.worldbank.org/curated/en/252071468167659875 /Determinants-of-farm-policies-in-the-United-States-1996-2008.

Orden, David, Robert Paarlberg, and Terry Roe. *Policy Reform in American Agriculture: Analysis and Prognosis*. Chicago: University of Chicago Press, 1999.

———. "What Is Happening to U.S. Farm Policy: A Chronology and Analysis of the 1995–96 Farm Bill Debate." Working Paper #96-4. St. Paul MN: International Agricultural Trade Research Consortium, 1996.

Orden, David, and Carl Zulauf. "Political Economy of the 2014 Farm Bill." *American Journal of Agricultural Economics*, 97, no. 5(June 11, 2015): 1298–1311.

Paarlberg, Robert, and Don Paarlberg. "Agricultural Policy in the Twentieth Century." *Agricultural History* 74, no. 2 (Spring 2000): 136–61.

Peek, George N., and Hugh S. Johnson. *Equality for Agriculture*. 2nd ed. Moline IL: H. W. Harrington, 1922.

Pennock, J. Roland. "Party and Constituency in Postwar Agricultural Price-Support Legislation." *Journal of Politics* 18, no. 2 (May 1956): 167–210.

Percival, Robert V. "Presidential Management of the Administrative State: The Not-So-Unitary Executive." *Duke Law Journal* 51, no. 3 (2001): 996–1013.

Perkins, Van L. *Crisis in Agriculture: The Agricultural Adjustment Administration and the New Deal, 1933*. Berkeley: University of California Press, 1969.

Peters, John G. "The 1977 Farm Bill: Coalitions in Congress." In *The New Politics of Food*, edited by Don F. Hadwiger and William P. Browne, 23–35. Lexington MA: Lexington Books, 1978.

———. "The 1981 Farm Bill." In *Food Policy and Farm Programs*. Vol. 34, no. 3, of *Proceedings of the Academy of Political Science*, edited by Don F. Hadwiger and Ross B. Talbot, 157–73. New York: Academy of Political Science, 1982.

Peterson, Jeffrey A. "The 1996 Farm Bill: What to (Re) Do in 2002." *Kansas Journal of Law and Public Policy* 11, no. 1 (2001): 65–88.

Pollan, Michael. *The Omnivore's Dilemma: A Natural History of Four Meals*. New York: Penguin Books, 2006.

Porter, Laurellen. "Congress and Agricultural Policy, 1977." In *The New Politics of Food*, edited by Don F. Hadwiger and William P. Browne, 15–22. Lexington MA: Lexington Books, 1978.

Rapp, David. *How the U.S. Got into Agriculture and Why It Can't Get Out*. Washington DC: Congressional Quarterly, 1988.

Rasmussen, Wayne D. "New Deal Agricultural Policies after Fifty Years." *Minnesota Law Review* 68, no. 2 (1983): 353–78.

Ritchie, Mark. "The 'De-Coupled' Approach to Agriculture." Minneapolis: Institute for Agricultural and Trade Policy, 1988. http://www.iatp.org /documents/the-de-coupled-approach-to-agriculture.

———. "Global Agricultural Trade Negotiations and Their Potential Impact on Minnesota." *Journal of the Minnesota Academy of Science* 54, no. 2 (1989): 1–9.

Robinson, Kenneth L. *Farm and Food Policies and Their Consequences*. Englewood Cliffs NJ: Prentice Hall, 1989.

Rosenbaum, Dorothy. "The Relationship between SNAP and Work Among Low-Income Households." Washington DC: Center on Budget and Policy Priorities, 2013. http://www.cbpp.org/research/the-relationship-between -snap-and-work-among-low-income-households.

———. "Ryan Budget Would Slash SNAP funding by $127 Billion over Ten Years." Washington DC: Center on Budget and Policy Priorities, 2011. http://www.cbpp.org/research/ryan-budget-would-slash-snap-funding -by-135-billion-over-ten-years?fa=view&id=3923.

Rosenbaum, Dorothy, and Stacy Dean. "House Agriculture Committee Farm Bill Would Throw 2 to 3 Million People off of SNAP." Washington DC: Center on Budget and Policy Priorities, 2012. http://www.cbpp.org /research/house-agriculture-committee-farm-bill-would-throw-2-to-3 -million-people-off-of-snap?fa=view&id=3800.

Saloutos, Theodore. "New Deal Agricultural Policy: An Evaluation." *Journal of American History* 61, no. 2 (September 1974): 394–416.

Santayana, George. *The Life of Reason: Introduction and Reason in Common Sense.* Edited by Marianne Sophie Wokeck. Cambridge MA: MIT Press, 2011.

Schapsmeier, Edward L., and Frederick H. Schapsmeier. "Eisenhower and Agricultural Reform: Ike's Farm Policy Legacy Appraised." *American Journal of Economics and Sociology* 1, no. 2 (April 1992): 147–60.

———. "Eisenhower and Ezra Taft Benson: Farm Policy in the 1950s." *Agricultural History* 44, no. 4, (Oct. 1970): 369–78.

———. *Ezra Taft Benson and the Politics of Agriculture: The Eisenhower Years, 1953–1961.* Danville IL: Interstate Printers and Publishers, 1975.

———. "Farm Policy from FDR to Eisenhower: Southern Democrats and the Politics of Agriculture." *Agricultural History, Southern Agriculture Since the Civil War: A Symposium* 53, no. 1 (January 1979): 353–71.

Schertz, Lyle P., and Otto C. Doering III. *The Making of the 1996 Farm Act.* Ames: Iowa State University Press, 1999.

Schick, Alan. "How the Budget Was Won and Lost." In *President and Congress: Assessing Reagan's First Year*, edited by Norman J. Ornstein. Washington DC: American Enterprise Institute, 1982.

Schnepf, Randy. "Background on the U.S.-Brazil WTO Cotton Subsidy Dispute." Rep. No. RL32571. Washington DC: Congressional Research Service, 2005.

———. "Brazil's WTO Case against the U.S. Cotton Program." Rep. No. RL32571. Washington DC: Congressional Research Service, 2011.

———. "Farm Safety-Net Payments under the 2014 Farm Bill: Comparison by Program Crop." Washington DC: Congressional Research Service, 2017.

———. "Measuring Equity in Farm Support Levels." Washington DC: Congressional Research Service, 2010.

———. "Status of the WTO Brazil-U.S. Cotton Case." Rep. No. R43336. Washington DC: Congressional Research Service, 2014.

———. "U.S. Peanut Program and Issues." Washington DC: Congressional Research Service, 2016.

Schnepf, Randy, and Bryan Just. *Rice: Background for 1995 Farm Legislation.* Agricultural Economic Report No. (AER-713). Washington DC: U.S. Department of Agriculture, Economic Research Service, 1995. https://www.ers.usda.gov/publications/pub-details/?pubid=40651.

Schnepf, Randy, and Brent D. Yacobucci. "Renewable Fuel Standard (RFS): Overview and Issues." Rep. No. R40155. Washington DC: Congressional Research Service, 2010.

Schudson, Michael. "Notes on Scandal and the Watergate Legacy." *American Behavioral Scientist* 74, no. 9 (May 2004): 1231–38.

Schuh, G. Edward. "The New Macroeconomics of Agriculture." Proceedings Issue, *American Journal of Agricultural Economics* 58, no. 5 (December 1976).

Scott, Robert. "Exported to Death: The Failure of Agriculture Deregulation." *Minnesota Journal of Global Trade* 9, no. 1 (2000): 87–102.

Shaw, John T. *Richard G. Lugar, Statesman of the Senate: Crafting Foreign Policy from Capitol Hill.* Bloomington: Indiana University Press, 2012.

Shumaker, Michael J. "Tearing the Fabric of the World Trade Organization: United States—Subsidies on Upland Cotton." *North Carolina Journal of International Law and Commercial Regulation* 32, no. 3 (2006–7): 547–604.

Sinclair, Barbara. "Agenda Control and Policy Success: Ronald Reagan and the 97th House." *Legislative Studies Quarterly* 10, no. 3 (1985): 291–314.

Slichter, Gertrude Almy. "Franklin D. Roosevelt and the Farm Problem, 1929–1932." *Mississippi Valley Historical Review* 43, no. 2 (September 1956): 238–58.

Smith, Jean Edward. *FDR.* New York: Random House, 2007.

Snyder, Robert E. "Huey Long and the Cotton-Holiday Plan of 1931." *Louisiana History: The Journal of the Louisiana Historical Association* 18, no. 2 (1977): 133–60.

Soth, Lauren. "Farm Policy: A Look Backward and Forward." *Social Research* 27 (1960): 127–38.

Spar, Karen. "Budget 'Sequestration' and Selected Program Exemptions and Special Rules." Rep. No. R42050. Washington DC: Congressional Research Service, 2013.

Spitze, Robert G. F. "The Agriculture and Food Act of 1981: Continued Policy Evolution." *North Central Journal of Agricultural Economics* 5, no. 2 (July 1983): 65–75.

———. "A Continuing Evolution in U.S. Agricultural and Food Policy—The 1990 Act." *Agricultural Economics* 77 (1992): 125–39.

———. "The Evolution and Implications of the U.S. Food Security Act of 1985." *Agricultural Economics* 1, no. 2 (1987): 175–90.

———. "Future Agricultural and Food Policy." *Southern Journal of Agricultural Economics* 13, no. 1 (July 1981): 11–19.

Stubbs, Megan. "Conservation Compliance and U.S. Farm Policy." Rep. No. R42459. Washington DC: Congressional Research Service, 2014. http://nationalaglawcenter.org/wp-content/uploads/assets/crs/r42459.pdf.

Sumner, Daniel A. "Implications of the US Farm Bill of 2002 for Agricultural Trade and Trade Negotiations." *Australian Journal of Agricultural and Resource Economics* 46, no. 3 (2003): 99–122.

Sumner, Daniel A., Julian M. Alston, and Joseph W. Glauber. "Evolution of the Economics of Agricultural Policy." *American Journal of Agricultural Economics* 92, no. 2 (January 2010): 403–23.

Tweeten, Luther G. *Foundations of Farm Policy.* 2nd. ed. Lincoln: University of Nebraska Press, 1979.

Tyner, Wallace E. "The US Ethanol and Biofuels Boom: Its Origins, Current Status, and Future Prospects." *BioScience* 58, no. 7 (2008): 646–53.

U.S. Department of Agriculture, Economic Research Service. "Farm Commodity Programs and Their Effects." *National Food Review* 13, no. 1 (January–March 1990). http://usda.mannlib.cornell.edu/usda/ers/NatlFoodReview//1990s/1990/NatlFoodReview-1990-Vol-13-01.pdf.

U.S. Government Accountability Office. "1983 Payment-in-Kind Program Overview: Its Design, Impact and Cost." Rep. No. GAO-RCED-85-89. Washington DC: U.S. Government Accountability Office, 1985.

———. "Basic Changes Needed to Avoid Abuse of the $50,000 Payment Limit." Rep. No. RCED 87–176. Washington DC: U.S. Government Accountability Office, 1987. http://www.gao.gov/products/rced-87-176.

———. "Crop Insurance: Reducing Subsidies for Highest Income Participants Could Save Federal Dollars with Minimal Effect on the Program." Rep. No. GAO-15-356. Washington DC: U.S. Government Accountability Office, 2015.

———. "Crop Insurance: Savings Would Result from Program Changes and Greater Use of Data Mining." Rep. No. GAO-12–256. Washington DC: U.S. Governmental Accountability Office, 2012. http://www.gao.gov/products/gao-12-256.

U.S. House of Representatives, Joint Committee on Taxation. "Description of the Administration's Payment-in-Kind (PIK) Program, Including Tax Issues Raised by the Program." Washington DC: U.S. Government Printing Office, 1983.

Uslaner, Eric M., and M. Margaret Conway. "The Responsible Congressional Electorate: Watergate, the Economy and Vote Choice in 1974." *American Political Science Review* 79, no. 3 (September 1985): 788–803.

Vance, Rupert B. "Human Factors in the South's Agricultural Readjustment." *Law and Contemporary Problems, Agricultural Readjustment in the South: Cotton and Tobacco: A Symposium* 1, no. 3 (June 1934): 259–74.

Waldron, Jeremy. "The Dignity of Legislation." *Maryland Law Review* 54, no. 2 (1995): 633–66.

———. "Legislation, Authority, and Voting." *Georgetown Law Journal* 84, no. 6 (1995–96): 2185–2214.

Walter, Alan S. "Impacts of the Congressional Budget Process on Agricultural Legislation." In *The New Politics of Food*, edited by Don F. Hadwiger and William P. Browne, 37–42. Lexington MA: Lexington Books, 1978.

Watson, Nathan. "Federal Farm Subsidies: A History of Governmental Control, Recent Attempts at a Free Market Approach, the Current Backlash, and Suggestions for Future Action." *Drake Journal of Agricultural Law* 9, no. 2 (2004): 279–98.

Weatherford, M. Stephen, and Lorraine M. McDonnell. "Ronald Reagan as Legislative Advocate: Passing the Reagan Revolution's Budgets in 1981 and 1982." *Congress and The Presidency* 32, no. 1 (2005): 1–29.

Whatley, Warren C. "Labor for the Picking: The New Deal in the South." *Journal of Economic History* 43, no. 4 (1983): 905–29.

Williams, Craig L. "Soil Conservation and Water Pollution Control: The Muddy Record of the United States Department of Agriculture." *Boston College Environmental Affairs Law Review* 7, no. 3 (1979): 365–422.

Winders, Bill. *The Politics of Food Supply: U.S. Agricultural Policy in the World Economy.* New Haven: Yale University Press, 2009.

———. "'Sowing the Seeds of Their Own Destruction': Southern Planters, State Policy and the Market, 1933–1973." *Journal of Agrarian Change* 6, no. 2 (April 2006): 143–66.

Woodman, Harold D. "Post–Civil War Southern Agriculture and the Law." *Agricultural History: Southern Agriculture since the Civil War: A Symposium* 53, no. 1 (January 1979): 319–37.

Young, C. Edwin, David W. Skully, Paul C. Westcott, and Linwood Hoffman. "Economic Analysis of Base Acre and Payment Yield Designations Under the 2002 U.S. Farm Act." Economic Research Report Number 12. Washington DC: U.S. Department of Agriculture, Economic Research Service, 2005.

Youngberg, G. "U.S. Agriculture Policy in the 1970's: Continuity and Change in an Uncertain World." *Policy Studies Journal* 4 (1975): 25–31.

INDEX

crop insurance (*continued*)
of, 60; Federal Crop Insurance Improvement Act (1980), 206; millennial farm
policy and, 205–8, 213, 233; MYA prices
(2000–2016), 326; policies and loss ratio
(2000–2016), 324; revenue-based, 206–7,
227, 228, 234, 287; revolution and reform
in 1990s and, 178–79, 195, 206; STAX, 247–
48, 258, 266, 267, 276
crop prices: Agricultural Act of 2014 and,
282–83; conservation program acreage
and, 299; evolutionary patterns in farm
policy development and, 289–95; fault
lines in farm policy drawn by, 286–89;
future policy predictions and, 310–11;
millennial farm policy, depressed crop
prices affecting, 205; productivity paradox
or farm problem (high production/
low crop prices) driving farm policy, 10;
revolution and reform in 1990s, crop price
decreases (from 1997) affecting, 194–96,
200, 204; revolution and reform in 1990s,
crop price increases (to 1997) affecting,
176, 184–85, 187, 188, 192, 200, 201–4; RFS,
increased crop prices due to, 224, 227,
229–30, 235, 294, 301; supply and demand
affecting, 286–87; technological advances
in production and, 286–87; volatility in
70s/80s (*See* volatile crop prices and farm
policy). *See also* decoupling of prices and
production; fixed prices and fixed-price
supports; price supports; target prices
crops. *See specific types, e.g.* wheat
CRP. *See* Conservation Reserve Program
CSP. *See* Conservation Stewardship Program
Cuban missile crisis, 130

dairy, 16, 56, 57, 59, 152, 155, 190, 193, 255, 257,
267–68, 338n15
D&PL (Delta and Pine Land Company), 30,
31, 41–42, 343–44nn122–123
Daschle, Tom, 177, 178, 189, 190, 192, 197,
213, 218
debt ceiling crisis, 240, 244
decoupling of prices and production:
Agricultural Act of 2014 and, 266, 277–78;
crop insurance and, 206; evaluation of
policy, 297, 299, 300; evolutionary patterns

in farm policy development and, 294;
millennial farm policy and, 205, 206, 211,
217, 221; revolution and reform in 1990s
and, 184, 193, 196, 198–202; WTO and
Brazil/U.S. cotton dispute, 221
deficiency payments, 150, 158, 161, 166, 183,
184, 185, 194, 200, 241, 292, 293
de la Garza, Kika, 150, 156, 174, 192, 403n129
Delta and Pine Land Company (D&PL), 30,
31, 41–42, 343–44nn122–123
Democrats: Agricultural Act of 2014 and,
7–8, 238–39, 246–48, 250, 252, 253, 255,
261–66, 269, 272, 280; mid-century farm
policy and, 97, 100, 102, 103, 104, 106,
109–11, 113–14, 116, 121–24, 126, 128, 133–34;
millennial farm policy and, 209, 210,
213–15, 217, 218, 222–26, 233; in New Deal
and WWII, 43, 64, 67, 68; origins of farm
policy (1909–1933) and, 15, 18, 20, 22; in
post-WWII era (1945–1949), 80–82, 84,
86, 89, 358n55; revolution and reform in
1990s and, 173, 175–77, 179–86, 188–90,
192–97, 200, 202, 203, 400n59; volatile
crop prices in 70s/80s and, 140, 144, 145,
148–57, 160, 162–64, 166, 167, 170. *See also*
southern Democrats
demographic issues: African American
northern migration, mid-century, 61, 115,
121, 133, 352n141, 352n150; black-operated
farms in South, decline in, 170; coalition between urban and rural interests
(1964–1965), 124–30; coalition between
urban and rural interests (Agricultural
Act of 2014), 240, 256, 262–63; decrease
in number of farmers and increase in size
of farms (1933–1945), 69, 296–97, 355n203;
mid-century shift of U.S. population to ur-
ban/suburban areas, 98, 115, 132–34, 308–9;
weakening of overall support for helping
farmers mid-century, 109, 134; WWII/
postwar employment boom and northern
migration of sharecroppers/tenants, 94, 133
demonstration farms, 11–12
Department of Agriculture. *See* U.S. Department of Agriculture
direct payments: Agricultural Act of 2014
eliminating, 239–41, 249, 250, 255, 258,
260, 269, 276–77; defined, 334; evolution-

free rider problem, 3, 10, 35, 161; future policy predictions, 310–12; as global issue, 3–4; history, importance of, 2–3; internal pressures on, 6; productivity paradox or farm problem (high production/low crop prices) driving, 1–4, 10; regional/commodity interests affecting, 8; risk, fundamental issues of, 3–5; self-governance, as window into, xii, 35, 39, 285, 302, 308–10; and yield, 243–44, 300

FarmPolicy.com, xi

Farm Security Administration, 358n55

Farm Security and Rural Investment Act (2002), 205, 208–19, 242, 270, 279, 311, 331

Farm Service Agency, ix

fault lines in farm policy, 5–8, 284, 285–312; crop prices and, 286–89; evaluating policies resulting from, 295–301; evolutionary patterns in farm policy development and, 289–95; food assistance programs and, 305–6, 309 (See also food stamps/food aid); future policy predictions, 310–12; historical context for, 285–86; legislative process and coalition-building, 301–10; regional/commodity interests, 8, 289, 338n15 (See also corn (Midwest); cotton (South); wheat (Great Plains))

FCIC (Federal Crop Insurance Corporation), 60

FDR. See Roosevelt, Franklin Delano

Federal Agriculture Improvement and Reform (FAIR) Act (1996), 183–94, 195, 196–204, 208, 232, 294, 311, 331

Federal Crop Insurance Corporation (FCIC), 60

Federal Crop Insurance Improvement Act (1980), 206

federal debt ceiling crisis, 240, 244

Federal Farm Board, 18

federal government, debates over role of, 308–10

Federalist Papers, 448n30

fencerow to fencerow planting, 146, 149

fertilizer and fertilization, 41, 56, 76, 96, 106, 301

Filburn, Wickard v., 66

Findley, Paul, 122, 125–26, 148

fiscal cliff, 255

Fite, Gilbert, 26

fixed loan rates, high, 64, 65, 68, 92, 102, 109, 179, 277, 294, 296

fixed payments: Agricultural Act of 2014 and, 275; millennial farm policy and, 211; revolution and reform in 1990s and, 184–85, 188, 189, 191–94, 196–200, 203

fixed prices and fixed-price supports, ix, 4, 7; Agricultural Act of 2014 and, 241–44, 247–49, 256–57, 259, 261, 262, 266, 269, 276–78; definition of fixed-price policy, 334; evaluation of policy, 296; high, fixed loan rates, 64, 65, 68, 92, 102, 109, 179, 277, 294, 296; mid-century farm policy and, 102, 106, 117; millennial farm policy and, 218; New Deal and WWII, 65, 66; origins of farm policy and, 24; in post-WWII period, 75, 79, 82, 87, 89, 91; revolution and reform in 90s and, 178; volatile crop prices of 70s/80s and, 137, 154. See also target prices

Foley, Tom, 145, 147, 150, 157, 382n72

Food, Agriculture, Conservation and Trade (FACT) Act (1990), 173, 175, 176–81, 188, 196, 197, 292, 293, 331

Food, Conservation, and Energy Act (2008), ix, xi, 223–32, 255, 294, 331

Food and Agricultural Act (1962), 117–23, 130, 133, 134, 270, 292, 330

Food and Agricultural Act (1965), 125–28, 129, 130, 136, 292, 330

Food and Agriculture Act (1977), 146–49, 292, 331

Food Control Act, 12

food costs and consumer interests in 1970s, 136, 139

Food for Peace (1954), 293

food movement, 1–2

Food Security Act (1985), 157–68, 172, 173, 175, 178, 292–93, 331

Food Stamp Act (1964), 124–25, 134, 306

food stamps/food aid: Agricultural Act of 2014 and, 237, 241, 245–47, 250, 252–57, 260–66, 269–75, 280–82; Butz on farm program spending versus, 426n43; compensatory payments to keep food prices low as alternative to, 83; fault lines in farm policy and, 305–6, 309; FDR's Food Stamp

olution and reform in 1990s and, 193, 201, 403n120; volatile crop prices of 70s/80s and, 152, 154–55, 157, 388n144, 391n190

Pearl Harbor (1941), 65, 68

Peek, George, 14, 37

Pelosi, Nancy, 225

penny sales, 25

Peruvian anchovy harvest, 139

Peterson, Collin, 224, 225, 226, 238, 249, 260–61, 262, 264, 268, 269

PFC (production flexibility contract) payments, 196, 197–98

PIK (payment in kind), 116–17, 158–59, 166, 172, 293, 333, 394n229

planting flexibility for farmers: Agricultural Act of 2014 and, 266–67; millennial farm policy continuing, 212; revolution and reform in 1990s introducing, 174, 176, 193, 406n159

PLC (price loss coverage), 267

Poage, W. R., 85, 100, 111, 116, 128–30, 136, 143–46

Politico.com, xi

pork/hogs, 26, 33, 76, 287

post-WWII farm policy (1945–1949), 73–94; abundance view and, 74; acreage reduction in, 74, 87–88, 92–94; Agricultural Act (1948), 77–82, 89, 90, 93, 330; Agricultural Act (1949), 82–89, 90, 94, 101, 130, 131, 188, 190, 197, 255, 266–67, 291, 330; Brannan Plan, 83–86, 94, 98, 141, 143; coalition, breakdown of, 73–77, 89–94; corn (Midwest), 76–77, 78, 86, 87–88, 89–94; cotton (South), 76–77, 78, 80, 81, 87–88, 89–94; Democrats and, 80–82, 84, 86, 89, 358n55; economic circumstances after war and, 74–75; House Agriculture Committee, 81, 85, 89; loans and loan rates, 74, 75, 81, 82, 83, 87, 88, 90–92, 94; Marshall Plan, 75; New Deal mentality, transition from, 70–71; parity system, 73, 75, 78–81, 84–93, 86, 356; Republicans and, 76, 77, 78, 80–82, 85, 89; Senate Agriculture Committee, 79, 81, 83, 86, 88; surplus commodities, accumulation of, 73, 82, 86, 87; technological advances in production, 75–76, 82, 355n11, 357n49; wheat (Great Plains), 87–88

potatoes, 84

premium subsidy, 334

price loss coverage (PLC), 267

prices for crops. See crop prices

price supports: AAA provisions, 23–24, 25–26; abundance viewpoint versus, 70, 71; definition of price support loan, 333; Korean War and, 96–97; in McNary-Haugen legislation, 15; mid-century farm policy and, 101, 104, 113; New Deal and, 57–58; in Peek-Johnson proposal, 14; in post-WWII era, 75–77, 79–80, 82, 83, 84–88, 89–91; production controls and, 58–59. See also fixed prices and fixed-price supports; loans and loan rates; parity system

production controls: AAA provisions, 24–25, 26, 33–35, 343n112; abundance viewpoint versus, 70; Agricultural Act of 1970 formally ending, 137; Bankhead Cotton Control Act (1934), instituting compulsory controls, 40–47; domestic allotment plan, 19–22, 48, 290; evaluation of policy, 296; farm bill (1965) ending many of, 127; grain companies opposing, 169; Great Depression, development of concept in, 19–22; mid-century farm policy and, 98; New Deal and, 54–59; in post-WWII era, 75–77; price supports and, 58–59; weather and, 34, 41, 53, 55, 56, 198. See also acreage reduction/diversion

production flexibility contract (PFC) payments, 196, 197–98

production incentives, 53, 75, 80, 142, 146, 175, 199, 203, 267, 291, 293, 295, 296, 300

Proxmire, William, 111

Pryor, David, 189–90

Quie, Albert, 111

racial issues and civil rights: anti-lynching legislation, 61, 352n141, 352n150; black-operated farms in South, decline in, 170; Civil Rights Act (1964), 128; farm coalition damaged by, 305; integrated recreational facilities, turning farmland into, 121–22; mid-century farm policy and, 114, 115, 121–22, 128, 130, 133–34; northern migration, mid-century, of African Americans, 61, 115, 121, 133, 352n141, 352n150; under Roosevelt

concept, 52; post-WWI economy and, 13; in post-WWII era, 87–88; prices received by farmers (1909–2014), 314; production levels (1933–2013), 316; regional/commodity interests affecting farm policy, 8; revolution and reform in 1990s and, 173, 178, 182, 198, 201; Soviet wheat deal (1972), 139, 141, 288; target prices and, 288; uses of (1950–2013), 321; volatile crop prices in 70s/80s and, 137, 138, 139, 141, 145–49, 155, 158, 163, 166, 383n88; weather and climate, vulnerability to, 60, 287, 289

White, E. D., 361n117

Whitten, Jamie, 126

Whittington, William, 63

Wickard, Claude, 65, 358n55

Wickard v. Filburn, 66

Winders, Bill, *The Politics of Food Supply: U.S. Agricultural Policy in the World Economy* (2009), x

World Trade Organization (wto): Brazil/U.S. cotton dispute, 220–21, 223, 228, 229, 232–36, 239, 241, 246–48, 266, 275, 276, 279, 283; formation of, 204, 219–20

World War I and postwar economy of 1920s, 9–10, 11–13, 78

World War II, 64–66, 67, 68–69

Wright, Jim, 382n72

yield-based crop insurance policy, 206–7

yield policy (crop insurance), 334

Young, Milton, 87, 100, 107, 126, 141, 147, 168, 227

Zulauf, Carl, x

Milton Keynes UK
Ingram Content Group UK Ltd.
UKHW031842070924
447867UK00002B/10